Advances in Mechanics and Mathematics

Volume 25

For more titles in this series, go to
http://www.springer.com/series/5613

Mikhail Z. Zgurovsky • Valery S. Mel'nik
Pavlo O. Kasyanov

Evolution Inclusions and Variation Inequalities for Earth Data Processing II

Differential-Operator Inclusions and Evolution Variation Inequalities for Earth Data Processing

Springer

Dr. Mikhail Z. Zgurovsky
Pavlo O. Kasyanov
Valery S. Mel'nik
National Technical University of Ukraine "Kyiv Polytechnic Institute"
Institute for Applied System Analysis
National Academy of Sciences of Ukraine
37, Peremogy Ave.
03056 Kyiv
Ukraine
zgurovsm@hotmail.com

ISSN 1571-8689 e-ISSN 1876-9896
ISBN 978-3-642-13877-5 e-ISBN 978-3-642-13878-2
DOI 10.1007/978-3-642-13878-2
Springer Heidelberg Dordrecht London New York

Library of Congress Control Number: 2010936816

Cover design: deblik, Berlin

Printed on acid-free paper

Springer is part of Springer Science+Business Media (www.springer.com)

Preface

By this book we continue the series of monographs devoted to investigation method for mathematical models of non-linear geophysical processes and fields. In the first volume the content of the second volume is announced.

We cheerfully express the gratitude to:

V.O. Gorban, I.N. Gorban, N.V. Gorban, V.I. Ivanenko, A.N. Novikov, V.I. Obuchovskii, N.A. Perestyuk, A.E. Shishkov for useful remarks;

O.P. Kogut and N.V. Zadoyanchuk for the exceptional diligence when preparing the electronic version of the book and friendly help in linguistic issues;

Kathleen Cass and Olena L. Poptsova for a technical support of our book;

editors of the "Springer" Publishing House who worked with our book and everybody who took part in preparation of the manuscript.

Kyiv, Ukraine
August 2010

Dr. Mikhail Z. Zgurovsky
Valery S. Mel'nik
Pavlo O. Kasyanov

Contents

Acronyms

a.e.	Almost everywhere
FG method	Faedo–Galerkin method
for a.e.	For almost each
l.s.c.	Lower semicontinuous
LTS	Linear topological space
N-s.b.v.	N-semibounded variation
N-sub-b.v.	N-subbounded variation
r.c.	Radial continuous
r.l.s.c.	Radial lower semicontinuous
r.s.c.	Radial semicontinuous
r.u.s.c.	Radial upper semicontinuous
s.b.v.	Semibounded variation
s.m.	Semimonotone
sub-m.	Submonotone
sub-b.v.	Subbounded variation
u.h.c.	Upper hemicontinuous
u.s.c.	Upper semicontinuous
u.s.b.v.	Uniform semibounded variation
V-s.b.v.	V-semibounded variation
w.l.s.c.	Weakly lower semicontinuous

Introduction

At an analysis and control of different geophysical and socio-economical processes it is often appears such problem: at a mathematical modelling of effects related to friction and viscocity, quantum effects, a description of different nature waves the existing "gap" between rather high degree of the mathematical theory of analysis and control for non-linear processes and fields and practice of its using in applied scientific investigations make us require rather stringent conditions for interaction functions. These conditions related to linearity, monotony, smoothness, continuity and can substantially have an influence on the adequacy of mathematical model. Let us consider for example some diffusion process. Its mathematical model has the next form:

$$\begin{cases} y_t - \Delta y + f(y) = g(t,x) \quad in \quad \Omega \times (\tau; T), \\ y\big|_{\partial\Omega} = 0, \\ y\big|_{\tau=T} = y_0, \end{cases} \tag{1}$$

here $n \geq 2$, $\Omega \subset \mathbb{R}^n$ is a bounded domain with a rather smooth boundary, $-\infty < \tau < T < +\infty$, $g : \Omega \times (\tau; T) \to \mathbb{R}$, $y_0 : \Omega \to \mathbb{R}$ are rather regular functions, $f : \mathbb{R} \to \mathbb{R}$ is an interaction function, $y : \Omega \times (\tau; T) \to \mathbb{R}$ is an unknown function. It is well known that if f is a rather smooth function and satisfies for example the next condition of no more than polynomial growth:

$$\exists p > 1, \quad \exists c > 0: \quad |f(s)| \leq c(1 + |s|^{p-1}) \quad \forall s \in \mathbb{R}, \tag{2}$$

then problem (1) has a unique rather regular solution. Let us consider the case when f is continuous and initial data and external forces are nonregular (for example $y_0 \in L_2(\Omega)$, $g \in L_2(\Omega \times (\tau; T))$). Then, as a rule, we consider the generalized setting of problem (1):

$$\begin{cases} y'(t) + A(y(t)) + B(y(t)) = g(t) \quad \text{for a.e.} \quad t \in (\tau; T), \\ y(\tau) = y_0, \end{cases} \tag{3}$$

here $A : V_1 \to V_1^*$ is an energetic extension of operator "$-\Delta$", $B : V_2 \to V_2^*$ is the Nemytskii operator for F, $V_1 = H_0^1(\Omega)$ is a real Sobolev space, $V_2 = L_p(\Omega)$, $V_1^* = H^{-1}(\Omega)$, $V_2^* = L_q(\Omega)$, q is the conjugated index, y' is a derivative of an element $y \in L_2(\tau, T; V_1) \cap L_p(\tau, T; V_2)$ and it is considered in the sense of the space $\mathscr{D}^*([\tau; T], V_1^* + V_2^*)$.

A solution of problem (3) in the class $W = \{y \in L_2(\tau, T; V_1) \cap L_p(\tau, T; V_2) \big| y' \in L_2(\tau, T; V_1^*) + L_q(\tau, T; V_2^*)\}$ refers to be the generalized solution of problem (1).

To prove the existence of solutions for problem (1) as a rule we need to add supplementary "signed condition" for an interaction function f, for example,

$$\exists \alpha, \beta > 0: \quad f(s)s \geq \alpha |s|^p - \beta \quad \forall s \in \mathbb{R}. \tag{4}$$

But we do not succeed in proving the uniqueness of the solution of such problem in the general case. Note that technical condition (4) provides a dissipation too. We remark also that different conditions for parameters of problem (1) provide corresponding conditions for generated mappings A and B.

Problem (3) is usually investigated in more general case:

$$\begin{cases} y' + \mathscr{A}(y) = g, \\ y(\tau) = y_0, \end{cases} \tag{5}$$

here $\mathscr{A} : X \to X^*$ is the Nemytskii operator for $A + B$,

$$\mathscr{A}(y)(t) = A(y(t)) + B(y(t)) \quad \text{for a.e.} \quad t \in (\tau; T),\ y \in X,$$

$$X = L_2(\tau, T; V_1) \cap L_p(\tau, T; V_2), \quad X^* = L_2(\tau, T; V_1^*) + L_q(\tau, T; V_2^*).$$

Solutions of problem (5) are also searched in the class $W = \{y \in X \big| y' \in X^*\}$.

In cases when the continuity of the interaction function f have an influence on the adequacy of mathematical model fundamentally then problem (1) is reduced to such problem:

$$\begin{cases} y_t - \Delta y + F(y) \ni g(x,t) \quad \text{in } Q = \Omega \times (\tau; T), \\ y\big|_{\partial\Omega} = 0, \\ y\big|_{\tau=T} = y_0, \end{cases} \tag{6}$$

here

$$F(s) = [\underline{f}(s), \overline{f}(s)], \quad \underline{f}(s) = \underline{\lim}_{t \to s} f(t), \quad \overline{f}(s) = \overline{\lim}_{t \to s} f(t), \quad s \in \mathbb{R},$$

$$[a, b] = \{\alpha a + (1 - \alpha) b \big| \alpha \in [0, 1]\},$$

$-\infty < a < b < +\infty$.

A solution of such differential-operator inclusion

$$\begin{cases} y' + \mathscr{A}(y) \ni g, \\ y(\tau) = y_0, \end{cases} \tag{7}$$

is usually thought to be the generalized solution of problem (6). Here $\mathscr{A} : X \rightrightarrows X^*$,

$$\mathscr{A}(u) = \{p \in X^* \big| p(t) \in A(u(t)) + B(u(t)), \text{ for a.e. } t \in (\tau; T)\}, \ u \in X,$$

$A : V_1 \to V_1^*$ is the energetic extension of "$-\Delta$" in $H_0^1(\Omega)$, $B : V_2 \to C_v(V_2^*)$ is the Nemytskii operator for F:

$$B(v) = \{z \in V_2^* \big| z(x) \in F(v(x)) \text{ for a.e. } x \in \Omega\}, \quad v \in V_2.$$

Taking into account all variety of classes of mathematical models for different nature geophysical processes and fields we propose rather general approach to investigation of them in this book. Further we will study classes of mathematical models in terms of general properties of generated mappings like $\mathscr{A}$.

This monograph is the continuation of [ZMK10]. Let us consider some denotations and results, that we will use in this book. Let X be a Banach space, X^* be its topologically adjoint,

$$\langle \cdot, \cdot \rangle_X : X^* \times X \to \mathbb{R}$$

be the canonical duality between X and X^*, 2^{X^*} be a family of all subsets of the space X^*, let $A : X \to 2^{X^*}$ be the multivalued map,

$$\text{graph} A = \{(\xi; y) \in X^* \times X \,|\xi \in A(y)\},$$

$$\text{Dom} A = \{y \in X \,|A(y) \neq \emptyset\}.$$

The multivalued map A is called strict if $\text{Dom}A = X$. Together with every multivalued map A we consider its upper

$$[A(y), \xi]_+ = \sup_{d \in A(y)} \langle d, \xi \rangle_X$$

and lower

$$[A(y), \xi]_- = \inf_{d \in A(y)} \langle d, \xi \rangle_X$$

support functions, where $y, \xi \in X$. Let also

$$\|A(y)\|_+ = \sup_{d \in A(y)} \|d\|_{X^*}, \quad \|A(y)\|_- = \inf_{d \in A(y)} \|d\|_{X^*}, \quad \|\emptyset\|_+ = \|\emptyset\|_- = 0.$$

For arbitrary sets $C, D \in 2^{X^*}$ we set

$$\text{dist}(C, D) = \sup_{e \in C} \inf_{d \in D} \|e - d\|_{X^*}, \quad d_H(C, D) = \max\{\text{dist}(C, D), \text{dist}(D, C)\}.$$

Then, obviously,

$$\|A(y)\|_+ = d_H(A(y), 0) = \text{dist}(A(y), 0), \quad \|A(y)\|_- = \text{dist}(0, A(y)).$$

Together with the operator $A : X \to 2^{X^*}$ let us consider the following maps

$$\text{co}A : X \to 2^{X^*} \quad \text{and} \quad \overset{*}{\overline{\text{co}}}\, A : X \to 2^{X^*},$$

defined by relations

$$(\text{co}A)(y) = \text{co}(A(y)) \quad \text{and} \quad (\overset{*}{\overline{\text{co}}}\, A(y)) = \overset{*}{\overline{\text{co}}}\, (A(y))$$

respectively, where $\overset{*}{\overline{\text{co}}}\, (A(y))$ is the weak star closure of the convex hull $\text{co}(A(y))$ for the set $A(y)$ in the space X^*. Besides for every $G \subset X$

$$(\text{co}A)(G) = \bigcup_{y \in G} (\text{co}A)(y), \quad (\overset{*}{\overline{\text{co}}}\, A)(G) = \bigcup_{y \in G} (\overset{*}{\overline{\text{co}}}\, A)(y).$$

Further we will denote the strong, weak and weak star convergence by $\to, \overset{w}{\to}, \overset{*}{\to}$ or $\to, \rightharpoonup, \rightharpoonup$ respectively. As $C_v(X^*)$ we consider the family of all nonempty convex closed bounded subsets from X^*.

Proposition 1. *[ZMK10, Proposition 1.2.1] Let $A, B, C : X \rightrightarrows X^*$. Then for all $y, v, v_1, v_2 \in X$ the following statements take place:*

1. *The functional $X \ni u \to [A(y), u]_+$ is convex, positively homogeneous and lower semicontinuous;*
2. $[A(y), v_1 + v_2]_+ \leq [A(y), v_1]_+ + [A(y), v_2]_+$,
   $[A(y), v_1 + v_2]_- \geq [A(y), v_1]_- + [A(y), v_2]_-$,
   $[A(y), v_1 + v_2]_+ \geq [A(y), v_1]_+ + [A(y), v_2]_-$,
   $[A(y), v_1 + v_2]_- \leq [A(y), v_1]_+ + [A(y), v_2]_-$;
3. $[A(y) + B(y), v]_+ = [A(y), v]_+ + [B(y), v]_+$,
   $[A(y) + B(y), v]_- = [A(y), v]_- + [B(y), v]_-$;
4. $[A(y), v]_+ \leq \|A(y)\|_+ \|v\|_X$,
   $[A(y), v]_- \leq \|A(y)\|_- \|v\|_X$;
5. $\|\overset{*}{\overline{\text{co}}}\, A(y)\|_+ = \|A(y)\|_+$, $\|\overset{*}{\overline{\text{co}}}\, A(y)\|_- = \|A(y)\|_-$,
   $[A(y), v]_+ = \left[\overset{*}{\overline{\text{co}}}\, A(y), v\right]_+$, $[A(y), v]_- = \left[\overset{*}{\overline{\text{co}}}\, A(y), v\right]_-$;
6. $\|A(y) - B(y)\|_+ \geq |\, \|A(y)\|_+ - \|B(y)\|_- \,|$,
   $\|A(y) - B(y)\|_- \geq \|A(y)\|_- - \|B(y)\|_+$;
7. $d \in \overset{*}{\overline{\text{co}}}\, A(y) \quad \Leftrightarrow \quad \forall \omega \in X \;\; [A(y), \omega]_+ \geq \langle d, w \rangle_X$;
8. $d_H(A(y), B(y)) \geq |\|A(y)\|_+ - \|B(y)\|_+|$,
   $d_H(A(y), B(y)) \geq |\|A(y)\|_+ - \|B(y)\|_+|$,
   where d_H is Hausdorff metric;
9. $\text{dist}(A(y) + B(y), C(y)) \leq \text{dist}(A(y), C(y)) + \text{dist}(B(y), 0)$,
   $\text{dist}(C(y), A(y) + B(y)) \leq \text{dist}(C(y), A(y)) + \text{dist}(0, B(y))$,
   $d_H(A(y) + B(y), C(y)) \leq d_H(A(y), C(y)) + d_H(B(y), 0)$;

10. for any $D \subset X^*$ and bounded $E \in C_v(X^*)$

$$\operatorname{dist}(D, E) = \operatorname{dist}(\overset{*}{\overline{\operatorname{co}}}\, D, E).$$

Proposition 2. *[ZMK10, Proposition 1.2.2] The inclusion $d \in \overset{*}{\overline{\operatorname{co}}}\, A(y)$ holds true if and only if one of the following relations takes place:*

$$\begin{aligned} \textit{either} \quad & [A(y), v]_+ \geq \langle d, v\rangle_X \quad \forall v \in X, \\ \textit{or} \quad & [A(y), v]_- \leq \langle d, v\rangle_X \quad \forall v \in X. \end{aligned}$$

Proposition 3. *[ZMK10, Proposition 1.2.3] Let $D \subset X$ and $a(\cdot\,,\cdot) : D \times X \to \mathbb{R}$. For each $y \in D$ the functional $X \ni w \mapsto a(y, w)$ is positively homogeneous, convex and lower semicontinuous if and only if there exists the multivalued map $A : X \to 2^{X^*}$ such that $D(A) = D$ and*

$$a(y, w) = [A(y), w]_+ \qquad \forall y \in D(A),\ w \in X.$$

Proposition 4. *[ZMK10, Proposition 1.2.4] The functional $\|\cdot\|_+ : C_v(X^*) \to \mathbb{R}_+$ satisfies the following properties:*

1. $\{\bar{0}\} = A \quad \Leftrightarrow \quad \|A\|_+ = 0,$
2. $\|\alpha A\|_+ = |\alpha|\|A\|_+, \quad \forall \alpha \in \mathbb{R},\ A \in C_v(X^*),$
3. $\|A + B\|_+ \leq \|A\|_+ + \|B\|_+ \quad \forall A, B \in C_v(X^*).$

Proposition 5. *[ZMK10, Proposition 1.2.5] The functional $\|\cdot\|_- : C_v(X^*) \to \mathbb{R}_+$ satisfies the following properties:*

1. $\bar{0} \in A \quad \Leftrightarrow \quad \|A\|_- = 0,$
2. $\|\alpha A\|_- = |\alpha|\|A\|_-, \quad \forall \alpha \in \mathbb{R},\ A \in C_v(X^*),$
3. $\|A + B\|_- \leq \|A\|_- + \|B\|_- \quad \forall A, B \in C_v(X^*).$

Let us consider some classes of multivalued maps that we introduced in [ZMK10]. As before, let X be a Banach space, X^* be its topologically adjoint,

$$\langle\cdot,\cdot\rangle_X : X^* \times X \to \mathbb{R}$$

be the duality form (Fig. 1).

We remind that the multivalued map $A : D(A) \subset X \to 2^{X^*}$ is called the monotone one if

$$\langle d_1 - d_2, y_1 - y_2\rangle_X \geq 0 \quad \forall y_1, y_2 \in D(A) \quad \forall d_1 \in A(y_1), \forall d_2 \in A(y_2).$$

Using the above mentioned brackets it is easy to see that the multivalued operator $A : D(A) \subset X \to 2^{X^*}$ is monotone if and only if

$$[A(y_1), y_1 - y_2]_- \geq [A(y_2), y_1 - y_2]_+ \qquad \forall y_1, y_2 \in D(A).$$

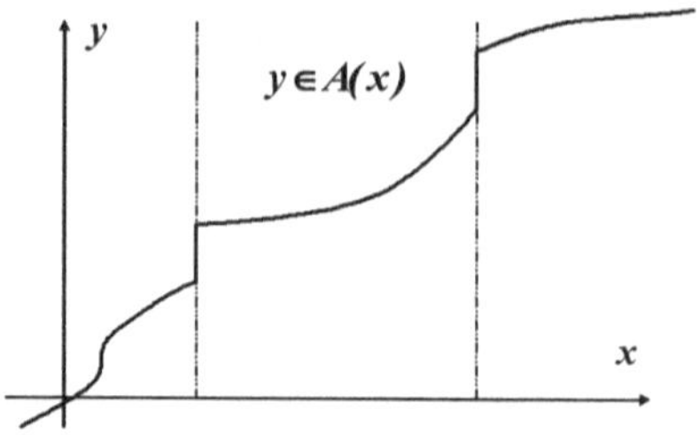

Fig. 1 The monotone multivalued map

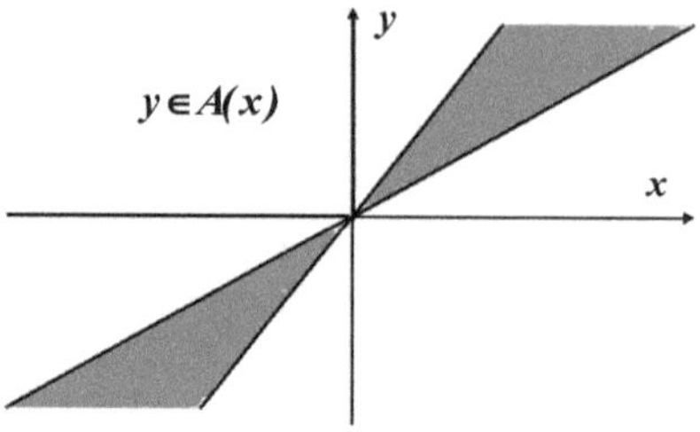

Fig. 2 The "−"-coercive multivalued map

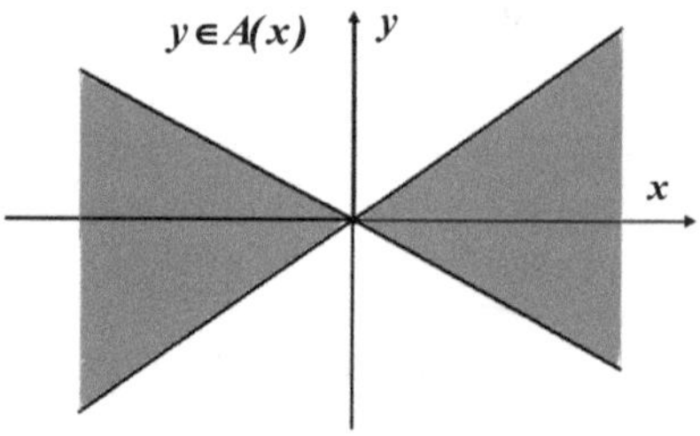

Fig. 3 The "+"-coercive multivalued map, but not "−"-coercive

In addition to the common monotony of multivalued maps we are interested in the following concepts (Figs. 2 and 3):

- *N-monotony*, namely

$$[A(y_1), y_1 - y_2]_- \geq [A(y_2), y_1 - y_2]_- \qquad \forall y_1, y_2 \in D(A);$$

- *V-monotony*, namely

$$[A(y_1), y_1 - y_2]_+ \geq [A(y_2), y_1 - y_2]_+ \qquad \forall y_1, y_2 \in D(A);$$

- *w-monotony*, namely

$$[A(y_1), y_1 - y_2]_+ \geq [A(y_2), y_1 - y_2]_- \qquad \forall y_1, y_2 \in D(A).$$

Definition 1. Let $D(A)$ be some subset. The multivalued map $A : D(A) \subset X \to 2^{X^*}$ is called:

- *Weakly +(-)-coercive*, if for each $f \in X^*$ there exists $R > 0$ such that

$$[A(y), y]_{+(-)} \geq \langle f, y \rangle_X \quad \forall y \in X \cap D(A) : \ \|y\|_X = R.$$

- *+(-)-coercive*, if

$$\frac{[A(y), y]_{+(-)}}{\|y\|_X} \to +\infty \quad \text{as} \quad \|y\|_X \to +\infty, \quad y \in D(A);$$

- *Uniformly +(−)-coercive* if for some $c > 0$

$$\frac{[A(y), y]_{+(-)} - c\|A(y)\|_{+(-)}}{\|y\|_X} \to +\infty \quad \text{as} \quad \|y\|_X \to +\infty, \quad y \in D(A);$$

- *Bounded* if for any $L > 0$ there exists $l > 0$ such that $\|A(y)\|_+ \leq l \ \forall y \in D(A)$ $\|y\|_X \leq L$;
- *Locally bounded*, if for an arbitrary fixed $y \in D(A)$ there exist constants $m > 0$ and $M > 0$ such that $\|A(\xi)\|_+ \leq M$ when $\|y - \xi\|_X \leq m, \xi \in D(A)$;
- *Finite-dimensionally locally bounded*, if for any finite-dimensional space $F \subset X$ the contraction of A on $F \cap D(A)$ is locally bounded.

Definition 2. A strict multivalued map $A : X \rightrightarrows X^*$ is called:

- *Radial lower semicontinuous (r.l.s.c.)* if $\forall y, \xi \in X$

$$\varliminf_{t \to 0+} [A(y + t\xi), \xi]_+ \geq [A(y), \xi]_-;$$

- *Radial upper semicontinuous (r.u.s.c.)* if the real function

$$[0, 1] \ni t \to [A(y + t\xi), \xi]_+$$

  is upper semicontinuous at the point $t = 0$ for any $y, \xi \in X$.
- *Radial semicontinuous (r.s.c.)* if $\forall y, \xi \in X$

$$\varliminf_{t \to 0+} [A(y - t\xi), \xi]_+ \geq [A(y), \xi]_-;$$

- *Radial continuous (r.c.)* if the real function

$$[0, 1] \ni t \to [A(y + t\xi), \xi]_-$$

  is continuous at the point $t = 0$ from the right for any $y, \xi \in X$;

- *(upper) Hemicontinuous (u.h.c.)* if the function

$$X \ni x \longmapsto [A(x), y]_+$$

is u.s.c. on X for any $y \in X$;
- *λ-pseudomonotone on* X if for any sequence $\{y_n\}_{n\geq 0} \subset X$ such that $y_n \rightharpoonup y_0$ in X as $n \to +\infty$ from the inequality

$$\overline{\lim_{n\to\infty}} \langle d_n, y_n - y_0 \rangle_X \leq 0, \tag{8}$$

where $d_n \in \overset{*}{\overline{\text{co}}}\, A(y_n)$ $\forall n \geq 1$ the existence of subsequences $\{y_{n_k}\}_{k\geq 1}$ from $\{y_n\}_{n\geq 1}$ and $\{d_{n_k}\}_{k\geq 1}$ from $\{d_n\}_{n\geq 1}$ follows for which the next relation holds true:

$$\underline{\lim_{k\to\infty}} \langle d_{n_k}, y_{n_k} - w \rangle_X \geq [A(y_0), y_0 - w]_- \quad \forall w \in X; \tag{9}$$

- *λ_0-pseudomonotone on* X, if for any sequence $\{y_n\}_{n\geq 0} \subset X$ such that $y_n \rightharpoonup y_0$ in X, $d_n \rightharpoonup d_0$ in X^* as $n \to +\infty$ where $d_n \in \overset{*}{\overline{\text{co}}}\, A(y_n)$ $\forall n \geq 1$ from the inequality (8) the existence of subsequences $\{y_{n_k}\}_{k\geq 1}$ from $\{y_n\}_{n\geq 1}$ and $\{d_{n_k}\}_{k\geq 1}$ from $\{d_n\}_{n\geq 1}$ follows for which the relation (9) holds true.

The above mentioned multivalued map satisfies:
- *Condition* $(\kappa)_{+(-)}$ if for each bounded set D in X there exists $c \in \mathbb{R}$ such that

$$[A(v), v]_{+(-)} \geq -c\|v\|_X \quad \forall v \in D \setminus \{\bar{0}\};$$

- *Condition* (Π) if for any $k > 0$, any bounded set $B \subset X$, any $y_0 \in X$ and for some selector $d \in A$ for which

$$\langle d(y), y - y_0 \rangle_X \leq k \quad \text{for all } y \in B, \tag{10}$$

there exists $C > 0$ such that

$$\|d(y)\|_{X^*} \leq C \quad \text{for all } y \in B.$$

Proposition 6. *[ZMK10, Proposition 1.2.6] If a multivalued operator $A : X \rightrightarrows X^*$ satisfies Condition (Π) then it satisfies Condition $(\kappa)_+$ as well.*

During the investigation of evolution inclusions and variation inequalities, that describe mathematical models of nonlinear geophysical processes, we will use some properties of multimaps, represented in [ZMK10] (Fig. 4).

Corollary 1. *[ZMK10, Corollary 1.2.2] Let $\varphi : X \to \mathbb{R}$ be a convex lower semicontinuous functional such that*

$$\frac{\varphi(y)}{\|y\|_X} \to +\infty \quad \textit{as} \quad \|y\|_X \to \infty.$$

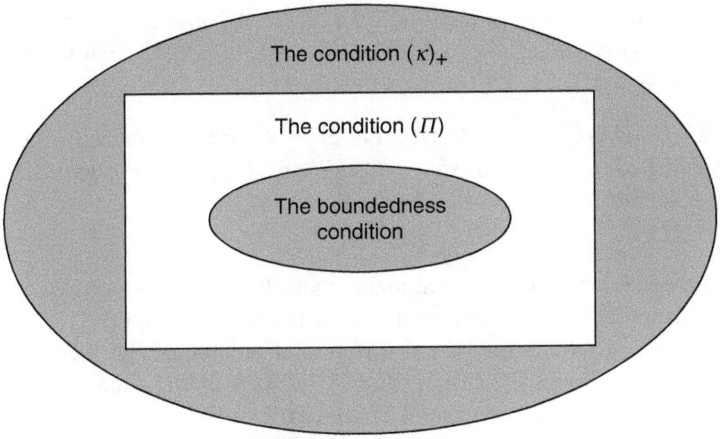

Fig. 4 Some properties for strict multivalued map in Banach space

Then its subdifferential map

$$\partial\varphi(y) = \{p \in X^* \mid \langle p, \omega - y\rangle_X \le \varphi(\omega) - \varphi(y) \quad \forall \omega \in X\} \ne \emptyset, \quad y \in X$$

is $+$*-coercive and hence* $-$*-coercive, uniformly* $-$*-coercive and uniformly* $+$*-coercive.*

Proposition 7. *[ZMK10, Proposition 1.2.30] Let a function* $\varphi : X \to \mathbb{R}$ *be convex, lower semicontinuous on* X*. Then the multivalued map* $B = \partial\varphi : X \to C_v(X^*)$ *is* λ_0*-pseudomonotone on* X *and it satisfies Condition* (Π)*.*

Now let X be a Banach space such that $X = X_1 \cap X_2$ where X_1, X_2 is the interpolation pair of reflexive Banach spaces [TRI78] which satisfies

$$X_1 \cap X_2 \quad \text{is dense in} \quad X_1, X_2. \tag{11}$$

Definition 3. A pair of multivalued maps $A : X_1 \to 2^{X_1^*}$ and $B : X_2 \to 2^{X_2^*}$ is called *s-mutually bounded* if for each $M > 0$ and a bounded set $B \subset X$ there exists a constant $K(M) > 0$ such that from

$$\|y\|_X \le M \quad \text{and} \quad \langle d_1(y), y\rangle_{X_1} + \langle d_2(y), y\rangle_{X_2} \le M \quad \forall y \in B$$

it follows that

$$\text{either} \quad \|d_1(y)\|_{X_1^*} \le K(M), \quad \text{or} \quad \|d_2(y)\|_{X_2^*} \le K(M) \quad \forall y \in B$$

for some selectors $d_1 \in A$ and $d_2 \in B$.

Remark 1. [ZMK10, Remark 1.2.20] If one of the maps $A : X_1 \to 2^{X_1^*}$ *or* $B : X_2 \to 2^{X_2^*}$ *is bounded then the pair* $(A; B)$ *is s-mutually bounded.*

Lemma 1. *[ZMK10, Lemma 1.2.9] Let $A : X_1 \rightrightarrows X_1^*$ and $B : X_2 \rightrightarrows X_2^*$ be some multivalued +(−)-coercive maps which satisfy Condition (κ). Then the multivalued map $C := A + B : X \rightrightarrows X^*$ is +(−)-coercive too.*

Lemma 2. *[ZMK10, Lemma 1.2.10] Let $A : X_1 \rightrightarrows X_1^*$ and $B : X_2 \rightrightarrows X_2^*$ be strict multivalued maps satisfying Condition (Π). Then the pair $(A; B)$ is s-mutually bounded and the multivalued map $C := A + B : X \rightrightarrows X^*$ satisfies Condition (Π).*

Now we consider subdifferential maps, that play an important role in the non-smooth analysis and the optimization theory [PSH80, AE84, DV81], in nonlinear boundary value problems for partial differential equations, the theory of control of the distributed systems [ZM99, LIO69], as well as the theory of differential games and mathematical economy [AF90, CHI97]. For basic properties of such maps we refer the reader to [AE84, DV81, IT79]. In [ZMK10] we generalized basic properties of subdifferentials and local subdifferentials known for Banach spaces to the case of Frechet spaces. We present some variations of obtained results, that we will use during the investigation of differential-operator inclusions and evolution variation inequalities for Earth Data Processing.

Let U be a convex subset in X, $F : X \mapsto \overline{\mathbb{R}} = \mathbb{R} \cup \{+\infty\}$ be a functional

$$\mathrm{dom} F = \{x \in X \,|\, F(x) \neq +\infty\}.$$

The set

$$\partial F(x_0; U) = \{p \in X^* \mid \langle p, x - x_0\rangle_X \leq F(x) - F(x_0)\ \forall x \in U\}$$

refers to a *local subdifferential* of a functional F in a point $x_0 \in U$. Observe that $\partial F(x_0; U_1) \supset \partial F(x_0; U_2)$, if $U_1 \subset U_2$. In particular, $\partial F(x_0; X) = \partial F(x_0) \subset \partial F(x_0; U)$. The last set is called the *subdifferential* of F at the point x_0 (Fig. 5).

Proposition 8. *[AE84, p.191] Let X be a Banach space. Then the norm $\|\cdot\|_X$ in X is subdifferentiable functional and*

$$\partial\|\cdot\|_X(x) = \{\, p \in X^* \,|\, \langle p, x\rangle_X = \|x\|_X,\ \|p\|_{X^*} = 1\,\} \quad \forall x \in X.$$

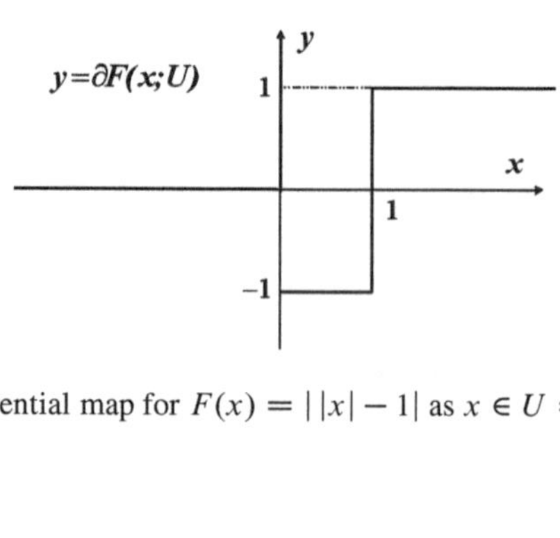

Fig. 5 The local subdifferential map for $F(x) = ||x| - 1|$ as $x \in U = \mathbb{R}_+$

In the case, when $\alpha > 1$

$$\partial\left(\frac{1}{\alpha}\|\cdot\|_X^\alpha\right)(x) = \|\cdot\|^{\alpha-1}\partial\|\cdot\|_X(x)$$
$$= \{\, p \in X^* \mid \langle p, x\rangle_X = \|x\|_X^\alpha,\ \|p\|_{X^*} = \|x\|_X^{\alpha-1} \,\} \quad \forall x \in X.$$

Theorem 1. *[ZMK10, Theorem 1.3.3] Let* U *be a convex body in* X, $F : X \mapsto \mathbb{R} \cup \{+\infty\}$ *be a convex functional on* U *and a lower semicontinuous functional on* int$U$ *(*int$U \subset$ domF*). Then for every* $x_0 \in$ int$U$ *and every* $h \in X$*, the quantity*

$$D_+F(x_0; h) = \lim_{t\to 0+} \frac{F(x_0 + th) - F(x_0)}{t} \tag{12}$$

is finite and the following statements hold true:

(i) There exists a counterbalanced (cf. [RUD73]) convex absorbing neighborhood of zero Θ *(*$x_0 + \Theta \subset$ int$U$*) such that for every* $h \in \Theta$

$$F(x_0) - F(x_0 - h) \le D_+F(x_0; h) \le F(x_0 + h) - F(x_0); \tag{13}$$

(ii) The functional int$U \times X \ni (x; h) \mapsto D_+F(x; h)$ *is upper semicontinuous.*
(iii) The functional $D_+F(x_0; \cdot) : X \mapsto \mathbb{R}$ *is positively homogeneous and semiadditive for every* $x_0 \in$ int$U$.
(iv) There exist a neighborhood $O(h_0)$ *and a constant* $c_1 > 0$ *such that for every* $x_0 \in$ int$U$ *and every* $h_0 \in X$,

$$|D_+F(x_0; h) - D_+F(x_0; h_0)| \le c_1 d(h, h_0) \quad \textit{for every } h \in O(h_0).$$

Definition 4. A multivalued map $A : X \rightrightarrows X^*$ is called:

(a) *-Upper semicontinuous (*-u.s.c.), if for any set B open in the $\sigma(X^*, X)$ topology the set $A_M^{-1}(B) = \{x \in X \mid A(x) \subset B\}$ is open in X.
(b) Upper hemicontinuous, if the function

$$X \ni x \mapsto [A(x), y]_+ = \sup_{d\in A(x)} \langle d, y\rangle_X$$

is upper semicontinuous for each $y \in X$.

Let us note that (b) follows from (a).

Theorem 2. *[ZMK10, Theorem 1.3.4] Let* U *be a convex body and* int$U \subset$ dom$F$*, where* $F : X \to \overline{\mathbb{R}}$ *is a convex functional on* U *and a semicontinuous function on* int$U$*. Then*

(i) $\partial F(x; U)$ *is a nonempty convex compact set for every* $x \in$ int$U$ *in the* $\sigma(X^*; X)$ *topology.*
(ii) $\partial F(\cdot; U) : U \rightrightarrows X^*$ *is a monotone map (on* U*).*

(iii) The map $\mathrm{int}U \ni x \mapsto \partial\varphi(x;U) \subset X^*$ *is *-upper semicontinuous (on* $\mathrm{int}\,U$*) and*

$$[\partial\varphi(x_0;U),h]_+ = D_+\varphi(x_0;h) \quad \textit{for all } h \in X \textit{ and } x_0 \in \mathrm{int}\,U. \tag{14}$$

Theorem 3. *[ZMK10, Theorem 1.3.5] Let* $F : X \mapsto \overline{\mathbb{R}}$ *be a convex on* U *lower semicontinuous on* $\mathrm{int}U$ *functional. Then for each* $x_0 \in \mathrm{intdom}\varphi$ *the set* $\partial\varphi(x_0)$ *is nonempty convex compact in* $\sigma(X^*;X)$*-topology, the map*

$$\mathrm{intdom}\,\varphi \ni x \longmapsto \partial\varphi(x) \subset X^*$$

*is *-u.s.c. and the following equality takes place*

$$[\partial\varphi(x_0),u]_+ = D_+\varphi(x_0;u) \qquad \forall u \in X. \tag{15}$$

Some proofs are based on the following Proposition which is a generalization of the Weierstrass Theorem on the case of locally bounded sets.

Lemma 3. *[ZMK10, Lemma 1.4.3] Suppose* W *is locally convex space,* W^* *is its topologically adjoint,* $E \subset W^*$ *is a set closed in the topology* $\tau(W^*;W)$*,* $L : E \to \widehat{R} = R \cup \{-\infty\}$ *is its upper semicontinuous main functional in the topology* $\tau(W^*;W)$*. Besides, let either the set* E *be equicontinuous in* W^**, or the following analogue of coercivity hold true: for an arbitrary set* $U \subset W^*$*, which is not equicontinuous and* $\lambda \subset R$ *there exists* $w_\lambda \in U$ *such that* $L(w_\lambda) \le \lambda$*.*

Then the functional L *is upper bounded on* E *and reaches on* E *its upper boundary* l*, and here the set* $\{w \in E | L(w) = l\}$ *is compact in the topology* $\tau(W^*;W)$*.*

The role of the classical acute angle Lemma [ZM04] in demonstration of solvability for nonlinear operator equations with monotone coercive maps in finite-dimensional space is well-known. In [ZMK10, Sect. 1.4] the minimax inequalities are investigated. By using of this apparatus multivalued analogues of "acute angle Lemma" are proved.

Corollary 2. *[ZMK10, Corollary 1.4.3] Suppose* Y *is finite-dimensional space,* $F : \bar{B}_r \to C_v(Y)$ *are strictly u.s.c. maps where*

$$\bar{B}_r = \{y \in Y | \, \|y\|_Y \le r\}.$$

If here

$$[F(y),y]_+ \ge 0 \qquad \forall y \in Y : \quad \|y\|_Y = r, \tag{16}$$

then there exists $\bar{x} \in \bar{B}_r$ *for which* $\bar{0} \in F(\bar{x})$*.*

By using this apparatus the new constructive solvability theorems for operator inclusions and variation inequalities are obtained in [ZMK10, Chap. 2]. More particular, suppose X is a reflexive Banach space, Y^* is a normalized space, $U \subset Y^*$ is a nonempty subset, $A : X \times U \to 2^{X^*}$ is a multivalued map, $f \in X^*$, $u \in U$ are some fixed elements,

$$K_{f,u} = \{y \in X \mid f \in A(y,u)\}.$$

In [ZMK10, Chap. 2] it is investigated some properties of the set $K_{f,u}$.

Definition 5. An element $y \in X$ is called a weak solution of the inclusion $f \in A(y,u)$ if

$$[A(y,u), w]_+ \geq \langle f, w \rangle_X \quad \forall w \in X. \tag{17}$$

Theorem 4. *[ZMK10, Theorem 2.2.1] Let for any $u \in U$ $A(\cdot,u) : X \to 2^{X^*}$ be a strict λ-pseudomonotone finite-dimensionally bounded map and for any $f \in X^*$ there exists $r > 0$ such that*

$$[A(y,u) - f, y]_+ \geq 0 \quad \forall y \in \partial B_r \subset X.$$

Then $\forall f \in X^*$, $u \in U$ there exists a weak solution of the operator inclusion $f \in A(y,u)$.

References

[AE84] Aubin J-P, Ekeland I (1984) Applied nonlinear analysis. Wiley-Interscience Publication, New York

[AF90] Aubin J-P, Frankowska H (1990) Set-valued analysis. Birkhauser, Boston

[CHI97] Chikrii AA (1997) Conflict-controlled processes. Kluwer Academic, Boston

[DV81] Dem'yanov VF, Vasiliev LV (1981) Non-differentiated optimization [in Russian]. Nauka, Moscow

[IT79] Ioffe AD, Tihomirov VM (1979) Theory of extremal problems. North Holland, Amsterdam

[LIO69] Lions JL (1969) Quelques methodes de resolution des problemes aux limites non lineaires. Dunod Gauthier-Villars, Paris

[PSH80] Pshenichniy BN (1980) Convex analysis and extremal problems [in Russian]. Nauka, Moscow

[RUD73] Rudin W (1973) Functional analysis. McGraw-Hill, New York

[TRI78] Triebel H (1978) Interpolation theory, function spaces, differential operators. North-Holland Publishing Company, Amsterdam

[ZM99] Zgurovsky MZ, Mel'nik VS (1999) Nonlinear analysis and control of infinite-dimentional systems [in Russian]. Naukova dumka, Kiev

[ZM04] Zgurovsky MZ, Mel'nik VS (2004) Nonlinear analysis and control of physical processes and fields. Springer, Berlin

[ZMK10] Zgurovsky MZ, Mel'nik VS, Kasyanov PO (2010) Evolution inclusions and variation inequalities for earth data processing volume I: operator inclusions and variation inequalities for earth data processing. Springer, Berlin

Chapter 1
Auxiliary Statements

Abstract In this chapter we consider the universal toolbar required for the investigation of evolutional Earth Data Processes which contains in particular partial differential equations with discontinuous or multivalued relationship between determinative parameters of the problem. In Sect. 1.1 we introduce classes of phase spaces and extended phase spaces of the generalized solutions for differential-operator inclusions and evolutional multivariational inequalities. We study the structured properties of such spaces, give the theorems of embedding for non-reflexive classes of spaces of distributions with integrable derivatives. We consider also the basis theory for such spaces. In the second section we study classes of energetic extensions and the Nemytskii operators for differential operators from evolutional mathematical models for geophysical processes and fields. In particular, we consider a class of multivalued variational calculus operators. We prove the completeness with respect to the sum for quasibounded w_{λ_0}-pseudomonotone weakly coercive maps. The introduced in this chapter results are used for the qualitative and constructive investigation of differential-operator inclusions and evolutional multi-variational inequalities in the next chapters more than once. Corresponding theorems contains these results. The convergence of the Faedo–Galerkin method for differential-operators inclusions with maps of pseudomonotone type is proved by the help of the obtained properties. The compact embedding theorems in non-reflexive spaces together with the results of Sect. 1.2 allow us to sweep a substantially wider class of evolutional problems and validate the penalty method for evolutional multivariational inequalities with maps of w_{λ_0}-pseudomonotone type. Since in the majority of cases the introduced properties of classes of infinite-dimensional distribution spaces have not been considered yet, so the exposition of results are made with detail proofs. The obtained in this section results have an independent value.

M.Z. Zgurovsky et al., *Evolution Inclusions and Variation Inequalities for Earth Data Processing II*, Advances in Mechanics and Mathematics 25,
DOI 10.1007/978-3-642-13878-2_1,

1.1 Functional Spaces: The Embedding and Approximation Theorems

If it is necessary to describe a nonstationary process, which evolve in some domain Ω from some finite dimensional space $\mathbb{R}^n$ during the time interval S, we may deal with state and time functions, i.e. with functions that put in correspondence for the each pair $\{t, x\} \in S \times \Omega$ the real number or vector $u(t, x)$. In virtue of this approach the time and the space variables are equivalent. But there is one more convenient approach to the mathematical description for nonstationary processes [GGZ74, LIO69]: for each point in time t it is mapped the state function $u(t, \cdot)$. (For example for each point of time we put the temperature distribution or velocity distribution in the domain Ω.)

geophysical process or field	$\Rightarrow$	mathematical model (partial differential equation)	$\Rightarrow$	differential-operator inclusion
		$y : \Omega \times [\tau, T] \to \mathbb{R}^m$		$y : [\tau, T] \to (\Omega \to \mathbb{R}^m)$ $y(\cdot, t) \in (\Omega \to \mathbb{R}^m)$, $t \in [\tau, T]$

Thus, we consider some functions, well-defined at the time interval S, with values from the state functions space (for example in the space $H_0^1(\Omega)$). Therefore, at studying some problems that depend on time it is rather natural to consider some function spaces, that acts from S into some infinite dimensional space X, in particular, it is natural to consider the spaces of integrable and differentiable functions. Further we will consider only **real** linear spaces.

In this section we introduce function spaces that will be used under investigation of differential-operator inclusions of such type in infinite dimensional spaces:

$$Lu + A(u) + B(u) \ni f, \quad u \in D(L), \tag{1.1}$$

where $A : X_1 \to 2^{X_1^*}$, $B : X_2 \to 2^{X_2^*}$ are strict multivalued maps of $D(L)_{\lambda_0}$-pseudomonotone type with nonempty, convex, closed, bounded values, X_1, X_2 are Banach spaces continuously embedded in some Hausdorff linear topological space, $X = X_1 \cap X_2$, $L : D(L) \subset X \to X^*$ is linear, monotone, closed, densely defined operator with a linear definitional domain $D(L)$. Moreover, we prove the important properties for this spaces. We consider constructions that guarantee the convergence of Faedo–Galerkin method for differential-operator inclusions with w_{λ_0}-pseudomonotone maps. The compact embedding Theorems for irreflexive spaces together with results from the Sect. 1.2 allow us to consider the appreciably wider class of problems. The outcomes reduced with detailed proofs, because the given classes of spaces, in most cases, were not considered yet.

In the following referring to Banach spaces X, Y, when we write

$$X \subset Y$$

we mean the embedding in the set-theory sense and in the topological sense.

For the very beginning let us consider ideas of the sum and intersection of Banach spaces required for studying of anisotropic problems.

For $n \geq 2$ let $\{X_i\}_{i=1}^n$ be some family of Banach spaces.

Definition 1.1. *The interpolation family* refers a family of Banach spaces $\{X_i\}_{i=1}^n$ such that for some linear topological space (LTS) Y we have

$$X_i \subset Y \quad \text{for all} \quad i = \overline{1,n}.$$

As $n = 2$ the interpolation family is called *the interpolation pair.*

Further let $\{X_i\}_{i=1}^n$ be some interpolation family. On the analogy of [GGZ74, p.23], in the linear variety $X = \cap_{i=1}^n X_i$ we consider the norm

$$\|x\|_X := \sum_{i=1}^n \|x\|_{X_i} \qquad \forall x \in X, \tag{1.2}$$

where $\|\cdot\|_{X_i}$ is the norm in X_i.

Proposition 1.1. *Let $\{X, Y, Z\}$ be an interpolation family. Then*

$$X \cap (Y \cap Z) = (X \cap Y) \cap Z = X \cap Y \cap Z, \quad X \cap Y = Y \cap X$$

both in the sense of equality of sets and in the sense of equality of norms.

We also consider the linear space

$$Z := \sum_{i=1}^n X_i = \left\{ \sum_{i=1}^n x_i \;\middle|\; x_i \in X_i,\ i = \overline{1,n} \right\}$$

with the norm

$$\|z\|_Z := \inf \left\{ \max_{i=\overline{1,n}} \|x_i\|_{X_i} \;\middle|\; x_i \in X_i,\ \sum_{i=1}^n x_i = z \right\} \qquad \forall z \in Z. \tag{1.3}$$

Proposition 1.2. *Let $\{X_i\}_{i=1}^n$ be an interpolation family. Then $X = \cap_{i=1}^n X_i$ and $Z = \sum_{i=1}^n X_i$ are Banach spaces and it results in*

$$X \subset X_i \subset Z \quad \textit{for all} \quad i = \overline{1,n}. \tag{1.4}$$

Proof. Since X is a linear space, from properties of $\|\cdot\|_{X_i}$ and from the definition of $\|\cdot\|_X$ on X it follows that $\|\cdot\|_X$ is the norm on X.

Let us prove the completeness of X. From the definition of $\|\cdot\|_X$ on X it follows that every Cauchy sequence $\{x_n\}_{n\geq 1}$ in X is fundamental, so it converges in X_i and in Y $\forall\, i=\overline{1,n}$, where Y is the LTS in the Definition 1.1. Hence, due to $\{X_i\}_{i=1}^n$ is the interpolation family and to the uniqueness of the limit of a sequence $\{x_n\}_{n\geq 1}$ in LTS Y it follows that for some $x\in X$ and for all $i=\overline{1,n}$

$$x_n \to x \quad \text{in} \quad X_i \quad \text{as} \quad n\to\infty.$$

So, $x_n \to x$ in X as $n\to\infty$.

Now let us check that $\|\cdot\|_Z$ is the norm on Z.

If $\|z\|_Z=0$, then thanks to (1.3) for each $m\geq 1$ there exists $x_{mi}\in X_i$ $(i=\overline{1,n})$ such that

$$z=\sum_{i=1}^n x_{mi}, \qquad \|x_{mi}\|_{X_i}<\frac{1}{n}.$$

For every $i=\overline{1,n}$ the sequence x_{mi} tends to $\bar{0}$ in X_i, and so in Y too. Thus $\sum_{i=1}^n x_{mi} \to \bar{0}$ in Y as $m\to+\infty$ and $z=\bar{0}$. On the other hand, let $z=\bar{0}$. Then $\|z\|_Z\leq \max_{i=\overline{1,n}}\|\bar{0}\|_{X_i}=0$.

The another norm properties for $\|\cdot\|_Z$ follow from the properties of inf, max and norms $\|\cdot\|_{X_i}$, $i=\overline{1,n}$.

Let us check that the linear space Z under the above norm is a Banach space. Let $\{z_m\}_{m\geq 1}$ be a Cauchy sequence in Z. It contains a subsequence $\{z_{m_k}\}_{k\geq 1}$ with the property

$$\|z_{m_k}-z_{m_{k-1}}\|_Z<2^{-k} \quad \text{for} \quad k\geq 2.$$

From (1.3) for every $k\geq 2$ there exists

$$z_{m_k}-z_{m_{k-1}}=\sum_{j=1}^n u_{kj},$$

where $u_{kj}\in X_j$, $\|u_{kj}\|_X<2^{1-k}$ for each $j=\overline{1,n}$ and $k\geq 2$. Further,

$$z_{m_1}=\sum_{j=1}^n u_{1j}, \quad u_{1j}\in X_j, \quad j=\overline{1,n}.$$

For every $k\geq 1$ let us set

$$x_{kj}=\sum_{i=1}^k u_{ij}, \quad j=\overline{1,n}.$$

Hence

$$z_{m_k} = \sum_{j=1}^{n} x_{kj} \qquad \forall k \geq 1.$$

For all $j = \overline{1,n}$ the sequence x_{kj} converges in X_j (according to its construction) to some $x_j \in X_j$. Let us set $z = \sum\limits_{j=1}^{n} x_j$. Then we have

$$\|z - z_{m_k}\|_Z \leq \max_{j=\overline{1,n}} \|x_j - x_{kj}\|_{X_j} \quad \forall k \geq 1.$$

From here it follows that z_{m_k} converges to z in Z as $k \to +\infty$. From the estimation

$$\|z - z_m\|_Z \leq \|z - z_{m_k}\|_Z + \|z_{m_k} - z_m\|_Z$$

and taking into account that the sequence $\{z_m\}_{m\geq 1}$ is fundamental we obtain

$$\lim_{m\to\infty} \|z - z_m\|_Z = 0.$$

The embedding (1.4) follows from the definition of Banach spaces $(X, \|\cdot\|_X)$ and $(Z, \|\cdot\|_Z)$. □

Remark 1.1. ([GGZ74, p.24]) Let Banach spaces X and Y satisfy the following conditions

$$X \subset Y, \qquad X \quad \text{is dense in} \quad Y,$$
$$\|x\|_Y \leq \gamma \|x\|_X \quad \forall x \in X, \qquad \gamma = const.$$

Then

$$Y^* \subset X^*, \qquad \|f\|_{X^*} \leq \gamma \|f\|_{Y^*} \quad \forall f \in Y^*.$$

Moreover, if X is reflexive, then Y^* is dense in X^*.

Let $\{X_i\}_{i=1}^n$ be an interpolation family such that the space $X := \cap_{i=1}^n X_i$ with the norm (1.2) is dense in X_i for all $i = \overline{1,n}$. Due to Remark 1.1 the space X_i^* may be considered as subspace of X^*. Thus we can construct $\sum_{i=1}^n X_i^*$ and

$$\sum_{i=1}^{n} X_i^* \subset \left(\bigcap_{i=1}^{n} X_i \right)^* . \tag{1.5}$$

Under the given assumptions X is dense in $Z := \sum\limits_{i=1}^{n} X_i$ for every $i = \overline{1,n}$. So X_i is dense in Z too. Thanks to Remark 1.1 we can consider space Z^* as a subspace of X_i^* for all $i = \overline{1,n}$, and also as a subspace of $\cap_{i=1}^n X_i^*$, i.e.

$$\left(\sum_{i=1}^{n} X_i \right)^* \subset \bigcap_{i=1}^{n} X_i^* . \tag{1.6}$$

Proposition 1.3. *Let $\{X_i\}_{i=1}^n$ be an interpolation family such that the space $X := \cap_{i=1}^n X_i$ with the norm (1.2) is dense in X_i for all $i = \overline{1,n}$. Then*

$$\sum_{i=1}^n X_i^* = \left(\bigcap_{i=1}^n X_i\right)^*$$

and

$$\left(\sum_{i=1}^n X_i\right)^* = \bigcap_{i=1}^n X_i^*$$

both in the sense of sets equality and in the sense of the equality of norms.

Proof. We consider the space $\mathscr{X} := \prod_{i=1}^n X_i$ with the norm

$$\|\{x_1, x_2, \dots, x_n\}\|_{\mathscr{X}} = \sum_{i=1}^n \|x_i\|_{X_i} \qquad \forall x = \{x_1, x_2, \dots, x_n\} \in \mathscr{X};$$

let $\mathscr{L}$ be the subspace of $\mathscr{X}$ defined by

$$\mathscr{L} = \{\{x, x, \dots, x\} \mid x \in X\}.$$

For a fixed $f \in X^*$ let us set

$$u(\{x, x, \dots, x\}) = f(x) \quad \forall x \in X.$$

Hence u is a linear functional on $\mathscr{L}$ with the norm $\|u\|_{\mathscr{L}^*} = \|f\|_{X^*}$. By the Hahn–Banach Theorem for the functional u there exists a linear functional v defined on $\mathscr{X}$ such that

$$\|v\|_{\mathscr{X}} = \|u\|_{\mathscr{L}^*} = \|f\|_{X^*}.$$

For every $i = \overline{1,n}$ we set

$$g_i(x) = v(\{\bar{0}, \dots, \bar{0}, x_i, \bar{0}, \dots, \bar{0}\}) \qquad \forall x_i \in X_i.$$

Hence it is clear that $g_i \in X_i^*$ for all $i = \overline{1,n}$ and

$$\max_{i=\overline{1,n}} \|g_i\|_{X_i^*} \le \|v\|_{Z^*} = \|f\|_{X^*}.$$

By the construction,

$$f(x) = \sum_{i=1}^n g_i(x) \qquad \forall x \in X,$$

i.e. $f = \sum_{i=1}^n g_i \in \sum_{i=1}^n X_i^*$. Thus it follows

$$\|f\|_{\sum_{i=1}^n X_i^*} \leq \max_{i=\overline{1,n}} \|g_i\|_{X_i^*} \leq \|f\|_{X^*}.$$

On the other hand $\|f\|_{X^*} = \sup_{\sum_{i=1}^n \|x\|_{X_i}=1} f(x)$

$$\leq \sup_{\sum_{i=1}^n \|x\|_{X_i}=1} \inf\left\{ \sum_{i=1}^n \|g_i\|_{X_i^*}\|x\|_{X_i} \,\middle|\, g_i \in X_i^*, \sum_{i=1}^n g_i = f \right\}$$

$$\leq \inf\left\{ \max_{i=\overline{1,n}} \|g_i\|_{X_i^*} \,\middle|\, g_i \in X_i^*, \sum_{i=1}^n g_i = f \right\} = \|f\|_{\sum_{i=1}^n X_i^*}.$$

The latest inequalities and (1.5) prove the first part of the Theorem.

Let us prove the remaining part.

Lemma 1.1. *Let $f \in \cap_{i=1}^n X_i^*$. Then for every $k = \overline{2,n}$ and $x_i,\ y_i \in X_i$ $(i = \overline{1,k})$ such that $\sum_{i=1}^k x_i = \sum_{i=1}^k y_i =: x$ we have*

$$\sum_{i=1}^k f(x_i) = \sum_{i=1}^k f(y_i) =: f(x). \tag{1.7}$$

Proof. We prove this Proposition arguing by induction.

Let $x_i, y_i \in X_i$ $(i = 1, 2)$ such that $x_1 + x_2 = y_1 + y_2 =: x$. Then $x_1 - y_1 = y_2 - x_2 \in X_1 \cap X_2$ and

$$f(x_1) - f(y_1) = f(x_1 - y_1) = f(y_2 - x_2) = f(y_2) - f(x_2).$$

From the last the necessary Proposition is follows.

Now we assume that for some $k = \overline{2, n-1}$ and for arbitrary $x_i,\ y_i \in X_i$ $(i = \overline{1,k})$ such that $\sum_{i=1}^k x_i = \sum_{i=1}^k y_i =: x$ equality (1.7) is valid.

Let $x_i,\ y_i \in X_i$ $(i = \overline{1, k+1})$ such that $\sum_{i=1}^{k+1} x_i = \sum_{i=1}^{k+1} y_i =: x$. Thus

$$x_{k+1} - y_{k+1} = \sum_{i=1}^k (y_i - x_i) \in \left(\sum_{i=1}^k X_i\right) \cap X_{k+1},$$

and so, by the induction assumption, we obtain

$$f(x_{k+1}) - f(y_{k+1}) = f(x_{k+1} - y_{k+1}) = f\left(\sum_{i=1}^k (y_i - x_i)\right) = \sum_{i=1}^k (f(y_i) - f(x_i))$$

and the Lemma follows. □

According to Lemma 1.1 let us continue any fixed functional $f \in \cap_{i=1}^n X_i^*$ to some functional on Z in such way:

for $z = \sum_{i=1}^n x_i$, where $x_i \in X_i\ \forall i = \overline{1,n}$

$$f(z) = \sum_{i=1}^n f(x_i).$$

From relation (1.7) it follows that the given definition is correct and does not depend on the representation of z as $\sum_{i=1}^n x_i$. Since

$$f(z) \le \inf\left\{ \sum_{i=1}^n \|f\|_{X_i^*}\|x_i\|_{X_i} \;\middle|\; x_i \in X_i,\ \sum_{i=1}^n x_i = z \right\} \le \left(\sum_{i=1}^n \|f\|_{X_i^*} \right) \|z\|_Z,$$

then $f \in Z^*$ and $\|f\|_{Z^*} \le \|f\|_{\cap_{i=1}^n X_i^*}$. Taking into account (1.6) we have $Z^* = \cap_{i=1}^n X_i^*$ as equality of the sets. In order to prove the equality of norms it is sufficient to show the inequality $\|f\|_{\cap_{i=1}^n X_i^*} \le \|f\|_{Z^*}$. For every $\varepsilon > 0$ there exists $x_i \in X_i$ such that

$$\|f\|_{X_i^*} \le f(x_i) + \varepsilon/n, \quad \|x_i\|_{X_i} = 1.$$

Hence

$$\begin{aligned}\|f\|_{\cap_{i=1}^n X_i^*} &= \sum_{i=1}^n \|f\|_{X_i^*} \le f\left(\sum_{i=1}^n x_i \right) + \varepsilon \le \|f\|_{Z^*} \left\| \sum_{i=1}^n x_i \right\|_Z + \varepsilon \\ &\le \|f\|_{Z^*} \max_{i=\overline{1,n}} \|x_i\|_{X_i} + \varepsilon = \|f\|_{Z^*} + \varepsilon\end{aligned}$$

and from here the delivered conclusion follows. □

Theorem 1.1. ***(The reflexivity criterium: [PET67, p. 57])*** *Banach space E is reflexive if and only if each bounded in E sequence contains the weakly convergent in E subsequence.*

Theorem 1.2. ***(Banach–Alaoglu: [GGZ74, p. 30])*** *In reflexive Banach space each bounded sequence contains the weakly convergent subsequence.*

Now let us consider classes of integrable by Bochner distributions with values in Banach space. Such extended phase spaces generally appear when studying evolutional geophysical processes and fields.

Now let Y be some Banach space, Y^* its topological adjoint space, S be some compact time interval. We consider the classes of functions defined on S and imagines in Y (or in Y^*).

The set $L_p(S;Y)$ of all measured by Bochner functions (see [GGZ74]) as $1 \le p \le +\infty$ with the natural linear operations with the norm

$$\|y\|_{L_p(S;Y)} = \left(\int_S \|y(t)\|_Y^p dt \right)^{1/p}$$

is a Banach space. As $p = +\infty$ $L_\infty(S;Y)$ with the norm

$$\|y\|_{L_\infty(S;Y)} = \operatorname{vrai\,max}_{t \in S} \|y(t)\|_Y$$

is a Banach space.

The next Theorem shows that under the assumption of reflexivity and separability of Y the adjoint with $L_p(S;Y)$, $1 \le p < +\infty$, space $(L_p(S;Y))^*$ may be identified with $L_q(S;Y^*)$, where q is such that $p^{-1} + q^{-1} = 1$.

Theorem 1.3. *If the space Y is reflexive and separable and $1 \le p < +\infty$, then each element $f \in (L_p(S;Y))^*$ has the unique representation*

$$f(y) = \int_S \langle \xi(t), y(t) \rangle_Y dt \quad \textit{for every } y \in L_p(S;Y)$$

with the function $\xi \in L_q(S;Y^*)$, $p^{-1} + q^{-1} = 1$. The correspondence $f \to \xi$, with $f \in (L_p(S;Y))^*$ is linear and

$$\|f\|_{(L_p(S;Y))^*} = \|\xi\|_{L_q(S;Y^*)}.$$

Now let us consider the reflexive separable Banach space V with the norm $\|\cdot\|_V$ and the Hilbert space $(H, (\cdot,\cdot)_H)$ with the norm $\|\cdot\|_H$, and for the given spaces let the next conditions be satisfied

$$\begin{gathered} V \subset H, \quad V \text{ is dense in } H, \\ \exists \gamma > 0 : \|v\|_H \le \gamma \|v\|_V \quad \forall v \in V. \end{gathered} \tag{1.8}$$

Due to Remark 1.1 under the given assumptions we may consider the adjoint with H space H^* as a subspace of V^* that is adjoint with V. As V is reflexive then H^* is dense in V^* and

$$\|f\|_{V^*} \le \gamma \|f\|_{H^*} \quad \forall f \in H^*,$$

where $\|\cdot\|_{V^*}$ and $\|\cdot\|_{H^*}$ are the norm in V^* and in H^*, respectively.

Further, we identify the spaces H and H^*. Then we obtain

$$V \subset H \subset V^*$$

with continuous and dense embedding.

Definition 1.2. The triple of spaces $(V; H; V^*)$, that satisfy the latter conditions will be called *the evolution triple*.

Let us point out that under identification H with H^* and H^* with some subspace of V^*, an element $y \in H$ is identified with some $f_y \in V^*$ and

$$(y, x) = \langle f_y, x \rangle_V \quad \forall x \in V,$$

where $\langle \cdot, \cdot \rangle_V$ is the canonical pairing between V^* and V. Since the element y and f_y are identified then, under condition (1.8), the pairing $\langle \cdot, \cdot \rangle_V$ will denote the inner product on H $(\cdot, \cdot)$.

We consider p_i, r_i, $i = 1, 2$ such that $1 < p_i \le r_i \le +\infty$, $p_i < +\infty$. Let $q_i \ge r_i' \ge 1$ well-defined by

$$p_i^{-1} + q_i^{-1} = r_i^{-1} + r_i'^{-1} = 1 \qquad \forall i = 1, 2.$$

Remark that $1/\infty = 0$.

Now we consider some Banach spaces that play an important role in the investigation of the differential-operator equations and evolution variation inequalities in nonreflexive Banach spaces.

Referring to evolution triples $(V_i; H; V_i^*)$ $(i = 1, 2)$ such that

$$\text{the set } V_1 \cap V_2 \text{ is dense in the spaces } V_1,\ V_2 \text{ and } H \tag{1.9}$$

we consider the functional Banach spaces (Proposition 1.2)

$$X_i = X_i(S) = L_{q_i}(S; V_i^*) + L_{r_i'}(S; H), \quad i = 1, 2$$

with norms

$$\|y\|_{X_i} = \inf\Big\{ \max\big\{ \|y_1\|_{L_{q_i}(S;V_i^*)};\ \|y_2\|_{L_{r_i'}(S;H)} \big\} \ \Big| \\ \Big|\ y_1 \in L_{q_i}(S; V_i^*), y_2 \in L_{r_i'}(S; H),\ y = y_1 + y_2 \Big\},$$

for all $y \in X_i$, and

$$X = X(S) = L_{q_1}(S; V_1^*) + L_{q_2}(S; V_2^*) + L_{r_2'}(S; H) + L_{r_1'}(S; H)$$

with

$$\|y\|_X = \inf\{ \max_{i=1,2} \{ \|y_{1i}\|_{L_{q_i}(S;V_i^*)};\ \|y_{2i}\|_{L_{r_i'}(S;H)} \} \mid y_{1i} \in L_{q_i}(S; V_i^*), \\ y_{2i} \in L_{r_i'}(S; H),\ i = 1, 2;\ y = y_{11} + y_{12} + y_{21} + y_{22} \},$$

for each $y \in X$. As $r_i < +\infty$, due to Theorem 1.3 and to Theorem 1.3 the space X_i is reflexive. Analogously, if $\max\{r_1, r_2\} < +\infty$, the space X is reflexive.

Under the latter Theorems we identify the adjoint with $X_i(S)$, $X_i^* = X_i^*(S)$, with $L_{r_i}(S; H) \cap L_{p_i}(S; V_i)$, where

$$\|y\|_{X_i^*} = \|y\|_{L_{r_i}(S;H)} + \|y\|_{L_{p_i}(S;V_i)} \quad \forall y \in X_i^*,$$

and, respectively, the adjoint with $X(S)$ space $X^* = X^*(S)$ we identify with

$$L_{r_1}(S; H) \cap L_{r_2}(S; H) \cap L_{p_1}(S; V_1) \cap L_{p_2}(S; V_2),$$

where

$$\|y\|_{X^*(S)} = \|y\|_{L_{r_1}(S;H)} + \|y\|_{L_{r_2}(S;H)} + \|y\|_{L_{p_1}(S;V_1)} + \|y\|_{L_{p_2}(S;V_2)} \quad \forall y \in X^*.$$

On $X(S) \times X^*(S)$ we denote by

$$\begin{aligned}\langle f, y\rangle = \langle f, y\rangle_S &= \int_S (f_{11}(\tau), y(\tau))_H \, d\tau + \int_S (f_{12}(\tau), y(\tau))_H \, d\tau \\ &+ \int_S \langle f_{21}(\tau), y(\tau)\rangle_{V_1} \, d\tau + \int_S \langle f_{22}(\tau), y(\tau)\rangle_{V_2} \, d\tau \\ &= \int_S (f(\tau), y(\tau)) \, d\tau \quad \forall f \in X, \ y \in X^*,\end{aligned}$$

where $f = f_{11} + f_{12} + f_{21} + f_{22}$, $f_{1i} \in L_{r_i'}(S; H)$, $f_{2i} \in L_{q_i}(S; V_i^*)$, $i = 1, 2$.

In that case when $\max\{r_1, r_2\} < +\infty$, with corresponding to the "standard" denotations [GGZ74, p. 171], the spaces X^*, X_1^* and X_2^*, further will denote as X, X_1 and X_2 respectively, and vice verse, X, X_1 and X_2 as X^*, X_1^* and X_2^* respectively. The given renotation is correct in virtue of the next Proposition, that is the direct corollary of Proposition 1.3 and of Theorem 1.3.

Proposition 1.4. *As* $\max\{r_1, r_2\} < +\infty$, *the spaces* X, X_1 *and* X_2 *are reflexive.*

Let $\mathscr{D}(S)$ be the space of the principal functions on S. For a Banach space X as $\mathscr{D}^*(S; X)$ we will denote the family of all linear continuous maps from $\mathscr{D}(S)$ into X, with the weak topology. The elements of the given space are called the distributions on S with values in X. For each $f \in \mathscr{D}^*(S; X)$ the generalized derivative is well-defined by the rule

$$f'(\varphi) = -f(\varphi') \quad \forall \varphi \in \mathscr{D}(S).$$

We remark that each locally integrable in Bochner sense function u we can identify with corresponding distribution $f_u \in \mathscr{D}^*(S; X)$ in such way:

$$f_u(\varphi) = u(\varphi) = \int_S u(t)\varphi(t)dt \quad \forall \varphi \in \mathscr{D}(S), \tag{1.10}$$

where the integral is regard in the Bochner sense. We will interpret the family of all locally Bochner integrable functions from $(S \to X)$ as subspace in $\mathscr{D}^*(S; X)$. Thus the distributions, that allow the representation (1.10), we will consider as functions from $(S \to X)$. We are also remark that the correspondence $\mathscr{D}^*(S;X) \ni f \to f' \in \mathscr{D}^*(S;X)$ is continuous [GGZ74, p.169].

Definition 1.3. [GGZ74, p.146] As $C^m(S;X)$, $m \geq 0$ we refer the family of all functions from $(S \to X)$, that have the continuous derivatives by the order m inclusively. In that case when S is a compact interval $C^m(S;X)$ is a Banach space with the norm

$$\|y\|_{C^m(S;X)} = \sum_{i=0}^{m} \sup_{t \in S} \left\| y^{(i)}(t) \right\|_X,$$

where $y^{(i)}(t)$ is the strong derivative from y at the point $t \in S$ by the order $i \geq 1$; $y^{(0)} \equiv y$.

Let $V = V_1 \cap V_2$. Then $V^* = V_1^* + V_2^*$. At studying the differential-operator inclusions and evolution variation inequalities together with the spaces X and X^* one more space, which we will denote as $W^* = W^*(S)$ plays the important role. Let us set

$$W^*(S) = \{y \in X^*(S) |\ y' \in X(S)\},$$

where the derivative y' from $y \in X^*$ is considered in the sense of scalar distributions space $\mathscr{D}^*(S;V^*)$.

The class W^* generally contains the generalized solutions of the first order differential-operator equations and inclusions with maps of pseudomonotone type. By analogy with Sobolev spaces it is required to study some structured properties, embedding and approximations theorems as well as some "rules of work" with elements of such spaces.

Theorem 1.4. *The set W^* with the natural operations and graph norm for y':*

$$\|y\|_{W^*} = \|y\|_{X^*} + \|y'\|_X \quad \forall y \in W^*$$

is Banach space.

Proof. The norm axioms for $\|\cdot\|_{W^*}$ directly follow from its definition. The completeness of W^* concerning the just defined norm follows from the next reasonings. Let $\{y_n\}_{n\geq 1}$ is the Cauchi sequence in W^*. Then, in virtue of the completeness of X and X^* it follows that for some $y \in X^*$ and $v \in X$

$$y_n \to y \text{ in } X^* \quad \text{and} \quad y_n' \to v \text{ in } X \quad \text{as } n \to +\infty.$$

Therefore from [GGZ74, Lemma IV.1.10] and from the continuous dependence of the derivative on the distribution in $\mathscr{D}^*(S;V^*)$ [GGZ74, p.169] it follows that $y' = v \in X$. □

Together with $W^* = W^*(S)$ we consider Banach space

$$W_i^* = W_i^*(S) = \{y \in L_{p_i}(S;V_i) |\ y' \in X(S)\}, \quad i = 1,2,$$

with the norm

$$\|y\|_{W_i^*} = \|y\|_{L_{p_i}(S;V_i)} + \|y'\|_X \quad \forall y \in W_i^*.$$

We also consider the space $W_0^* = W_0^*(S) = W_1^*(S) \cap W_2^*(S)$ with the norm

$$\|y\|_{W_0^*} = \|y\|_{L_{p_1}(S;V_1)} + \|y\|_{L_{p_2}(S;V_2)} + \|y'\|_X \quad \forall y \in W_0^*.$$

The space W^* is continuously embedded in W_i^* for $i = \overline{0,2}$.

Theorem 1.5. *It results in $W_i^* \subset C(S;V^*)$ with continuous embedding for $i = \overline{0,2}$.*

Proof. Let $i = 1,2$ be fixed, $y \in W_i^*$ and $\forall t_0, t \in S$ we set $\xi(t) = \int\limits_{t_0}^{t} y'(\tau)d\tau$ which has sense in the virtue of the local integrability of y'. It is obvious that

$$\|\xi(t) - \xi(s)\|_{V^*} \le \int\limits_t^s \|y'(\tau)\|_{V^*} d\tau \ \ \forall\, s \ge t$$

from which follows $\xi \in C(S;V^*)$. Then $\xi' = y'$, it means that $y(t) = \xi(t) + z$ for a.e. $t \in S$ and some $z \in V^*$. Therefore, the function y also belongs to $C(S;V^*)$. Note, that S is compact. Then in virtue of $X \subset L_1(S;V^*)$ we obtain

$$\|\xi(t)\|_{V^*} \le \int\limits_S \|y'(\tau)\|_{V^*} d\tau \le k\|y'\|_X \quad \forall t \in S. \tag{1.11}$$

Then due to the continuity of embedding $V_i \subset V^*$ we have

$$\begin{aligned} \|z\|_{V^*}(\text{mes}(S))^{1/p_i} = \Big(\int\limits_S \|z\|_{V^*}^{p_i} ds\Big)^{1/p_i} &= \|y - \xi\|_{L_{p_i}(S;V^*)} \\ &\le k_1\Big(\|y\|_{L_{p_i}(S;V^*)} + \|\xi\|_{C(S;V^*)}\Big) \\ &\le k_2\Big(\|y\|_{L_{p_i}(S;V_i)} + \|y'\|_X\Big), \end{aligned} \tag{1.12}$$

where k_2 does not depend on $y \in W_i^*$.

Now let $y \in W_0^* \subset C(S;V^*)$. In virtue of (1.11)–(1.12) for $i = 1,2$ there exists $k_3 \ge 0$ that $\|y\|_{C(S;V^*)} \le k_3\|y\|_{W_0}$ for all $y \in W_0^*$. □

Remark 1.2. From the definition of norms in the spaces W^* and W_0^* we obtain $W^* \subset C(S;V^*)$ with continuous embedding for the compact S in the natural topology of the space W^*.

Theorem 1.6. *The set $C^1(S;V)\cap W_0^*$ is dense in W_0^*.*

Proof. We prove this Proposition for more general case. At the beginning we suppose $S=\mathbb{R}$. Let us choose such a function $K\in C_0^\infty(S)$ that $\int\limits_S K(\tau)d\tau=1$ and use the Sobolev mid-value method. Let us set for definiteness

$$K(\tau)=\begin{cases}\mu\exp\left\{-\frac{\tau^2}{\tau^2-1}\right\} & \text{for } |\tau|\le 1,\\ 0 & \text{for } |\tau|>1,\end{cases}$$

where μ is the constant of normalization and suppose $K_n(\tau)=nK(n\tau)$ for every $\tau\in S$ and $n\ge 1$. It is obvious that $K_n\in C_0^\infty(S)$ and

$$\int\limits_S K_n(\tau)d\tau=1\ \ \forall\, n\ge 1.$$

For every $y\in W_0^*$ let us define the sequence of functions

$$y_n(t)=\int\limits_S K_n(t-\tau)y(\tau)d\tau. \tag{1.13}$$

It is easy to check that $y_n\in C^1(S;V)$ and

$$y_n'=\int\limits_S K_n'(t-\tau)y(\tau)d\tau=\int\limits_S K_n(t-\tau)y'(\tau)d\tau. \tag{1.14}$$

Besides $y_n\in L_{p_i}(S;V_i)$ and $y_n\to y$ in $L_{p_i}(S;V_i)$ for $(i=1,2)$. The last follows from the inequality $\|y_n\|_{L_{p_i}(S;V_i)}\le\|K\|_{L_1(S)}\|y\|_{L_{p_i}(S;V_i)}$ and from following estimations:

$$\begin{aligned}\|y_n-y\|^{p_i}_{L_{p_i}(S;V_i)}&=\int\limits_S\left\|\int\limits_{t-1/n}^{t+1/n}K_n(t-\tau)(y(\tau)-y(t))d\tau\right\|^{p_i}_{V_i}dt\\
&\le\int\limits_S\left(\int\limits_{-1/n}^{1/n}|K_n(s)|\,\|y(t+s)-y(t)\|_{V_i}ds\right)^{p_i}dt\\
&\le\int\limits_S\left\{\left(\int\limits_{-1/n}^{1/n}|K_n(s)|^{q_i}ds\right)^{p_i/q_i}\int\limits_{-1/n}^{1/n}\|y(t+s)-y(t)\|^{p_i}_{V_i}ds\right\}dt\\
&\le\frac{n}{2}(2\mu)^{p_i}\int\limits_{-1/n}^{1/n}\left(\int\limits_S\|y(t+s)-y(t)\|^{p_i}_{V_i}dt\right)ds.\end{aligned}$$

Pointing out that for arbitrary $y \in L_{p_i}(S; V_i)$ $(1 \le p_i < \infty)$ and for every h the function

$$y_h(t) = \begin{cases} y(t+h) & \text{for } t+h \in S, \\ 0 & \text{for } t+h \notin S \end{cases}$$

belongs to $L_{p_i}(S; V_i)$ and $\|y_h - y\|_{L_{p_i}(S;V_i)} \to 0$ as $h \to 0$ (see [GGZ74, Lemma IV.1.5]), then

$$\lim_{n\to\infty} \|y_n - y\|^{p_i}_{L_{p_i}(S;V_i)} \le \lim_{n\to\infty} (2\mu)^{p_i} \sup_{|s|\le 1/n} \|y_s - y\|^{p_i}_{L_{p_i}(S;V_i)} = 0 \text{ for } i = 1, 2.$$

Now we prove the convergence of derivatives. Let $y' \in X$ and $y' = \xi_1 + \xi_2 + \eta_1 + \eta_2$ where $\xi_i \in L_{q_i}(S; V_i^*)$, $\eta_i \in L_{r_i'}(S; H)$, $i = 1, 2$. By the analogy with (1.13) we suppose

$$\xi_{n,i}(t) = \int_S K_n(t-\tau)\xi_i(\tau)d\tau, \quad \eta_{n,i}(t) = \int_S K_n(t-\tau)\eta_i(\tau)d\tau \text{ for } i = 1, 2.$$

Then in virtue of (1.14) by the analogy to the previous case, $y_n' = \xi_{n,1} + \xi_{n,2} + \eta_{n,1} + \eta_{n,2}$ and besides $\xi_{n,i} \to \xi_i$ in $L_{q_i}(S; V_i^*)$ and $\eta_{n,i} \to \eta_i$ in $L_{r_i'}(S; H)$ for $i = 1, 2$. By definition of $\|\cdot\|_X$, it follows

$$\lim_{n\to\infty} \|y_n' - y'\|_X \le \lim_{n\to\infty} \max\{\|\xi_{n,1} - \xi\|_{L_{q_1}(S;V_1^*)}; \|\xi_{n,2} - \xi\|_{L_{q_2}(S;V_2^*)};$$
$$\|\eta_{n,1} - \eta\|_{L_{r_1'}(S;H)}; \ \|\eta_{n,2} - \eta\|_{L_{r_2'}(S;H)}\} = 0.$$

From here we conclude that for every $n \ge 1$ $y_n \in C^1(S; V) \cap W_0^*$ and the sequence $\{y_n\}_{n\ge1}$ converges to $y \in W_0^*$ in W_0^*.

Now let us consider the case when S is semibounded. Without loss of generality we suppose $S = [0, \infty)$. For $y \in W_0^* = W_0^*(S)$ we put $y_h(t) = y(t+h)$ for every $h > 0$. Then, in virtue of [GGZ74, Lemma IV.1.5] it is easy to show that for $i = 1, 2$ $y_h \to y$ in $L_{p_i}(S; V_i)$ and $y_h' \to y'$ in X as $h \to 0+$. Remark that $y_h \in W_0^*$. To complete the proof it is sufficient to show that for every fixed $y \in W_0^*(S)$ and for $h > 0$ the element $y_h \in W_0^*$ can be sufficiently exactly approximated by the functions from $C^1(S; V) \cap W_0^*$.

For some $y \in W_0^*(S)$ and $h > 0$ let us consider the function

$$\xi(t) = \begin{cases} \varphi(t)y(t+h) & \text{for } t \ge -h, \\ 0 & \text{for } t < -h, \end{cases}$$

where $\varphi \in C^1(\mathbb{R})$, $\varphi(t) = 1$ if $t \ge -\frac{h}{2}$ and $\varphi(t) = 0$ if $t < -h$. Then for every $t \ge 0$ $\xi(t) = y_h(t)$ and due to the definition of derivative in sense of scalar distribution space $D^*(S; V^*)$ it follows that

$$\xi'(t) = \begin{cases} \varphi'(t)y(t+h) + \varphi(t)y'(t+h) & \text{for } t \ge -h, \\ 0 & \text{for } t < -h. \end{cases}$$

Let us prove that $\xi \in W_0^*(\mathbb{R})$. Since $y_h \in W_0^*(S)$ we have $\xi|_{[0,\infty)} \in X^*(S)$. Because of $\xi|_{(-\infty;-h)} = 0$ it remains to consider the section $[-h, 0)$.

From $\sup\limits_{s\in[-h,0)} |\varphi(s)| = 1$ we have

$$\int_{-h}^{0} \|\xi(s)\|_{V_i}^{p_i} ds \leq \int_{-h}^{0} |\varphi(s)|^{p_i} \|y(s+h)\|_{V_i}^{p_i} ds$$
$$\leq \int_{-h}^{0} \|y(s+h)\|_{V_i}^{p_i} ds = \int_{0}^{h} \|y(\tau)\|_{V_i}^{p_i} d\tau \quad (i = 1, 2).$$

Thus, $\xi|_{[-h,0)} \in L_{p_i}([-h,0); V_i)$ for $i = 1, 2$. Similarly we can prove that $\xi' \in X(\mathbb{R})$. So, $\xi \in W_0^*(\mathbb{R})$ and in virtue of the previous case there exists a sequence of elements $\xi_n \in C^1(\mathbb{R}; V) \cap W_0^*(\mathbb{R})$ that converges to ξ in $W_0^*(\mathbb{R})$.

Now we set $\zeta_n = \xi_n|_S \in C^1(S; V) \cap W_0^*(S)$ for every $n \geq 1$. Here $\zeta_n \to y_h$ in $W_0^*(S)$ as $n \to \infty$, because $\xi|_S = y_h$.

Let us consider, at last, the case when S is bounded. For every $y \in W_0^*(S)$ (where $S = [a, b]$, $a < b$) we put

$$\xi(t) = \begin{cases} \varphi(t)y(t) & \text{for } t \in [a,b], \\ 0 & \text{for } t > b, \end{cases}$$

$$\eta(t) = \begin{cases} (1-\varphi(t))y(t) & \text{for } t \in [a,b], \\ 0 & \text{for } t < a. \end{cases}$$

Let φ be such function from $C^1(S)$ that $\varphi(t) = 0$ in some neighborhood of the point b and $\varphi(t) = 1$ in some neighborhood of the point a. Note that $y(t) = \xi(t) + \eta(t)$ for all $t \in S$. It is easy to check that $\xi \in W_0^*([a,\infty))$ and $\eta \in W_0^*((-\infty, b])$. Therefore, due to the previous case, there exist such sequences

$$\{\xi_n\}_{n\geq 1} \subset C^1([a,\infty); V) \cap W_0^*([a,\infty))$$

and

$$\{\eta_n\}_{n\geq 1} \subset C^1((-\infty,b); V) \cap W_0^*((-\infty,b)),$$

that

$$\xi_n \to \xi \text{ in } W_0^*([a,\infty)) \text{ and } \eta_n \to \eta \text{ in } W_0^*((-\infty,b)) \text{ as } n \to \infty.$$

So, $(\xi_n + \eta_n)|_S \to y$ in $W_0^*(S)$.

The Theorem is proved. □

Theorem 1.7. *$W_0^* \subset C(S; H)$ with continuous embedding. Moreover, for every $y, \xi \in W_0^*$ and $s, t \in S$ the next formula of integration by parts takes place*

$$\big(y(t),\xi(t)\big)-\big(y(s),\xi(s)\big)=\int_s^t\Big\{(y'(\tau),\xi(\tau))+(y(\tau),\xi'(\tau))\Big\}d\tau. \tag{1.15}$$

In particular, when $y=\xi$ *we have:*

$$\frac{1}{2}\Big(\|y(t)\|_H^2-\|y(s)\|_H^2\Big)=\int_s^t(y'(\tau),y(\tau))d\tau. \tag{1.16}$$

Proof. To simplify the proof we consider $S=[a,b]$ for some

$$-\infty<a<b<+\infty.$$

The validity of formula (1.15) for $y,\xi\in C^1(S;V)$ is checked by direct calculation. Now let $\varphi\in C^1(S)$ be such fixed that $\varphi(a)=0$ and $\varphi(b)=1$. Moreover, for $y\in C^1(S;V)$ let $\xi=\varphi y$ and $\eta=y-\varphi y$. Then, due to (1.15):

$$\big(\xi(t),y(t)\big)=\int_a^t\Big\{\varphi'(s)\big(y(s),y(s)\big)+2\varphi(s)(y'(s),y(s))\Big\}ds,$$

$$-\big(\eta(t),y(t)\big)=\int_t^b\Big\{-\varphi'(s)\big(y(s),y(s)\big)+2(1-\varphi(s))(y'(s),y(s))\Big\}ds,$$

from here for $\xi_i\in L_{q_i}(S;V_i^*)$ and $\eta_i\in L_{r_i'}(S;H)$ $(i=1,2)$ such that $y'=\xi_1+\xi_2+\eta_1+\eta_2$ it follows:

$$\begin{aligned}
\|y(t)\|_H^2&=\int_t^b\Big\{\varphi'(s)\big(y(s),y(s)\big)+2\varphi(s)(y'(s),y(s))\Big\}ds-2\int_t^b(y'(s),y(s))ds\\
&\le\max_{s\in S}|\varphi'(s)|\cdot\|y\|_{C(S;V^*)}\cdot\|y\|_{L_1(S;V)}+2\int_S(\varphi(s)-1)(y'(s),y(s))ds\\
&\le\max_{s\in S}|\varphi'(s)|\cdot\|y\|_{C(S;V^*)}\cdot\|y\|_{L_1(S;V)}+2\max_{s\in S}|\varphi(s)-1|\\
&\quad\cdot\Big(\|\xi_1\|_{L_{q_1}(S;V_1^*)}\|y\|_{L_{p_1}(S;V_1)}+\|\xi_2\|_{L_{q_2}(S;V_2^*)}\|y\|_{L_{p_2}(S;V_2)}\\
&\quad+\|\eta_1\|_{L_{r_1'}(S;H)}\|y\|_{L_{r_1}(S;H)}+\|\eta_2\|_{L_{r_2'}(S;H)}\|y\|_{L_{r_2}(S;H)}\Big)\\
&\le\max_{s\in S}|\varphi'(s)|\cdot\|y\|_{C(S;V^*)}\cdot\Big(\|y\|_{L_{p_1}(S;V_1)}\mathrm{mes}(S)^{1/q_1}\\
&\quad+|y\|_{L_{p_2}(S;V_2)}\mathrm{mes}(S)^{1/q_2}\Big)+2\max_{s\in S}|\varphi(s)-1|
\end{aligned}$$

$$\cdot \left(\|\xi_1\|_{L_{q_1}(S;V_1^*)} + \|\xi_2\|_{L_{q_2}(S;V_2^*)} + \|\eta_1\|_{L_{r_1'}(S;H)} + \|\eta_2\|_{L_{r_2'}(S;H)} \right)$$
$$\times \left(\|y\|_{L_{p_1}(S;V_1)} + \|y\|_{L_{p_2}(S;V_2)} + \|y\|_{C(S;H)} \text{mes}(S)^{1/r_1} \right.$$
$$\left. + \|y\|_{C(S;H)} \text{mes}(S)^{1/r_2} \right).$$

Hence, due to Theorem 1.5, definition of $\|\cdot\|_X$, if we take in last inequality $\varphi(t) = \frac{t-a}{b-a}$ for all $t \in S$ we obtain

$$\|y\|^2_{C(S;H)} \leq C_2 \cdot \|y\|^2_{W_0^*} + C_3 \cdot \|y\|_{W_0^*} \cdot \|y\|_{C(S;H)}, \tag{1.17}$$

where C_1 is the constant from inequality $\|y\|_{C(S;V^*)} \leq C_1 \cdot \|y\|_{W_0^*}$ for every $y \in W_0^*$,

$$C_2 = 2 + \frac{C_1}{\min\{\text{mes}(S)^{1/p_1}, \text{mes}(S)^{1/p_2}\}}, \quad C_3 = 2 \cdot \max\left\{ \text{mes}(S)^{1/\min\{r_1, r_2\}}, 1 \right\}.$$

Remark that $\frac{1}{+\infty} = 0$, $C_2, C_3 > 0$. From (1.17) it obviously follows that

$$\|y\|_{C(S;H)} \leq C_4 \cdot \|y\|_{W_0^*} \quad \text{for all } y \in C^1(S;V), \tag{1.18}$$

where $C_4 = \frac{C_3 + \sqrt{C_3^2 + 4C_2}}{2}$ does not depend on y.

Now let us apply Theorem 1.6. For arbitrary $y \in W_0^*$ let $\{y_n\}_{n \geq 1}$ be a sequence of elements from $C^1(S;V)$ converging to y in W_0^*. Then in virtue of relation (1.18) we have

$$\|y_n - y_k\|_{C(S;H)} \leq C_4 \|y_n - y_k\|_{W_0^*} \to 0,$$

therefore, the sequence $\{y_n\}_{n \geq 1}$ converges in $C(S;H)$ and it has only limit $\chi \in C(S;H)$ such that for a.e. $t \in S$ $\chi(t) = y(t)$. So, we have $y \in C(S;H)$ and now the embedding $W_0^* \subset C(S;H)$ is proved. If we pass to limit in (1.18) with $y = y_n$ as $n \to \infty$ we obtain the validity of the given estimation $\forall y \in W_0^*$. It proves the continuity of the embedding W^* into $C(S;H)$.

Now let us prove formula (1.15). For every $y, \xi \in W_0^*$ and for corresponding approximating sequences $\{y_n, \xi_n\}_{n \geq 1} \subset C^1(S;V)$ we pass to the limit in (1.15) with $y = y_n$, $\xi = \xi_n$ as $n \to \infty$. In virtue of Lebesgue's Theorem and $W_0^* \subset C(S;V^*)$ with continuous embedding formula (1.15) is true for every $y \in W_0^*$.

The Theorem is proved. □

In virtue of $W^* \subset W_0^*$ with continuous embedding and due to the latter Theorem the next Proposition is true.

Corollary 1.1. *$W^* \subset C(S;H)$ with continuous embedding. Moreover, for every $y, \xi \in W^*$ and $s, t \in S$ formula (1.15) takes place.*

Remark 1.3. When $\max\{r_1, r_2\} < +\infty$, due to the standard denotations [GGZ74, p. 173], we will denote the space W^* as W; "$*$" will direct on nonreflexivity of the spaces X and W.

Referring to evolution triples $(V_i; H; V_i^*)$ $(i = 1, 2)$ such that

$$\text{the set } V_1 \cap V_2 \text{ is dense in the spaces } V_1,\ V_2 \text{ and } H$$

we conclude that the space $V = V_1 \cap V_2 \subset H$ with continuous and dense embedding. Since V is a separable Banach space, then there exists a complete in V and consequently in H countable vectors system $\{h_i\}_{i\geq 1} \subset V$.

Let us introduce some useful constructions which can be used for example for studying of differential-operator inclusions in infinite-dimensional spaces by the Faedo–Galerkin method.

Let for each $n \geq 1$

$$H_n = \operatorname{span}\{h_i\}_{i=1}^n,$$

on which we consider the inner product induced from H that we again denote by $(\cdot,\cdot)$; $P_n : H \to H_n \subset H$ the operator of orthogonal projection from H on H_n, i.e.

$$\forall h \in H \quad P_n h = \operatorname*{argmin}_{h_n \in H_n} \|h - h_n\|_H.$$

Definition 1.4. We say that the triple $(\{h_i\}_{i\geq 1}; V; H)$ satisfies *Condition* (γ) if $\sup\limits_{n\geq 1} \|P_n\|_{\mathscr{L}(V,V)} < +\infty$, i.e. there exists $C \geq 1$ such that

$$\forall v \in V,\ \forall n \geq 1 \quad \|P_n v\|_V \leq C \cdot \|v\|_V. \tag{1.19}$$

Remark 1.4. When the vectors system $\{h_i\}_{i\geq 1} \subset V$ is orthogonal in H, Condition (γ) means that the given system is a Schauder basis in Banach space V.

Remark 1.5. Since $P_n \in \mathscr{L}(V, V)$, its adjoint operator $P_n^* \in \mathscr{L}(V^*, V^*)$ and

$$\|P_n\|_{\mathscr{L}(V,V)} = \|P_n^*\|_{\mathscr{L}(V^*,V^*)}.$$

It is clear that, for each $h \in H$, $P_n h = P_n^* h$. Hence, we identify P_n with P_n^*. Then Condition (γ) means that for each $v \in V$ and $n \geq 1$

$$\|P_n v\|_{V^*} \leq C \cdot \|v\|_{V^*}.$$

Due to the equivalence of H^* and H it follows that $H_n^* \equiv H_n$.

Let us consider latter introduced Banach spaces:

$$X = L_{q_1}(S; V_1^*) + L_{q_2}(S; V_2^*) + L_{r_2'}(S; H) + L_{r_1'}(S; H),$$

$$X^* = L_{r_1}(S; H) \cap L_{r_2}(S; H) \cap L_{p_1}(S; V_1) \cap L_{p_2}(S; V_2),$$

$$W^* = \{y \in X \mid y' \in X^*\}$$

with corresponding norms.

For each $n \geq 1$ we consider Banach spaces

$$X_n = X_n(S) = L_{q_0}(S; H_n) \subset X, \quad X_n^* = X_n^*(S) = L_{p_0}(S; H_n) \subset X^*,$$

where $p_0 := \max\{r_1, r_2\}$, $q_0^{-1} + p_0^{-1} = 1$ with the natural norms. The space $L_{p_0}(S; H_n)$ is isometrically isomorphic to the adjoint space X_n^* of X_n, moreover, the map

$$X_n \times X_n^* \ni f, x \to \int_S (f(\tau), x(\tau))_{H_n} d\tau = \int_S (f(\tau), x(\tau)) d\tau = \langle f, x \rangle_{X_n}$$

is the duality form on $X_n \times X_n^*$. This Proposition is correct due to

$$L_{q_0}(S; H_n) \subset L_{q_0}(S; H) \subset L_{r_1'}(S; H) + L_{r_2'}(S; H) + L_{q_1}(S; V_1^*) + L_{q_2}(S; V_2^*).$$

Let us remark that $\langle \cdot, \cdot \rangle_S \big|_{X_n(S) \times X_n^*(S)} = \langle \cdot, \cdot \rangle_{X_n(S)}$.

Proposition 1.5. *For every $n \geq 1$ $X_n = P_n X$, i.e.*

$$X_n = \{P_n f(\cdot) \mid f(\cdot) \in X\}.$$

Moreover, if the triple $(\{h_j\}_{j \geq 1}; V_i; H)$, $i = 1, 2$ satisfies Condition (γ) with $C = C_i$, then

$$\text{for every } f \in X \text{ and } n \geq 1 \qquad \|P_n f\|_X \leq max\{C_1, C_2\} \cdot \|f\|_X.$$

Proof. Let us fix an arbitrary number $n \geq 1$. For every $y \in X$ let $y_n(\cdot) := P_n y(\cdot)$, i.e. $y_n(t) = P_n y(t)$ for almost all $t \in S$. Since P_n is linear and continuous on V_1^*, on V_2^* and on H we have that $y_n \in X_n \subset X$. It is immediate that the inverse inclusion is valid.

Now let us prove the second part of this Proposition. We suppose that Condition (γ) holds, let us fix $f \in X$ and $n \geq 1$. Then from Condition (γ) it follows that for every $f_{1i} \in L_{r_i'}(S; H)$ and $f_{2i} \in L_{q_i}(S; V_i^*)$ such that $f = f_{11} + f_{12} + f_{21} + f_{22}$ we have:

$$\|P_n f_{11}\|_{L_{r_1'}(S;H)} + \|P_n f_{12}\|_{L_{r_2'}(S;H)} + \|P_n f_{21}\|_{L_{q_1}(S;V_1^*)} + \|P_n f_{22}\|_{L_{q_2}(S;V_2^*)}$$

$$= \left(\int_S \|P_n f_{11}(\tau)\|_H^{r_1'} d\tau \right)^{\frac{1}{r_1'}} + \left(\int_S \|P_n f_{12}(\tau)\|_H^{r_2'} d\tau \right)^{\frac{1}{r_2'}}$$

$$+\left(\int\limits_S \|P_n f_{21}(\tau)\|_{V_1^*}^{q_1} d\tau\right)^{\frac{1}{q_1}} + \left(\int\limits_S \|P_n f_2(\tau)\|_{V_2^*}^{q_2} d\tau\right)^{\frac{1}{q_2}}$$

$$\le \left(\int\limits_S \|f_{11}(\tau)\|_H^{r_1'} d\tau\right)^{\frac{1}{r_1'}} + \left(\int\limits_S \|f_{12}(\tau)\|_H^{r_2'} d\tau\right)^{\frac{1}{r_2'}}$$

$$+ C_1\left(\int\limits_S \|f_{21}(\tau)\|_{V_1^*}^{q_1} d\tau\right)^{\frac{1}{q_1}} + C_2\left(\int\limits_S \|f_{22}(\tau)\|_{V_2^*}^{q_2} d\tau\right)^{\frac{1}{q_2}}$$

$$\le \max\{C_1, C_2\} \cdot \Big(\|f_{11}\|_{L_{r_1'}(S;H)} + \|f_{12}\|_{L_{r_2'}(S;H)}$$

$$+ \|f_{21}\|_{L_{q_1}(S;V_1^*)} + \|f_{22}\|_{L_{q_2}(S;V_2^*)}\Big),$$

because $C_1, C_2 \ge 1$. Therefore, due to the definition of norm in X we complete the proof. □

Proposition 1.6. *For every $n \ge 1$ it results in $X_n^* = P_n X^*$, i.e.*

$$X_n^* = \{P_n y(\cdot) \mid y(\cdot) \in X^*\},$$

and

$$\langle f, P_n y\rangle = \langle f, y\rangle \quad \forall y \in X^* \text{ and } f \in X_n.$$

Furthermore, if the triple $\big(\{h_j\}_{j\ge 1}; V_i; H\big)$, $i = 1,2$ satisfies Condition (γ) with $C = C_i$, then we get

$$\|P_n y\|_{X^*} \le max\{C_1, C_2\} \cdot \|y\|_{X^*} \quad \forall y \in X^* \text{ and } n \ge 1.$$

Proof. For every $y \in X^*$ we set $y_n(\cdot) := P_n y(\cdot)$, i.e. $y_n(t) = P_n y(t)$ for a.e. $t \in S$. As the operator P_n is linear and continuous on V_1, on V_2 and on H we have that $y_n \in X^*{}_n \subset X^*$. The inverse inclusion is obvious.

Due to Condition (γ) and to the definition of $\|\cdot\|_{L_{p_i}(S;V_i)}$ and $\|\cdot\|_{L_{r_i}(S;H)}$ it follows that

$$\|y_n\|_{L_{p_i}(S;V_i)} \le C_i \cdot \|y\|_{L_{p_i}(S;V_i)} \quad \text{and} \quad \|y_n\|_{L_{r_i}(S;H)} \le \|y\|_{L_{r_i}(S;H)}.$$

Thus $\|y_n\|_{X^*} \le max\{C_1, C_2\} \cdot \|y\|_{X^*}$.

Now let us show that for every $f \in X_n$

$$\langle f, y_n\rangle = \langle f, y\rangle.$$

As $f \in L_{p_0}(S; H_n)$, then we have

$$\langle f, y\rangle = \int_S (f(\tau), y(\tau))d\tau = \int_S (f(\tau), P_n y(\tau))d\tau$$
$$= \int_S (f(\tau), y_n(\tau))d\tau = \langle f, y_n\rangle,$$

because for every $n \geq 1$, $h \in H$ and $v \in H_n$ it results in

$$(h - P_n h, v) = (h - P_n h, v)_H = 0.$$

The Proposition is proved. □

When $p_0 < +\infty$ for each $n \geq 1$ we denote by I_n the canonical embedding of X_n^* in X^* ($\forall x \in X_n^*$ $I_n x = x$), $I_n^* : X \to X_n$ its adjoint operator (since X_n and X are reflexive).

Proposition 1.7. *When $p_0 < +\infty$ for each $n \geq 1$ and $f \in X$ $(I_n^* f)(t) = P_n f(t)$ for a.e. $t \in S$. Moreover, if the triples $(\{h_j\}_{j\geq 1}; V_i; H)$ as $i = 1, 2$ satisfy Condition (γ) with $C = C_i$, then*

$$\text{for each } f \in X \text{ and } n \geq 1 \qquad \|I_n^* f\|_X \leq \max\{C_1, C_2\} \cdot \|f\|_X,$$

i.e.

$$\sup_{n\geq 1} \|I_n^*\|_{\mathscr{L}(X;X)} \leq \max\{C_1, C_2\}.$$

Proof. Let $n \geq 1$ and $f \in X$ be fixed. Let us show that for a.e. $t \in S$ $(I_n^* f)(t) = P_n f(t)$. Since Remark 1.5 it follows that for each $x \in X_n^*$

$$\langle I_n^* f, x\rangle = \langle f, x\rangle = \int_S (f(\tau), x(\tau))d\tau = \int_S (f(\tau) - P_n f(\tau), x(\tau))d\tau$$
$$+ \int_S (P_n f(\tau), x(\tau))d\tau = \int_S (P_n f(\tau), x(\tau))d\tau,$$

since for all $u \in H_n$ and $v \in V^*$, $(v - P_n v, u) = 0$.

So, we are under conditions of Proposition 1.5.

The Proposition is proved. □

From the last Propositions and the properties of I_n^* it immediately follows the next

Corollary 1.2. *When $p_0 < +\infty$ for each $n \geq 1$ $X_n = P_n X = I_n^* X$, i.e.*

$$X_n = \{P_n f(\cdot) \mid f(\cdot) \in X\} = \{I_n^* f \mid f \in X\}.$$

Let us show the separation property (in some sense) for the space X^*.

Proposition 1.8. *Under the condition* $\max\{r_1, r_2\} < +\infty$ *the set* $\bigcup_{n\geq 1} X_n^*$ *is dense in* $(X^*, \|\cdot\|_{X^*})$.

Proof. (a) At first we prove that the set $L_\infty(S; V)$ is dense in space

$$(X^*, \|\cdot\|_{X^*}).$$

Let us fix $x \in X^*$. Then for every $n \geq 1$ we consider

$$x_n(t) := \begin{cases} x(t), \ \|x(t)\|_V \leq n \\ \quad 0, \text{ elsewhere.} \end{cases} \tag{1.20}$$

Obviously $\forall n \geq 1 \ x_n \in L_\infty(S; V)$. The continuous embedding of V into H assures the existence of some positive γ such that for $i = 1, 2$ and a.e. $t \in S$ we have

$$\begin{aligned} &\|x_n(t) - x(t)\|_H \leq \gamma \|x_n(t) - x(t)\|_V \to 0, \\ &\|x_n(t) - x(t)\|_{V_i} \leq \|x_n(t) - x(t)\|_V \to 0 \quad \text{as} \quad n \to \infty, \end{aligned} \tag{1.21}$$

$$\|x_n(t)\|_H \leq \|x(t)\|_H, \ \ \|x_n(t)\|_{V_i} \leq \|x(t)\|_{V_i}. \tag{1.22}$$

Further let us set

$$\phi_H^n(t) = \|x_n(t) - x(t)\|_H^{p_0}, \quad \phi_{V_i}^n(t) = \|x_n(t) - x(t)\|_{V_i}^{p_i}.$$

So, from (1.21) and (1.22) we obtain:

$$\phi_H^n(t) \to 0, \quad \phi_{V_i}^n(t) \to 0 \ \text{as}\ n \to \infty \quad \text{for a.e. } t \in S \tag{1.23}$$

and for almost every $t \in S$

$$|\phi_H^n(t)| \leq 2^{p_0} \|x(t)\|_H^{p_0} =: \phi_H(t), \quad |\phi_{V_i}^n(t)| \leq 2^{p_i} \|x(t)\|_{V_i}^{p_i} =: \phi_{V_i}(t). \tag{1.24}$$

Since $x \in X^*$, then $\phi_H, \phi_{V_1}, \phi_{V_2} \in L_1(S)$. Thus, due to (1.23) and (1.24), we can apply the Lebesgue Theorem with integrable majorants ϕ_H, ϕ_{V_1} and ϕ_{V_2} respectively. Hence it follows that $\phi_H^n \to \bar{0}$ and $\phi_{V_i}^n \to \bar{0}$ in $L_1(S)$ as $i = 1, 2$. Consequently $\|x_n - x\|_{X^*} \to 0$ as $n \to \infty$.

(b) Further, for some linear variety L from V we set

$$\Upsilon(L) := \{y \in (S \to L) \mid y \text{ is a simple function}\}$$

(see [GGZ74, p.152]). Let us show that set $\Upsilon(V)$ is dense in the normalized space $(L_\infty(S, V), \|\cdot\|_{X^*})$. Let x be fixed in $L_\infty(S, V)$; so, there exists a sequence $\{x_n\}_{n\geq 1} \subset \Upsilon(V)$ such that

$$x_n(t) \to x(t) \text{ in } V \quad \text{as}\ n \to \infty \quad \text{for a.e. } t \in S. \tag{1.25}$$

Since $x \in L_\infty(S, V)$ we have $\operatorname*{ess\,sup}_{t \in S} \|x(t)\|_V =: c < +\infty$. For every $n \geq 1$ let us introduce

$$y_n(t) := \begin{cases} x_n(t), \ \|x_n(t)\|_V \leq 2c \\ \bar{0}, \text{ else.} \end{cases} \tag{1.26}$$

From (1.25) and (1.26) it follows that $y_n \in \Upsilon(V)$ as $n \geq 1$ and moreover,

$$y_n(t) \to x(t) \text{ in } V \quad \text{as } n \to \infty \quad \text{for a.e. } t \in S.$$

Hence, taking into account $V \subset H$, as $i = 1, 2$ and for a.e. $t \in S$ we obtain the following convergences

$$y_n(t) \to x(t) \text{ in } H, \quad y_n(t) \to x(t) \text{ in } V_1, \quad y_n(t) \to x(t) \text{ in } V_2 \quad \text{as } n \to \infty.$$

As in (a), assuming

$$\phi_H \equiv \phi_{V_1} \equiv \phi_{V_2} \equiv \max\{(3c)^{p_1}, (3c)^{p_2}, (3c\gamma)^{p_0}\} \in L_1(S))$$

we obtain that $y_n \to x$ in X^* as $n \to \infty$. So, $\Upsilon(V)$ is dense in

$$(L_\infty(S, V), \|\cdot\|_{X^*}).$$

(c) Since the set $\operatorname{span}\{h_n\}_{n \geq 1} = \bigcup_{n \geq 1} H_n$ is dense in $(V, \|\cdot\|_V)$ and $V \subset H$ with continuous embedding it follows that the set

$$\Upsilon\left(\bigcup_{n \geq 1} H_n\right) = \bigcup_{n \geq 1} \Upsilon(H_n)$$

is dense in $(\Upsilon(V), \|\cdot\|_{X^*})$. In order to complete the proof we point out that for every $n \geq 1$ $\Upsilon(H_n) \subset X_n^*$.

The Proposition is proved. □

For every $n \geq 1$ let us define Banach space $W_n^* = \{y \in X_n^* | \ y' \in X_n\}$ with the norm

$$\|y\|_{W_n^*} = \|y\|_{X_n^*} + \|y'\|_{X_n},$$

where the derivative y' is considered in sense of scalar distributions space $\mathscr{D}^*(S; H_n)$. As far as

$$\mathscr{D}^*(S; H_n) = \mathscr{L}(\mathscr{D}(S); H_n) \subset \mathscr{L}(\mathscr{D}(S); V_\omega^*) = \mathscr{D}^*(S; V^*)$$

it is possible to consider the derivative of an element $y \in X_n^*$ in the sense of $\mathscr{D}^*(S; V^*)$. Remark that for every $n \geq 1$ $W_n^* \subset W_{n+1}^* \subset W^*$.

Proposition 1.9. *For every $y \in X^*$ and $n \geq 1$ $P_n y' = (P_n y)'$, where derivative of element $x \in X^*$ is in the sense of the scalar distributions space $\mathscr{D}^*(S; V^*)$.*

Remark 1.6. We pay our attention to the fact that in virtue of the previous assumptions the derivatives of an element $x \in X_n^*$ in the sense of $\mathscr{D}(S; V^*)$ and in the sense of $\mathscr{D}(S; H_n)$ coincide.

Proof. It is sufficient to show that for every $\varphi \in D(S)$ $P_n y'(\varphi) = (P_n y)'(\varphi)$. In virtue of definition of derivative in sense of $\mathscr{D}^*(S; V^*)$ we have

$$\forall \varphi \in D(S) \quad P_n y'(\varphi) = -P_n y(\varphi') = -P_n \int_S y(\tau)\varphi'(\tau)d\tau$$
$$= -\int_S P_n y(\tau)\varphi'(\tau)d\tau = -(P_n y)(\varphi') = (P_n y)'(\varphi).$$

The Proposition is proved. □

Due to Propositions 1.6, 1.5, 1.9 it follows the next

Proposition 1.10. *For every $n \geq 1$ $W_n^* = P_n W^*$, i.e.*

$$W_n^* = \{P_n y(\cdot) \mid y(\cdot) \in W^*\}.$$

Moreover, if the triple $(\{h_i\}_{i\geq 1}; V_j; H)$, $j = 1, 2$ satisfies Condition (γ) with $C = C_j$. Then for every $y \in W^*$ and $n \geq 1$

$$\|P_n y(\cdot)\|_{W^*} \leq \max\{C_1, C_2\} \cdot \|y(\cdot)\|_{W^*}.$$

Proof. If $y \in W_n^*$, then

$$y \in X_n^* \subset X^* \cap (S \to H_n) \quad \text{and} \quad y' \in X_n \subset X.$$

Thus, $y(\cdot) = P_n y(\cdot) \in W^*$. On the other hand, let $y \in W^*$. Then in virtue of Proposition 1.6 $y_n(\cdot) := P_n y(\cdot) \in X_n^*$, from Proposition 1.9 it follows that $(y_n)' = P_n y' \in X_n^*$. It means that, $y_n \in W_n$.

Let us assume that the triple $(\{h_i\}_{i\geq 1}; V_j; H)$, $j = 1, 2$ satisfies Condition (γ) with $C = C_j$. Then from Propositions 1.6 and 1.5 it follows that for each $y \in W$:

$$\begin{aligned}\|P_n y(\cdot)\|_W &= \|P_n y(\cdot)\|_X + \|P_n y'(\cdot)\|_{X^*} \\ &\leq \max\{C_1, C_2\} \left(\|y(\cdot)\|_X + \|y'(\cdot)\|_{X^*}\right) = \max\{C_1, C_2\}\|y(\cdot)\|_W.\end{aligned}$$

The Proposition is proved. □

Theorem 1.8. *Let the triple $(\{h_i\}_{i\geq 1}; V_j; H)$, $j = 1, 2$ satisfy Condition (γ) with $C = C_j$. We consider bounded in X^* set $D \subset X^*$ and $E \subset X$ that is bounded in X. For every $n \geq 1$ let us consider*

$$D_n := \{y_n \in X_n^* \mid y_n \in D \text{ and } y_n' \in P_n E\} \subset W_n^*.$$

Then

$$\|y_n\|_{W^*} \leq \|D\|_+ + C \cdot \|E\|_+ \quad \text{for all } n \geq 1 \text{ and } y_n \in D_n, \tag{1.27}$$

where $C = \max\{C_1, C_2\}$, $\|D\|_+ = \sup_{y \in D} \|y\|_{X^*}$ *and* $\|E\|_+ = \sup_{f \in E} \|f\|_X$.

Remark 1.7. Due to Proposition 1.5 D_n is well-defined and $D_n \subset W_n^*$ is true.

Proof. Due to Proposition 1.5 for every $n \geq 1$ and $y_n \in D_n$

$$\|y_n\|_{W^*} = \|y_n\|_{X^*} + \|y_n'\|_X \leq \|D\|_+ + \|P_n E\|_+ \leq \|D\|_+ + \max\{C_1, C_2\} \cdot \|E\|_+.$$

The Theorem is proved. □

Now let us give some generalizations of well-known compact embedding theorems for classes of infinite-dimensional distribution spaces with integrable derivatives which can be used when investigating differential-operator inclusions and evolutional multivariational inequalities.

Further, let B_0, B_1, B_2 be some Banach spaces such, that

$$B_0, B_2 \text{ are reflexive, } B_0 \subset B_1 \text{ with compact embedding} \tag{1.28}$$

$$B_0 \subset B_1 \subset B_2 \text{ with continuous embedding} \tag{1.29}$$

Lemma 1.2. *([LIO69, Lemma 1.5.1, p.71]) Under the assumptions (1.28)–(1.29) for an arbitrary* $\eta > 0$ *there exists* $C_\eta > 0$ *such that*

$$\|x\|_{B_1} \leq \eta \|x\|_{B_0} + C_\eta \|x\|_{B_2} \qquad \forall x \in B_0.$$

Corollary 1.3. *Let the assumptions (1.28)–(1.29) for Banach spaces* B_0, B_1 *and* B_2 *are verified,* $p_1 \in [1; +\infty]$, $S = [0, T]$ *and the set* $K \subset L_{p_1}(S; B_0)$ *such that*

(a) K *is precompact set in* $L_{p_1}(S; B_2)$.
(b) K *is bounded set in* $L_{p_1}(S; B_0)$.

Then K *is precompact set in* $L_{p_1}(S; B_1)$.

Proof. Due to Lemma 1.2 and to the norm definition in $L_{p_1}(S; B_i)$, $i = \overline{0,2}$ it follows that for an arbitrary $\eta > 0$ there exists such $C_\eta > 0$ that

$$\|y\|_{L_{p_1}(S;B_1)} \leq 2\eta \|y\|_{L_{p_1}(S;B_0)} + 2C_\eta \|y\|_{L_{p_1}(S;B_2)} \quad \forall y \in L_{p_1}(S; B_0). \tag{1.30}$$

Let us check inequality (1.30), when $p_1 \in [0, +\infty)$ (the case $p_1 = +\infty$ is direct corollary of Lemma 1.2):

$$\|y\|^{p_1}_{L_{p_1}(S;B_1)} = \int\limits_S \|y(t)\|^{p_1}_{B_1} dt \le \int\limits_S [\eta\|y(t)\|_{B_0} + C_\eta\|y(t)\|_{B_2}]^{p_1} dt$$

$$\le 2^{p_1-1}\left[\eta^{p_1}\int\limits_S \|y(t)\|^{p_1}_{B_0} dt + C_\eta^{p_1}\int\limits_S \|y(t)\|^{p_1}_{B_2} dt\right]$$

$$= 2^{p_1-1}\left[\eta^{p_1}\|y\|^{p_1}_{L_{p_1}(S;B_0)} + C_\eta^{p_1}\|y\|^{p_1}_{L_{p_1}(S;B_2)}\right]$$

$$\le 2^{p_1}\left[\eta\|y\|_{L_{p_1}(S;B_0)} + C_\eta\|y\|_{L_{p_1}(S;B_2)}\right]^{p_1} \quad \forall y \in L_{p_1}(S;B_0).$$

The last inequality follows from

$$\frac{a^{p_1}+b^{p_1}}{2} \le (a+b)^{p_1} \le 2^{p_1-1}(a^{p_1}+b^{p_1}) \quad \forall a,b \ge 0.$$

Now let $\{y_n\}_{n\ge1}$ be an arbitrary sequence from K. Then by the conditions of the given Proposition there exists $\{y_{n_k}\}_{k\ge1} \subset \{y_n\}_{n\ge1}$ that is a Cauchy subsequence in the space $L_{p_1}(S;B_2)$. So, thanks to inequality (1.30) for every $k,m \ge 1$

$$\|y_{n_k} - y_{n_m}\|_{L_{p_1}(S;B_1)} \le 2\eta\|y_{n_k} - y_{n_m}\|_{L_{p_1}(S;B_0)} + 2C_\eta\|y_{n_k} - y_{n_m}\|_{L_{p_1}(S;B_2)}$$
$$\le \eta C + 2C_\eta\|y_{n_k} - y_{n_m}\|_{L_{p_1}(S;B_2)},$$

where $C > 0$ is a constant that does not depend on m,k,η. Therefore, for every $\varepsilon > 0$ we can choose $\eta > 0$ and $N \ge 1$ such that:

$$\eta C < \varepsilon/2 \quad \text{and} \quad 2C_\eta\|y_{n_k} - y_{n_m}\|_{L_{p_1}(S;B_2)} < \varepsilon/2 \quad \forall m,k \ge N.$$

Thus,

$$\forall\varepsilon > 0 \quad \exists N \ge 1: \quad \|y_{n_k} - y_{n_m}\|_{L_{p_1}(S;B_1)} < \varepsilon \quad \forall m,k \ge N.$$

This fact means, that $\{y_{n_k}\}_{k\ge1}$ converges in $L_{p_1}(S;B_1)$. The Corollary is proved. □

Theorem 1.9. *([LIO69, p.71]) Under conditions (1.28)–(1.29), for all* $p_0, p_1 \in (1;+\infty)$ *Banach space*

$$W = \left\{y \in L_{p_1}(S;B_0) \mid y' \in L_{p_0}(S;B_2)\right\}$$

with the norm

$$\|y\|_W = \|y\|_{L_{p_1}(S;B_0)} + \|y'\|_{L_{p_0}(S;B_2)} \quad \forall y \in W,$$

where the derivative y' *of an element* $y \in L_{p_1}(S;B_0)$ *is considered in the sense of the scalar distribution space* $\mathscr{D}(S;B_2)$*, is compactly embedded in* $L_{p_1}(S;B_1)$.

Theorem 1.10. *Let conditions (1.28)–(1.29) for* B_0, B_1, B_2 *are satisfied,* $p_0, p_1 \in [1; +\infty)$, S *be a finite time interval and* $K \subset L_{p_1}(S; B_0)$ *be such, that*

(a) K *is bounded in* $L_{p_1}(S; B_0)$.

(b) For every $\varepsilon > 0$ *there exists such* $\delta > 0$ *that from* $0 < h < \delta$ *it results in*

$$\int_S \|u(\tau) - u(\tau + h)\|_{B_2}^{p_0} d\tau < \varepsilon \quad \forall u \in K. \tag{1.31}$$

Then K *is precompact in* $L_{\min\{p_0;p_1\}}(S; B_1)$.

Furthermore, if for some $q > 1$ K *is bounded in* $L_q(S; B_1)$, *then* K *is precompact in* $L_p(S; B_1)$ *for every* $p \in [1, q)$.

Remark 1.8. Further we consider that every element $x \in (S \to B_i)$ is equal to $\bar{0}$ out of the interval S.

Proof. At the beginning we consider the first case. For our goal it is enough to show, that it is possible to choose a Cauchy subsequence from every sequence $\{y_n\}_{n \geq 1} \subset K$ in $L_{\min\{p_0;p_1\}}(S; B_1)$. Due to Corollary 1.3 it is sufficient to prove this Proposition for $L_{\min\{p_0;p_1\}}(S; B_2)$.

For every $x \in K$ $\forall h > 0$ $\forall t \in S$ we put

$$x_h(t) := \frac{1}{h} \int_t^{t+h} x(\tau) d\tau,$$

where the integral is regarded in the sense of Bochner integral. We point out that $\forall h > 0$ $x_h \in C(S; B_0) \subset C(S; B_2)$.

Fixing a positive number ε, we construct for a set

$$K \subset L_{\min\{p_0;p_1\}}(S; B_0) \subset L_{\min\{p_0;p_1\}}(S; B_2)$$

a final ε-web in $L_{\min\{p_0;p_1\}}(S; B_2)$. For $\varepsilon > 0$ we choose $\delta > 0$ from (1.31). Then for every fixed h $(0 < h < \delta)$ we have:

$$\begin{aligned}
\|x_h(t + u) - x_h(t)\|_{B_2} &= \frac{1}{h} \left\| \int_{t+u}^{t+u+h} x(\tau) d\tau - \int_t^{t+h} x(\tau) d\tau \right\|_{B_2} \\
&= \frac{1}{h} \left\| \int_t^{t+h} x(\tau + u) d\tau - \int_t^{t+h} x(\tau) d\tau \right\|_{B_2} \\
&\leq \frac{1}{h} \int_t^{t+h} \|x(\tau + u) - x(\tau)\|_{B_2} d\tau
\end{aligned}$$

Moreover, from Hölder's inequality we obtain

$$\frac{1}{h}\int_t^{t+h}\|x(\tau+u)-x(\tau)\|_{B_2}d\tau \le \left(\frac{1}{h}\right)^{\frac{1}{p_0}}\left(\int_t^{t+h}\|x(\tau+u)-x(\tau)\|_{B_2}^{p_0}d\tau\right)^{\frac{1}{p_0}}$$
$$\le \left(\frac{1}{h}\right)^{\frac{1}{p_0}}\left(\int_0^{T}\|x(\tau+u)-x(\tau)\|_{B_2}^{p_0}d\tau\right)^{\frac{1}{p_0}}$$
$$\le \left(\frac{\varepsilon}{h}\right)^{\frac{1}{p_0}} \qquad \forall x\in K\ \forall\, 0<u<\delta\ \forall t\in S.$$

Therefore the family of functions $\{x_h\}_{x\in K}$ is equicontinuous.

Since $\forall x\in K\ \forall t\in S$ it results in

$$\|x_h(t)\|_{B_2} = \frac{1}{h}\left\|\int_t^{t+h}x(\tau)d\tau\right\|_{B_2} \le \frac{1}{h}\int_t^{t+h}\|x(\tau)\|_{B_2}d\tau$$
$$\le \left(\frac{1}{h}\right)^{\frac{1}{p_1}}\left(\int_t^{t+h}\|x(\tau)\|_{B_2}^{p_1}d\tau\right)^{\frac{1}{p_1}}$$
$$\le \left(\frac{1}{h}\right)^{\frac{1}{p_1}}\left(\int_0^{T}\|x(\tau)\|_{B_2}^{p_1}d\tau\right)^{\frac{1}{p_1}} \le \left(\frac{C}{h}\right)^{\frac{1}{p_1}},$$

the family of functions $\{x_h\}_{x\in K}$ is uniformly bounded, because of the constant $C\ge 0$ does not depend from $x\in K$. Hence, $\forall h:\ 0<h<\delta$ the family of functions $\{x_h\}_{x\in K}$ is precompact in $C(S;B_2)$, so in $L_{\min\{p_0,p_1\}}(S;B_2)$ too.

On the other hand, $\forall\, 0<h<\delta\ \forall x\in K\ \forall t\in S$

$$\|x(t)-x_h(t)\|_{B_2} \le \frac{1}{h}\int_t^{t+h}\|x(t)-x(\tau)\|_{B_2}d\tau$$
$$\le \frac{1}{h}\int_0^{h}\|x(t)-x(t+\tau)\|_{B_2}d\tau$$
$$\le \left(\frac{1}{h}\right)^{\frac{1}{p_0}}\left(\int_0^{h}\|x(t)-x(t+\tau)\|_{B_2}^{p_0}d\tau\right)^{\frac{1}{p_0}}.$$

From here, taking into account inequality (1.31) we receive:

$$\left(\int_0^T \|x(t)-x_h(t)\|_{B_2}^{p_0}dt\right)^{\frac{1}{p_0}} \le \left(\int_0^T \frac{1}{h}\int_0^h \|x(t)-x(t+\tau)\|_{B_2}^{p_0}d\tau dt\right)^{\frac{1}{p_0}}$$
$$= \left(\frac{1}{h}\int_0^h\int_0^T \|x(t)-x(t+\tau)\|_{B_2}^{p_0}dtd\tau\right)^{\frac{1}{p_0}}$$
$$< \left(\frac{1}{h}\int_0^h \varepsilon\, d\tau\right)^{\frac{1}{p_0}} = \varepsilon^{\frac{1}{p_0}}.$$

Hence, by virtue of the precompactness of system $\{x_h\}_{x\in K}$ in $L_{\min\{p_0,p_1\}}(S;B_2)$ $\forall\, 0<h<\delta$ we have that K is a precompact set in $L_{\min\{p_0,p_1\}}(S;B_2)$.

Let us consider the second case. Assume that for some $q>1$ the set K is bounded in $L_q(S;B_1)$. Similarly to the previous case, it is enough to show that for every $p\in[1;q)$ and $\{y_n\}_{n\ge1}\subset K$ there exists a subsequence $\{y_{n_k}\}_{k\ge1}\subset\{y_n\}_{n\ge1}$ and $y\in L_p(S;B_1)$ so that

$$y_{n_k}\to y \quad \text{in} \quad L_p(S;B_1) \quad \text{as} \quad k\to\infty.$$

Because of $y_n\to y$ in $L_{\min\{p_0,p_1\}}(S;B_1)$, up to a subsequence, as $n\to\infty$, we have $\exists\{y_{n_k}\}_{k\ge1}\subset\{y_n\}_{n\ge1}$ such that $\lambda(B_{n_k})\to 0$ as $k\to\infty$, where

$$B_n := \{t\in S \mid \|y_n(t)-y(t)\|_{B_1}\ge 1\}$$

for every $n\ge1$, λ is the Lebesgue measure on S. Then for every $k\ge1$

$$\int_S \|y_{n_k}(s)-y(s)\|_{B_1}^p ds = \int_{A_{n_k}} \|y_{n_k}(s)-y(s)\|_{B_1}^p ds + \int_{B_{n_k}} \|y_{n_k}(s)-y(s)\|_{B_1}^p ds$$
$$\le \int_{A_{n_k}} \|y_{n_k}(s)-y(s)\|_{B_1}^p ds$$
$$+\left(\int_S \|y_{n_k}(s)-y(s)\|_{B_1}^q ds\right)^{\frac{p}{q}} (\lambda(B_{n_k}))^{\frac{q-p}{q}}$$
$$=: I_{n_k}+J_{n_k},$$

where $A_n = S\setminus B_n$ for every $n\ge1$.

It is clear that $J_{n_k}\to0$ as $k\to\infty$. Let us consider I_{n_k}. Since $\{y_{n_k}\}_{k\ge1}$ is precompact in $L_{\min\{p_0,p_1\}}(S;B_1)$, there exists such $\{y_{m_k}\}_{k\ge1}\subset\{y_{n_k}\}_{k\ge1}$ that $y_{m_k}(t)\to y(t)$ in B_1 as $k\to\infty$ almost everywhere in S. Setting

$$\forall k \geq 1 \quad \forall t \in S \qquad \varphi_{m_k}(t) := \begin{cases} \|y_{m_k}(t) - y(t)\|_{B_1}^p, \ t \in A_n \\ \qquad\qquad 0, \text{ otherwice,} \end{cases}$$

using definition of A_{m_k}, sequence $\{\varphi_{m_k}\}_{k\geq 1}$ satisfies the conditions of the Lebesgue Theorem with the integrable majorant $\phi \equiv 1$. So $\varphi_{m_k} \to \bar{0}$ in $L_1(S)$ as $k \to \infty$. Thus, within to a subsequence, $y_n \to y$ in $L_q(S; B_1)$.

The Theorem is proved. □

For evolution triples $(V_i; H; V_i^*)$, $i = 1, 2$ that satisfy (1.9) let us consider latter introduced Banach spaces:

$$\begin{aligned} X &= L_{q_1}(S; V_1^*) + L_{q_2}(S; V_2^*) + L_{r_2'}(S; H) + L_{r_1'}(S; H), \\ X^* &= L_{r_1}(S; H) \cap L_{r_2}(S; H) \cap L_{p_1}(S; V_1) \cap L_{p_2}(S; V_2), \\ W_0^* &= \{y \in L_{p_1}(S; V_1) \cap L_{p_2}(S; V_2) \mid y' \in X\} \end{aligned}$$

with corresponding norms. The derivative y' of an element $y \in L_{p_1}(S; B_0)$ is considered in the sense of the scalar distribution space $\mathscr{D}^*(S; V^*)$, where $V^* = V_1^* + V_2^*$, $S = [0, T]$ – is the finite time interval.

The next Theorem is also new when $V_1 = V_2$, $r_1 = r_2$, $p_1 = p_2$. It plays an important role in the investigation of the differential-operator equations in non-reflexive Banach spaces with nonlinear operators with $(X^*; W_0^*)$-semibounded variation. The given result is the some generalization of one embedding Theorem from [LIO69, SIM86].

Theorem 1.11. *If one of the following conditions is true:*
or $V_1 \subset H$ with the compact embedding,
or $V_2 \subset H$ with the compact embedding,
then $W_0^* \subset L_p(S; H)$ with the compact embedding for any $p \in [1, +\infty)$.

Proof. At first let us prove the compact embedding $W_0^* \subset L_1(S; V^*)$, where $V^* = V_1^* + V_2^*$. For each $y \in W_0^*$ and for each $h \in \mathbb{R}$ we set

$$y_h(t) = \begin{cases} y(t+h), & \text{if } t + h \in S, \\ \bar{0}, & \text{else.} \end{cases}$$

Due to Theorem 1.7 the given definition is correct.

Lemma 1.3. *For an arbitrary $y \in W_0^*$ and for each $h \in \mathbb{R}$ the following estimation is true:*

$$\|y - y_h\|_{L_1(S;V^*)} \leq \left[T h^{1/p_1} + T h^{1/p_2} + \gamma T^{1/r_1} h + \gamma T^{1/r_2} h\right] \|y'\|_X, \quad (1.32)$$

where $\gamma \equiv$ const from the inequality:

$$\|v\|_{V^*} \leq \gamma \|v\|_H \quad \forall h \in H.$$

Proof. Let $y \in W_0^*$ be an arbitrary fixed element. As $y \in X$, then for all $d_{1i} \in L_{q_i}(S; V_i^*)$ and $d_{2i} \in L_{r_i'}(S; H)$ such that

$$y' = d_{11} + d_{12} + d_{21} + d_{22}, \tag{1.33}$$

we have

$$\begin{aligned}\|y - y_h\|_{L_1(S;V^*)} &= \int_S \|y(t+h) - y(t)\|_{V^*} dt = \int_S \| \int_t^{t+h} y'(\tau)d\tau\|_{V^*} dt \\ &\le \int_S \| \int_t^{t+h} d_{11}(\tau)d\tau\|_{V^*} dt + \int_S \| \int_t^{t+h} d_{12}(\tau)d\tau\|_{V^*} dt \\ &\quad + \gamma \int_S \| \int_t^{t+h} d_{21}(\tau)d\tau\|_H dt + \gamma \int_S \| \int_t^{t+h} d_{22}(\tau)d\tau\|_H dt.\end{aligned} \tag{1.34}$$

Let us estimate every pair separately. From the definition of the norm in $V^* = V_1^* + V_2^*$, due to [GGZ74, Lemma IV.1.8, p.169], [GGZ74, Theorem IV.1.6, p.153] and to the Hölder inequality it follows as $i = 1, 2$ that

$$\begin{aligned}&\int_S \| \int_t^{t+h} d_{1i}(\tau)d\tau\|_{V^*} dt \\ &\le \int_S \| \int_t^{t+h} d_{1i}(\tau)d\tau\|_{V_i^*} dt \\ &\le \int_S | \int_t^{t+h} \|d_{1i}(\tau)\|_{V_i^*} d\tau | dt = \int_S \int_S \chi_{[t;t+h]}(\tau)\|d_{1i}(\tau)\|_{V_i^*} d\tau dt \\ &\le \int_S \left[\left(\int_S \chi_{[t;t+h]}^{p_i}(\tau)d\tau\right)^{1/p_i} \cdot \left(\int_S \|d_{1i}(\tau)\|_{V_i^*}^{q_i} d\tau\right)^{1/q_i} \right] dt \\ &= h^{1/p_i} \int_S \left(\int_S \|d_{1i}(\tau)\|_{V_i^*}^{q_i} d\tau\right)^{1/q_i} dt = T \cdot h^{1/p_i} \cdot \|d_{1i}\|_{L_{q_i}(S;V_i^*)},\end{aligned} \tag{1.35}$$

where for each $t, \tau \in S$ and $h \ge 0$

$$\chi_{[t;t+h]}(\tau) = \begin{cases} 1, \text{ if } \tau \in S \cap < t, t+h >, \\ 0, \text{ else}, \end{cases}$$

$$< t, t+h > = \left[\min\{t, t+h\}, \max\{t, t+h\}\right].$$

Now let us estimate the second pair. Set

$$g_i(t) = \int_t^{t+h} d_{2i}(\tau)d\tau \quad \forall t \in S \text{ as } i = 1, 2,$$

pointed out that $g_i \in C(S;H)$, for each $t, t_0 \in S$ we have

$$\|g_i(t) - g_i(t_0)\|_H = \left\| \int_{t_0}^{t} d_{2i}(\tau)d\tau - \int_{t_0+h}^{t+h} d_{2i}(\tau)d\tau \right\|_H$$

$$\leq \left| \int_{t_0}^{t} \|d_{2i}(\tau)\|_H d\tau \right| + \left| \int_{t_0+h}^{t+h} \|d_{2i}(\tau)\|_H d\tau \right| \to 0 \quad \text{as } t \to t_0,$$

because $d_{2i} \in L_1(S;H)$. Since S is a compact interval, it follows that $g_i \in L_1(S;H)$. Thus, due to Proposition 8 with $X = L_1(S;H)$ it follows the existence of $h_{g_i} \in L_\infty(S;H) \equiv X^*$ such that

$$\int_S \|g_i(t)\|_H dt = \int_S (g_i(t), h_{g_i}(t))_H dt \quad \text{and} \quad \|h_{g_i}\|_{L_\infty(S;H)} = 1.$$

Therefore as $i = 1,2$

$$\int_S \left\| \int_t^{t+h} d_{2i}(\tau)d\tau \right\|_H dt = \int_S \left(\int_t^{t+h} d_{2i}(\tau)d\tau, h_{g_i}(t) \right) dt$$

$$= \int_S \int_t^{t+h} (d_{2i}(\tau), h_{g_i}(t))d\tau dt = \int_S \int_{\tau-h}^{\tau} (d_{2i}(\tau), h_{g_i}(t))dt d\tau$$

$$= \int_S \left(d_{2i}(\tau), \int_{\tau-h}^{\tau} h_{g_i}(t)dt \right) d\tau \leq \int_S \|d_{2i}(\tau)\|_H \cdot h \cdot \operatorname*{ess\,sup}_{t\in S} \|h_{g_i}(t)\|_H d\tau$$

$$= h \cdot \|d_{2i}\|_{L_1(S;H)} \leq T^{1/r_i} \cdot h \cdot \|d_{2i}\|_{L_{r_i'}(S;H)}.$$

Having in mind that $\frac{1}{+\infty} := 0$, due to relations (1.34) and (1.35), it results in

$$\|y - y_h\|_{L_1(S;V^*)} \leq T \cdot h^{1/p_1} \cdot \|d_{11}\|_{L_{q_1}(S;V_1^*)} + T \cdot h^{1/p_2} \cdot \|d_{12}\|_{L_{q_2}(S;V_2^*)}$$

$$+ \gamma T^{1/r_1} \cdot h \cdot \|d_{21}\|_{L_{r_1'}(S;H)} + \gamma T^{1/r_2} \cdot h \cdot \|d_{22}\|_{L_{r_2'}(S;H)}$$

$$\leq \max_{i=1,2} \left\{ \|d_{1i}\|_{L_{q_i}(S;V_i^*)}; \|d_{2i}\|_{L_{r_i'}(S;H)} \right\}$$

$$\times \left[T \cdot h^{1/p_1} + T \cdot h^{1/p_2} + \gamma T^{1/r_1} \cdot h + \gamma T^{1/r_2} \cdot h \right].$$

If we take the infimum by the all representations (1.33) we obtain the necessary estimation. □

Let us continue the proof of the given Theorem. We consider the bounded set $K \subset W_0^*$. Then for some $C > 0$

$$\|y\|_{L_{p_1}(S;V_1)} + \|y\|_{L_{p_2}(S;V_2)} \leq C, \quad \|y'\|_X \leq C \quad \forall y \in K. \tag{1.36}$$

Without loss of generality we assume $V_1 \subset H$ with the compact embedding. To prove the precompactness of K in $L_1(S; H)$ we apply Theorem 1.10 with $V = V_1$, $H = H$, $B = V^*$, $p_0 = 1$, $p_1 = p_1$. The last relations get the estimations (1.36) and (1.32). Moreover, from the estimation (1.36) and from Theorem 1.7 it follows the boundness of K in $L_q(S; H)$ $\forall q \geq 1$. So, in virtue of Theorem 1.10 it follows the precompactness of K in $L_q(S; H)$ $\forall q \geq 1$. □

Corollary 1.4. *Let $V_1 \subset V_2$ with the compact embedding. Then*

$$W_0^* \subset L_{p_1}(S; V_2)$$

with the compact embedding.

Proof. From the compact embedding for $V_1 \subset H$ it follows that $V_1 \subset H$ with the compact embedding. Hence, due to Theorem 1.11, $W_0^* \subset L_q(S; H)$ with the compact embedding for each $q \geq 1$, in particular, for $q = p_1$.

Now if we consider an arbitrary bounded set $K \subset W_0^*$ we obtain that K is a precompact set in $L_{p_1}(S; H)$. Thus, in virtue of Corollary 1.3, the set K is precompact in $L_{p_1}(S; V_2)$.

The Corollary is proved. □

Let Banach spaces B_0, B_1, B_2 satisfy all assumptions (1.28)–(1.29), $p_0, p_1 \in [1; +\infty)$ be arbitrary numbers. We consider the set with the natural operations

$$W = \{v \in L_{p_0}(S; B_0) \mid v' \in L_{p_1}(S; B_2)\},$$

where the derivative v' of an element $v \in L_{p_0}(S; B_0)$ is considered in the sense of the scalar distribution space $\mathscr{D}(S; B_2)$. It is clear, that

$$W \subset L_{p_0}(S; B_0).$$

Theorem 1.12. *The set W with the natural operations and the graph norm*

$$\|v\|_W = \|v\|_{L_{p_0}(S;B_0)} + \|v'\|_{L_{p_1}(S;B_2)}$$

is a Banach space.

Proof. The executing of the norm properties for $\|\cdot\|_W$ immediately follows from its definition. Now we consider the completeness of W referring to just defined norm. Let $\{v_n\}_{n\geq 1}$ be a Cauchy sequence in W. Hence, due to the completeness of $L_{p_0}(S;B_0)$ and $L_{p_1}(S;B_2)$ it follows that for some $y \in L_{p_0}(S;B_0)$ and $v \in L_{p_1}(S;B_2)$

$$y_n \to y \text{ in } L_{p_0}(S;B_0) \quad \text{and} \quad y_n' \to v \text{ in } L_{p_1}(S;B_2) \quad \text{as } n \to +\infty.$$

Due to [GGZ74, Lemma IV.1.10] and in virtue of continuous dependence of the derivative by the distribution in $\mathscr{D}^*(S;B_2)$ (see [GGZ74, p.169]) it follows, that $y' = v \in L_{p_1}(S;B_2)$.

The Theorem is proved. □

Theorem 1.13. *Under conditions (1.28)–(1.29) $W \subset C(S;B_2)$ with the continuous embedding.*

Proof. For a fixed $y \in W$ let us show that $y \in C(S;B_2)$. Let us put

$$\xi(t) = \int_{t_0}^{t} y'(\tau)d\tau \quad \forall t_0, t \in S.$$

The integral is well-defined because $y' \in L_1(S;B_2)$. On the other hand, from the inequality [GGZ74, p.153]

$$\|\xi(t) - \xi(s)\|_{B_2} \leq \int_t^s \|y'(\tau)\|_{B_2} d\tau \quad \forall s \geq t,\ s \in S$$

it follows that $\xi \in C(S;B_2)$. Due to [GGZ74, Lemma IV.1.8] $\xi' = y'$, so from [GGZ74, Lemma IV.1.9] it follows that

$$y(t) = \xi(t) + z \quad \text{for a.e.} \quad t \in S$$

for some fixed $z \in B_2$. Thus the function y also lies in $C(S;B_2)$.

In virtue of the continuous embedding of $L_{p_1}(S;B_2)$ in $L_1(S;B_2)$ we have that for some constant $k > 0$, which does not depend on y,

$$\|\xi(t)\|_{B_2} \leq \int_S \|y'(\tau)\|_{B_2} d\tau \leq k\|y'\|_{L_{p_1}(S;B_2)} \quad \forall t \in S.$$

From here, due to the continuous embedding $B_0 \subset B_2$, we have

$$\|z\|_{B_2}(\text{mes}(S))^{1/p_1} = \left(\int_S \|z\|_{B_2}^{p_1} ds\right)^{1/p_1} = \|y - \xi\|_{L_{p_1}(S;B_2)}$$
$$\leq k_1 \left(\|y\|_{L_{p_1}(S;B_2)} + \|\xi\|_{C(S;B_2)}\right)$$
$$\leq k_2 \left(\|y\|_{L_{p_0}(S;B_0)} + \|y'\|_{L_{p_1}(S;B_2)}\right),$$

where mes(S) is the "length" (the measure) of S, $k_2 > 0$ is a constant that does not depend on $y \in W$. Therefore, from the last two relations there exists $k_3 \geq 0$ such that

$$\|y\|_{C(S;B_2)} \leq k_3 \|y\|_W \quad \forall y \in W.$$

The Theorem is proved. □

The next result represents a generalization of the classical compactness Lemma [LIO69, Theorem 1.5.1, p.70] into the case $p_0, p_1 \in [1; +\infty)$. The similar proposition formulated in [SIM86] at the page 89 (see (10.6)) without valid proof. The Author remarks that the proof is based on Theorem 7 (the analogue of Theorem 1.10) and on Lemma 4 (the analogue of Lemma 1.4). The proof of the given lemma Author based on the inequalities (1.3)–(1.5), that represented in his paper without substantiation (the citation on the paper [SI3] is baseless). We remark that the proof for the analogue of (1.37) in the spaces $L_p(S; B_2)$ with $p > 1$ is easer and essentially differs from the proof for the case $p = 1$. We remark also that we are not assume the execution of (10.1) from [SIM86, p. 87]. So, we try to give the formal proof for some analogue (10.6) represented in paper [SIM86].

Theorem 1.14. *Under conditions (1.28)–(1.29), for all $p_0, p_1 \in [1; +\infty)$ Banach space W is compactly embedded in $L_{p_0}(S; B_1)$.*

Proof. At the beginning we prove the compact embedding of W in $L_1(S; B_2)$.
For every $y \in W$ and $h \in \mathbb{R}$ let us take

$$y_h(t) = \begin{cases} y(t+h), & \text{if } t+h \in S, \\ \bar{0}, & \text{otherwice.} \end{cases}$$

In virtue of Theorem 1.13 the given definition is correct.

Lemma 1.4. *For every $y \in W$ and $h \in \mathbb{R}$*

$$\|y - y_h\|_{L_1(S;B_2)} \leq h \cdot \|y'\|_{L_1(S;B_2)}. \tag{1.37}$$

Proof. Let $y \in W$ be fixed. Then

$$\|y - y_h\|_{L_1(S;B_2)} = \int_S \|y(t+h) - y(t)\|_{B_2} dt = \int_S \|\int_t^{t+h} y'(\tau)d\tau\|_{B_2} dt.$$

Let us put

$$g_y(t) = \int_t^{t+h} y'(\tau)d\tau = y(t+h) - y(t) \quad \forall t \in S,\ i = 1, 2.$$

Due to Theorem 1.13 the element $g_y \in C(S; B_2)$. So, as S is a compact set, we have that $g_y \in L_1(S; B_2)$. Therefore, due to Proposition 8 with $X = L_1(S; B_2)$ and to Theorem 1.3 it follows the existence of $h_y \in L_\infty(S; B_2^*) \equiv X^*$ such that

$$\int_S \|g_y(t)\|_{B_2} dt = \int_S \langle h_y(t), g_y(t)\rangle_{B_2} dt \quad \text{and} \quad \|h_y\|_{L_\infty(S;B_2^*)} = 1.$$

Hence,

$$\begin{aligned}
&\int_S \left\| \int_t^{t+h} y'(\tau)d\tau \right\|_{B_2} dt \\
&= \int_S \|g_y(t)\|_{B_2} dt = \int_S \langle h_y(t), g_y(t)\rangle_{B_2} dt \\
&= \int_S \left\langle h_y(t), \int_t^{t+h} y'(\tau)d\tau \right\rangle_{B_2} dt = \int_S \int_t^{t+h} \langle h_y(t), y'(\tau)\rangle_{B_2} d\tau dt \\
&= \int_S \int_{\tau-h}^{\tau} \langle h_y(t), y'(\tau)\rangle_{B_2} dt d\tau = \int_S \left\langle \int_{\tau-h}^{\tau} h_y(t)dt, y'(\tau) \right\rangle_{B_2} d\tau \\
&\le \operatorname*{ess\,sup}_{t\in S} \|h_y(t)\|_{B_2^*} \cdot h \cdot \int_S \|y'(\tau)\|_{B_2} d\tau \le h \cdot \|y'\|_{L_1(S;B_2)}.
\end{aligned}$$

So, we have obtained necessary estimation (1.37).

The Lemma is proved. □

Let us continue the proof of the given Theorem. Let $K \subset W$ be an arbitrary bounded set. Then for some $C > 0$

$$\|y\|_{L_{p_0}(S;B_0)} \le C, \quad \|y'\|_{L_{p_1}(S;B_2)} \le C \quad \forall y \in K. \tag{1.38}$$

In order to prove the precompactness of K in $L_1(S; B_1)$ let us apply Theorem 1.10 with $B_0 = B_0$, $B_1 = B_1$, $B_2 = B_2$, $p_0 = 1$, $p_1 = p_1$. Due to estimates (1.37) and (1.38) the all conditions of the given Theorem hold. So, the set K is precompact in $L_1(S; B_1)$ and hence in $L_1(S; B_2)$. In virtue of Theorem 1.13 and the Lebesgue

Theorem it follows that the set K is precompact in $L_{p_0}(S; B_0)$. Hence, due to Corollary 1.3 we obtain the necessary Proposition.

The Theorem is proved. □

Proposition 1.11. *Let Banach spaces B_0, B_1, B_2 satisfy conditions (1.28)–(1.29), $p_0, p_1 \in [1; +\infty)$, $\{u_h\}_{h\in I} \subset L_{p_1}(S; B_0)$, where $I = (0, \delta) \subset \mathbb{R}_+$, $S = [a, b]$ such that*

(a) $\{u_h\}_{h\in I}$ is bounded in $L_{p_1}(S; B_0)$.
(b) there exists such $c : I \to \mathbb{R}_+$ that $\lim\limits_{n\to\infty} c(\frac{b-a}{2^n}) = 0$ and

$$\forall h \in I \qquad \int_S \|u_h(t) - u_h(t+h)\|_{B_2}^{p_0} dt \leq c(h)h^{p_0}.$$

Then there exists $\{h_n\}_{n\geq 1} \subset I$ ($h_n \searrow 0+$ as $n \to \infty$) so that $\{u_{h_n}\}_{n\geq 1}$ converges in $L_{\min\{p_0,p_1\}}(S; B_1)$.

Remark 1.9. We assume $u_h(t) = \overline{0}$ when $t > b$.

Remark 1.10. Without loss of generality let us consider $S = [0, 1]$.

Proof. At first we prove this Proposition for $L_{p_0}(S; B_2)$. In virtue of Minkowski inequality for every $h = \frac{1}{2^N} \in I$ and $k \geq 1$

$$\left(\int_0^1 \|u_h(t) - u_{\frac{h}{2^k}}(t)\|_{B_2}^{p_0} dt\right)^{\frac{1}{p_0}}$$

$$\leq \left(\int_0^1 \|u_h(t) - u_h(t+h)\|_{B_2}^{p_0} dt\right)^{\frac{1}{p_0}} + \left(\int_0^1 \|u_h(t+h) - u_{\frac{h}{2^k}}(t+h)\|_{B_2}^{p_0} dt\right)^{\frac{1}{p_0}}$$

$$+ \left(\int_0^1 \|u_{\frac{h}{2^k}}(t+h) - u_{\frac{h}{2^k}}(t)\|_{B_2}^{p_0} dt\right)^{\frac{1}{p_0}}$$

$$\leq c^{\frac{1}{p_0}}(h)h + \left(\int_h^1 \|u_h(t) - u_{\frac{h}{2^k}}(t)\|_{B_2}^{p_0} dt\right)^{\frac{1}{p_0}}$$

$$+ \sum_{i=0}^{2^k-1} \left(\int_0^1 \left\| u_{\frac{h}{2^k}}\left(t + \frac{i+1}{2^k}h\right) - u_{\frac{h}{2^k}}\left(t + \frac{i}{2^k}h\right)\right\|_{B_2}^{p_0} dt\right)^{\frac{1}{p_0}}$$

$$\leq c^{\frac{1}{p_0}}(h)h + 2^k \frac{h}{2^k} c^{\frac{1}{p_0}}(h/2^k)$$

$$+\left(\int_h^1 \|u_h(t) - u_{\frac{h}{2^k}}(t)\|_{B_2}^{p_0} dt\right)^{\frac{1}{p_0}} \le h\left(c^{\frac{1}{p_0}}(h) + c^{\frac{1}{p_0}}(h/2^k)\right)$$

$$+\left(\int_h^1 \|u_h(t) - u_h(t+h)\|_{B_2}^{p_0} dt\right)^{\frac{1}{p_0}}$$

$$+\left(\int_h^1 \|u_h(t+h) - u_{\frac{h}{2^k}}(t+h)\|_{B_2}^{p_0} dt\right)^{\frac{1}{p_0}}$$

$$+\left(\int_h^1 \|u_{\frac{h}{2^k}}(t+h) - u_{\frac{h}{2^k}}(t)\|_{B_2}^{p_0} dt\right)^{\frac{1}{p_0}} \le \ldots \le 2h\left(c^{\frac{1}{p_0}}(h) + c^{\frac{1}{p_0}}(h/2^k)\right)$$

$$+\left(\int_{2h}^1 \|u_h(t) - u_h(t+h)\|_{B_2}^{p_0} dt\right)^{\frac{1}{p_0}} \le \ldots \le 2^N h\left(c^{\frac{1}{p_0}}(h) + c^{\frac{1}{p_0}}(h/2^k)\right)$$

$$= c^{\frac{1}{p_0}}(h) + c^{\frac{1}{p_0}}(h/2^k).$$

So, for every $N \ge 1$ and $k \ge 1$ it results in

$$\left(\int_0^1 \|u_{1/2^N}(t) - u_{1/2^{N+k}}(t)\|_{B_2}^{p_0} dt\right)^{\frac{1}{p_0}} \le c^{\frac{1}{p_0}}\left(\frac{1}{2^N}\right) + c^{\frac{1}{p_0}}\left(\frac{1}{2^{N+k}}\right).$$

In virtue of assumption (b) we can choose $\{h_n\}_{n\ge1} \subset \left\{\frac{1}{2^m}\right\}_{m\ge1} \bigcap I$ such that $c(h_n) \to 0$ as $n \to \infty$. So, the sequence $\{u_{h_n}\}_{n\ge1}$ is fundamental in $L_{p_0}(S; B_2)$. Because of $B_0 \subset B_1$ with compact embedding, the sequence $\{u_{h_n}\}_{n\ge1}$ is bounded in $L_{\min\{p_0,p_1\}}(S; B_0)$; due to Corollary 1.3 it follows that $\{u_{h_n}\}_{n\ge1}$ is fundamental in $L_{\min\{p_0,p_1\}}(S; B_1)$.

The Proposition is proved. □

Now we combine all results to obtain the necessary a priori estimates for the Faedo–Galerkin method.

Theorem 1.15. *Let all conditions of Theorem 1.8 are satisfied and $V \subset H$ with compact embedding. Then (1.27) be true and the set*

$$\bigcup_{n\ge1} D_n \textit{ is bounded in } C(S; H) \textit{ and precompact in } L_p(S; H)$$

for every $p \ge 1$.

Proof. Estimation (1.27) follows from Theorem 1.8. Now we use compactness Theorem 1.14 with $B_0 = V$, $B_1 = H$, $B_2 = V^*$, $p_0 = 1$, $p_1 = 1$. Remark that $X^* \subset L_1(S; V)$ and $X \subset L_1(S; V^*)$ with continuous embedding. Hence, the set

$$\bigcup_{n\geq 1} D_n \text{ is precompact in } L_1(S; H).$$

In virtue of (1.27) and Theorem 1.7 on continuous embedding of W^* in $C(S; H)$, it follows that the set

$$\bigcup_{n\geq 1} D_n \text{ is bounded in } C(S; H).$$

Further, by using standard conclusions and the Lebesgue Theorem we obtain the necessary Proposition.

The Theorem is proved. □

Now we consider $p_i, r_i, i = 1, 2$ such that $1 < p_i \leq r_i \leq +\infty$, $p_i < +\infty$. Let $q_i \geq r_i' \geq 1$ well-defined by

$$p_i^{-1} + q_i^{-1} = r_i^{-1} + r_i'^{-1} = 1 \qquad \forall i = 1, 2.$$

Remark that $1/\infty = 0$.

Now we consider some Banach spaces that play an important role in the investigation about differential-operator equations and evolution variation inequalities in nonreflexive Banach spaces.

For evolution triples $(V_i; H; V_i^*)$ $(i = 1, 2)$ such that satisfy (1.9) and for some compact time interval we consider the functional Banach spaces

$$X_i(I) = L_{q_i}(I; V_i^*) + L_{r_i'}(I; H), \quad i = 1, 2$$

and

$$X(I) = L_{q_1}(I; V_1^*) + L_{q_2}(I; V_2^*) + L_{r_2'}(I; H) + L_{r_1'}(I; H)$$

with corresponding norms. We remark that if $r_i < +\infty$ then the space $X_i(I)$ is reflexive. Analogously, if $\max\{r_1, r_2\} < +\infty$, then the space $X(I)$ is reflexive.

We identify $X_i^*(I)$, adjoint with $X_i(I)$, with $L_{r_i}(I; H) \cap L_{p_i}(I; V_i)$ and $X^*(I)$, adjoint with $X(I)$, with

$$L_{r_1}(I; H) \cap L_{r_2}(I; H) \cap L_{p_1}(I; V_1) \cap L_{p_2}(I; V_2).$$

On $X(I) \times X^*(I)$ we denote the duality form by the rule:

$$
\begin{aligned}
\langle f, y\rangle_I &= \int_I (f_{11}(\tau), y(\tau))_H\, d\tau + \int_I (f_{12}(\tau), y(\tau))_H\, d\tau \\
&\quad + \int_I \langle f_{21}(\tau), y(\tau)\rangle_{V_1}\, d\tau + \int_I \langle f_{22}(\tau), y(\tau)\rangle_{V_2}\, d\tau \\
&= \int_I (f(\tau), y(\tau))\, d\tau \quad \forall f \in X,\ y \in X^*,
\end{aligned}
$$

where $f = f_{11} + f_{12} + f_{21} + f_{22}$, $f_{1i} \in L_{r_i'}(I; H)$, $f_{2i} \in L_{q_i}(I; V_i^*)$, $i = 1, 2$.

Let $V = V_1 \cap V_2$, $\mathscr{F}(V)$ be a filter of all finite-dimensional subspaces from V. As V is separable, there exists a countable monotone increasing system of subspaces $\{H_i\}_{i\geq 1} \subset \mathscr{F}(V)$ complete in V, and consequently in H. On H_n we consider inner product induced from H, that we denote again as $(\cdot,\cdot)$. Moreover, let $P_n : H \to H_n \subset H$ be the operator of orthogonal projection from H on H_n:

$$
\text{for every } h \in H \qquad P_n h = \arg\min_{h_n \in H_n} \|h - h_n\|_H .
$$

Definition 1.5. We say that the triple $(\{H_i\}_{i\geq 1}; V; H)$ satisfies *Condition* (γ), if $\sup\limits_{n\geq 1} \|P_n\|_{\mathscr{L}(V,V)} < +\infty$, i.e. there exists such $C \geq 1$ that for every $v \in V$ and $n \geq 1$

$$
\|P_n v\|_V \leq C \cdot \|v\|_V . \tag{1.39}
$$

Some constructions that satisfy the above condition were presented in [KMP06].

Remark 1.11. It is easy to check that there exists such complete orthonormal in H vector system $\{h_i\}_{i\geq 1} \subset V$ such that for every $n \geq 1$ H_n is a linear capsule stretched on $\{h_i\}_{i=1}^n$. Then Condition (γ) means that the system is a Schauder basis in the space V (see [JAM72, p. 403]).

Remark 1.12. From the identification between H^* and H it follows that H_n^* and H_n are also identified.

Remark 1.13. In virtue of $P_n \in \mathscr{L}(V, V)$ for every $n \geq 1$ the adjoint operator $P_n^* \in \mathscr{L}(V^*, V^*)$ and $\|P_n\|_{\mathscr{L}(V,V)} = \|P_n^*\|_{\mathscr{L}(V^*,V^*)}$. It is obvious that for every $h \in H$ $P_n h = P_n^* h$. Hence, we identify P_n with its adjoint P_n^* for every $n \geq 1$. Then, Condition (γ) will mean that for every $v \in V$ and $n \geq 1$

$$
\|P_n v\|_V \leq C \cdot \|v\|_V \ \text{ and } \ \|P_n v\|_{V^*} \leq C \cdot \|v\|_{V^*} . \tag{1.40}
$$

Let us denote by S a subset of a real line which can be presented as no more than numerable join of convex sets in $\mathbb{R}$. We denote by $BC(S) = \{I_\alpha\}_{\alpha\in\Theta}$ the family of all convex bounded sets from S, distinct from a point.

Remark 1.14. Notice that $\Theta = \Theta(S)$, i.e. the set of indexes depends on the set S. Further, Θ will mean $\Theta(S)$.

Furthermore we set

$$X^{loc} = \{y : S \to V^* \mid \forall \alpha \in \Theta \quad y|_{I_\alpha} \in X(I_\alpha)\},$$

where $V^* = V_1^* + V_2^*$. In this space the local base of topology is the following:

$$\mathscr{B}_X := \left\{ \bigcap_{k=1}^{n} V(\alpha_k, \varepsilon_k) \;\middle|\; \varepsilon_k > 0,\ \alpha_k \in \Theta,\ k = \overline{1,n}, \quad n \geq 1 \right\},$$

where for every $\alpha \in \Theta$ and $\varepsilon > 0$:

$$V(\alpha, \varepsilon) = \left\{u \in X^{loc} \;\middle|\; \|u\|_{X(I_\alpha)} < \varepsilon\right\}$$

Lemma 1.5. *$\mathscr{B}_X$ is local base of some topology τ_X in X^{loc}, which converts the given space in a separable locally convex linear topological space and, moreover,*

(a) τ_X is compatible with the set of seminorms

$$\{\rho_\alpha(\cdot) = \|\cdot\|_{X(I_\alpha)}\}_{\alpha\in\Theta} \text{ on } X^{loc}; \tag{1.41}$$

(b) A set $E \subset X^{loc}$ is bounded only when $\forall \alpha \in \Theta$ ρ_α is bounded on E.

Proof. We prove the system of seminorms $\{\rho_\alpha\}_{\alpha\in\Theta}$ divides points on X^{loc}. Let $u \in X^{loc} \setminus \{\overline{0}\}$, then $\lambda(t \in S \mid u(t) \neq \overline{0}) > 0$, where λ is Lebesgue measure on $\mathbb{R}$. Because of S is a subset of a real line which can be presented as no more than numerable join of convex sets in $\mathbb{R}$, we have $\exists \alpha_0 \in \Theta : \|u\|_{X(I_{\alpha_0})} > 0$. From here it follows $\rho_{\alpha_0}(u) > 0$.

From [RUD73, Theorem 1.37] it follows, that the system of seminorms $\{\rho_\alpha\}_{\alpha\in\Theta}$ generates some locally convex topology τ_X on X^{loc}, which converts the given space in locally convex linear topological space, which local base we obtain by final intersections of such sets:

$$\left\{V(I_\alpha, \varepsilon) = \left\{u \in X^{loc} \mid \rho_\alpha(u) < \varepsilon\right\} \;\middle|\; \alpha \in \Theta,\ \varepsilon > 0\right\}.$$

The Proposition (b) follows from the same Theorem.

The Lemma is proved. □

Let

$$X^*_{loc} = \{y : S \to V^* \mid \forall \alpha \in \Theta \quad y|_{I_\alpha} \in X^*(I_\alpha)\}.$$

In this space the local base of topology is the following:

$$\mathscr{B}_{X^*} := \left\{ \bigcap_{k=1}^{n} V(\alpha_k, \varepsilon_k) \;\middle|\; \varepsilon_k > 0,\ \alpha_k \in \Theta,\ k = \overline{1,n}, \quad n \geq 1 \right\},$$

where for every $\alpha \in \Theta$ and $\varepsilon > 0$:

$$V(\alpha,\varepsilon) = \left\{u \in X^*_{loc} \;\middle|\; \|u\|_{X^*(I_\alpha)} < \varepsilon\right\}$$

Lemma 1.6. *$\mathscr{B}_{X^*}$ is local base of some topology τ_{X^*} in X^*_{loc}, which converts the given space in a separable locally convex linear topological space and, moreover,*

(a) τ_{X^*} is compatible with the set of seminorms

$$\{\rho_\alpha(\cdot) = \|\cdot\|_{X^*(I_\alpha)}\}_{\alpha\in\Theta} \text{ on } X^*_{loc}; \tag{1.42}$$

(b) a set $E \subset X^*_{loc}$ is bounded only when $\forall \alpha \in \Theta$ ρ_α is bounded on E.

Proof. As well as in Lemma 1.5 it is enough to show that system of seminorms $\{\rho_\alpha\}_{\alpha\in\Theta}$ divides points on X^*_{loc}. Let $u \in X^*_{loc} \setminus \{\bar{0}\}$, then $\lambda(t \in S \mid u(t) \neq \bar{0}) > 0$. Because of S is a subset of a real line which can be presented as no more than numerable join of convex sets in $\mathbb{R}$, we have $\exists \alpha_0 \in \Theta : \|u\|_{X^*(I_{\alpha_0})} > 0$. From here it follows $\rho_{\alpha_0}(u) > 0$.

From [RUD73, Theorem 1.37] it follows, that the given system of seminorms $\{\rho_\alpha\}_{\alpha\in\Theta}$ generates some locally convex topology τ_{X^*} on X^*_{loc}, which converts the given space in locally convex linear topological space, which local base we obtain by find intersections of such sets:

$$\left\{V(I_\alpha,\varepsilon) = \{u \in X^*_{loc} \mid \rho_\alpha(u) < \varepsilon\} \;\middle|\; \alpha \in \Theta,\ \varepsilon > 0\right\}.$$

The Proposition (b) follows from same Theorem.

The Lemma is proved. □

Remark 1.15. Let us note that for every $I \in BC(S)$ the space $X^*(I)$ is topologically adjoint with $X(I)$, but X^*_{loc} is not topologically adjoint with X^{loc}.

For every $n \geq 1$ and $I \in BC(S)$ we consider Banach spaces

$$X_n(I) = L_{q_0}(I; H_n) \subset X(I), \quad X^*_n(I) = L_{p_0}(I; H_n) \subset X^*(I),$$

where $p_0 := \max\{r_1, r_2\}$, $q_0^{-1} + p_0^{-1} = 1$ with the natural norms. The space $L_{q_0}(I; H_n)$ is isometrically isomorphic to $X^*_n(I)$, the adjoint space of $X_n(I)$, moreover, the map

$$X_n(I) \times X^*_n(I) \ni (f,x) \to \int_I (f(\tau), x(\tau))_{H_n} d\tau = \int_I (f(\tau), x(\tau)) d\tau = \langle f, x\rangle_{X_n(I)}$$

is the duality form on $X_n(I) \times X^*_n(I)$. This statement is correct in virtue of

$$L_{q_0}(I; H_n) \subset L_{q_0}(I; H) \subset L_{r_1'}(I; H) + L_{r_2'}(I; H) + L_{q_1}(I; V_1^*) + L_{q_2}(I; V_2^*).$$

Let us point out that $\langle \cdot,\cdot\rangle_I\big|_{X_n(I)\times X^*_n(I)} = \langle\cdot,\cdot\rangle_{X_n(I)}$.

Let us also consider the space

$$X_n^{loc} = \{y : S \to H_n \mid \forall \alpha \in \Theta \quad y|_{I_\alpha} \in X_n(I_\alpha)\}$$

which topology is compatible with the set of seminorms $\{\|\cdot\|_{X_n(I_\alpha)}\}_{\alpha \in \Theta}$, and

$$X_{n\,loc}^* = \{y : S \to H_n \mid \forall \alpha \in \Theta \quad y|_{I_\alpha} \in X_n^*(I_\alpha)\},$$

which topology is compatible with the set of seminorms $\{\|\cdot\|_{X_n^*(I_\alpha)}\}_{\alpha \in \Theta}$.

Proposition 1.12. *For every $n \geq 1$ we have $X_n^{loc} = P_n X^{loc}$, i.e.*

$$X_n^{loc} = \{P_n f(\cdot) \mid f(\cdot) \in X^{loc}\}.$$

Moreover, if the triple $(\{H_j\}_{j\geq 1}; V_i; H)$, $i = 1,2$ satisfies Condition (γ) with $C = C_i$, then for each $f \in X_{loc}$, $n \geq 1$ and $I \in BC(S)$ it results in

$$\|P_n f\|_{X(I)} \leq \max\{C_1, C_2\} \cdot \|f\|_{X(I)}.$$

Proof. To prove this Proposition we will use Proposition 1.5. Now we consider the first part.

"$\subset$" Let $x \in X_n^{loc}$ be arbitrary fixed. Then for almost all $t \in S$ $P_n x(t) = x(t)$. Moreover, for every $I \in BC(S)$ $x\big|_I \in X_n(I) \subset X(I)$. Thus, $x \in P_n X^{loc}$.

"$\supset$" Let $x \in P_n X^{loc}$ be arbitrary fixed. Then for some $y \in X^{loc}$ $P_n y(t) = x(t)$ for almost all $t \in S$. In virtue of Proposition 1.5 and the definition of X^{loc} it follows that for every $I \in BC(S)$ $x\big|_I = P_n y\big|_I \in X_n(I)$. Thus, $x \in X_n^{loc}$.

The second part of the given Proposition is the direct corollary of Proposition 1.5. This completes the proof. □

Proposition 1.13. *For every $n \geq 1$ we have $X_{n\,loc}^* = P_n X_{loc}^*$, i.e.*

$$X_{n\,loc}^* = \{P_n y(\cdot) \mid y(\cdot) \in X_{loc}^*\},$$

and

$$\langle f, P_n y\rangle_I = \langle f, y\rangle_I \quad \forall I \in BC(S),\ y \in X_{loc}^* \text{ and } f \in X_n^{loc}.$$

Furthermore, if the triple $(\{H_j\}_{j\geq 1}; V_i; H)$, $i = 1,2$ satisfies Condition (γ) with $C = C_i$, then it results in

$$\|P_n y\|_{X^*(I)} \leq \max\{C_1, C_2\} \cdot \|y\|_{X^*(I)} \quad \forall I \in BC(S),\ y \in X_{loc}^* \text{ and } n \geq 1.$$

Proof. To prove this Proposition we use Proposition 1.6. Now we consider the first part.

"$\subset$" Let $f \in X_{n\,loc}^*$ be arbitrary fixed. Then for almost all $t \in S$ $P_n f(t) = f(t)$. Moreover, for every $I \in BC(S)$ $f\big|_I \in X_n^*(I) \subset X^*(I)$. Thus, $f \in P_n X_{loc}^*$.

"$\supset$" Let $f \in P_n X^*_{loc}$ be arbitrary fixed. Then for some $g \in X^*_{loc}$ $P_n g(t) = g(t)$ for almost all $t \in S$. In virtue of Proposition 1.6 and the definition of X^*_{loc} it follows that for every $I \in BC(S)$ $f\big|_I = P_n f\big|_I \in X^*_n(I)$. Thus, $x \in X^*_{n\,loc}$.

The last statements of the Proposition is direct corollary of Proposition 1.6.

Proposition is proved. □

Proposition 1.14. *Under the condition* $\max\{r_1, r_2\} < +\infty$, *the set* $\bigcup_{n\geq 1} X^*_{n\,loc}$ *is dense in* X^*_{loc}.

Proof. Arguing by contradiction, let us assume that for some $f \in X^*_{loc}$ there is an open set from the base of topology of the locally convex linear topological space X^*_{loc}

$$\mathcal{O} = \bigcap_{k=1}^{n} V(\alpha_k, \varepsilon_k),$$

where $n \geq 1$ $\varepsilon_k > 0, \alpha_k \in \Theta, k = \overline{1,n}$,

$$V(\alpha, \varepsilon) = \left\{u \in X^*_{loc} \;\middle|\; \|u\|_{X^*(I_\alpha)} < \varepsilon\right\}, \quad \alpha \in \Theta,\ \varepsilon > 0,$$

such that

$$\left(\bigcup_{n\geq 1} X^*_{n\,loc}\right) \bigcap (f + \mathcal{O}) = \emptyset.$$

Thus

$$\left(\bigcup_{n\geq 1} X^*_{n\,loc}\right) \bigcap (f + \mathcal{O}) \supset \left(\bigcup_{n\geq 1} X^*_{n\,loc}\right) \bigcap f + V(\alpha_0, \varepsilon_0)) = \emptyset, \tag{1.43}$$

where

$$\varepsilon_0 = \frac{1}{n} \min_{k=\overline{1,n}} \varepsilon_k > 0, \quad \alpha_0 \in \Theta(\mathbb{R}) : \ BC(\mathbb{R}) \ni I_{\alpha_0} \supset \bigcup_{k=\overline{1,n}} I_{\alpha_k}.$$

Because of the set

$$(f + V(\alpha_0, \varepsilon_0))\big|_{I_{\alpha_0}} = \left\{f\big|_{I_{\alpha_0}} + g\big|_{I_{\alpha_0}} \;\middle|\; g \in V(\alpha_0, \varepsilon_0)\right\}$$

is open in $X^*(I_{\alpha_0})$, due to Proposition 1.8 the set

$$\bigcup_{n\geq 1} \left(X^*_{n\,loc}\big|_{I_{\alpha_0}}\right) = \bigcup_{n\geq 1} X^*_n(I_{\alpha_0})$$

is dense in $\left(X^*(I_{\alpha_0}), \|\cdot\|_{X^*(I_{\alpha_0})}\right)$ and from (1.43) we obtain the contradiction.

The proof is concluded. □

Now for an arbitrary $I \in BC(S)$ we consider Banach space

$$W^*(I) = \{y \in X(I) |\ y' \in X^*(I)\}$$

with the norm

$$\|y\|_{W(I)^*} = \|y\|_{X(I)^*} + \|y'\|_{X(I)},$$

where we mean the derivative y' of an element $y \in X(I)^*$ in the sense of the scalar distribution space $\mathscr{D}^*(I; V^*) = \mathscr{L}(\mathscr{D}(I); V_w^*)$, where V_w^* is equal to V^* with topology $\sigma(V^*; V)$ [RS80].

Together with $W(I)^*$ we consider Banach space

$$W_i^*(I) = \{y \in L_{p_i}(I; V_i) |\ y' \in X(I)\}, \quad i = 1, 2,$$

with the norm

$$\|y\|_{W_i^*(I)} = \|y\|_{L_{p_i}(I;V_i)} + \|y'\|_{X(I)} \quad \forall y \in W_i^*(I).$$

Also we consider the space $W_0^*(I) = W_1^*(I) \cap W_2^*(I)$ with the norm

$$\|y\|_{W_0^*(I)} = \|y\|_{L_{p_1}(I;V_1)} + \|y\|_{L_{p_2}(I;V_2)} + \|y'\|_{X(I)} \quad \forall y \in W_0^*(I).$$

Notice that the space $W^*(I)$ is continuously embedded in $W_i^*(I)$ for $i = \overline{0, 2}$.

Let us set

$$W_{0\,loc}^* = \{y \in L_{p_1}^{loc}(S; V_1) \cap L_{p_2}^{loc}(S; V_2) \mid y' \in X^{loc}\},$$

where the derivative y' of an element $y \in L_{p_1}^{loc}(S; V_1) \cap L_{p_2}^{loc}(S; V_2)$ is regarded in the sense of space of distributions $\mathscr{D}^*(S; V^*)$ and in this space a subbase of topology σ is assigned through the following sets:

$$\mathscr{C} = \Big\{U(\alpha, \varepsilon) = \big\{u \in W_{0\,loc}^* \mid \|u\|_{L_{p_1}(I_\alpha;V_1)} + \|u\|_{L_{p_2}(I_\alpha;V_2)} + \|u'\|_{X(I_\alpha)} < \varepsilon\big\} \;\Big|\; \alpha \in \Theta,\ \varepsilon > 0\Big\}.$$

Lemma 1.7. *$\mathscr{C}$ is a subbase of some topology σ in $W_{0\,loc}^*$, which turns the given space into separable locally convex linear topological space and, moreover:*

(a) σ is compatible with the set of seminorms

$$\{p_\alpha(u) = \|u\|_{L_{p_1}(I_\alpha;V_1)} + \|u\|_{L_{p_2}(I_\alpha;V_2)} + \|u'\|_{X(I_\alpha)}\}_{\alpha\in\Theta},$$

divides points on $W_{0\,loc}^*$.

(b) A set $E \subset W_{0\,loc}^*$ is bounded only when for every $\alpha \in \Theta$ p_α is bounded on E.

Proof. As well as in Lemma 1.5 it is enough to show that system of seminorms $\{\rho_\alpha\}_{\alpha\in\Theta}$ divides points on $W^*_{0\,loc}$. Let $u \in W^*_{0\,loc}\setminus\{\bar{0}\}$, then $\lambda(t \in S \mid u(t) \neq \bar{0}) > 0$. Because of S is a subset of a real line which can be presented as no more than numerable join of convex sets in $\mathbb{R}$, we have $\exists \alpha_0 \in \Theta : \|u\|_{W_0^*(I_{\alpha_0})} > 0$. From here it follows $\rho_{\alpha_0}(u) > 0$, as it was to be shown.

From [RUD73, Theorem 1.37] it follows that the system of seminorms $\{\rho_\alpha\}_{\alpha\in\Theta}$ generates some locally convex topology σ on $W^*_{0\,loc}$, which converts the given space in locally convex linear topological space, which local base we obtain by final intersections of such sets:

$$\left\{V(I_\alpha,\varepsilon) = \left\{u \in W^*_{0\,loc} \mid \rho_\alpha(u) < \varepsilon\right\} \,\middle|\, \alpha \in \Theta,\ \varepsilon > 0\right\}.$$

The Proposition (b) follows from the same Theorem.

The Lemma is proved. □

For a subset of a real line S which can be presented as no more than numerable join of convex sets in $\mathbb{R}$, distinct from a point, let us denote the family of all convex compact sets from S, distinct from a point by $BCC(S) = \{I_\alpha\}_{\alpha\in\Delta}$. We notice that the family of all subset of a real line S which can be presented as no more than numerable join of convex sets in $\mathbb{R}$, distinct from a point, coincides with the family of all subset of a real line S which can be presented as no more than numerable join of convex **compact** sets in $\mathbb{R}$, distinct from a point.

Let us also consider the space

$$C^{loc}(S;H) = \{y : S \to H \mid \forall \alpha \in \Delta \quad y|_{I_\alpha} \in C(I_\alpha;H)\}.$$

which topology is compatible with the set of seminorms $\{\|\cdot\|_{C(I_\alpha;H)}\}_{\alpha\in\Delta}$.

Theorem 1.16. *It results in $W^*_{0\,loc} \subset C^{loc}(S;H)$ with continuous embedding. Moreover, for every $y,\xi \in W_0^*$ and $s,t \in S$: $s < t$ and $(s,t) \in BC(S)$, the next formula of integration by parts takes place*

$$\big(y(t),\xi(t)\big) - \big(y(s),\xi(s)\big) = \int_s^t \Big\{(y'(\tau),\xi(\tau)) + (y(\tau),\xi'(\tau))\Big\}d\tau. \tag{1.44}$$

In particular, when $y = \xi$ we have:

$$\frac{1}{2}\Big(\|y(t)\|_H^2 - \|y(s)\|_H^2\Big) = \int_s^t (y'(\tau),y(\tau))d\tau.$$

Proof. At first let us prove the embedding $W^*_{0\,loc} \subset C^{loc}(S;H)$ in the sense of the set theory. Let $y \in W^*_{0\,loc}$ be fixed. Then for every $t \in S$, due to the set S can be presented as no more than numerable join of convex compact sets in $\mathbb{R}$, distinct from a point, there exists $I \in BCC(S)$ such that $t \in I$. Moreover, we can consider that t is an interior point of I in the space $(S, |\cdot|)$. Hence, due to the definition of $W^*_{0\,loc}$ and Theorem 1.7 it follows that $y\big|_I \in W^*_0(I) \subset C(I;H)$. Thus the function $y : S \to H$ is continuous in the point t. The necessary statement follows from the arbitrary of $t \in S$.

Now let us prove the continuous embedding $W^*_{0\,loc} \subset C^{loc}(S;H)$. Since the set S can be presented as no more than numerable join of convex compact sets in $\mathbb{R}$, distinct from a point, there exists $\Xi \subset \Delta$ (card $\Xi \leq \aleph_0$) such that

$$\bigcup_{\alpha\in\Xi} I_\alpha = S.$$

So, it is enough to show that for every $\alpha \in \Xi$ there is a continuous seminorm $\mu_\alpha : C^{loc}(S;H) \to \mathbb{R}$ and a constant $C_\alpha > 0$ such that

$$\|y\|_{W^*_0(I_\alpha)} \leq C_\alpha \mu_\alpha(u) \quad \forall u \in W^*_{0\,loc}.$$

This fact follows from Theorem 1.7, because for every $\alpha \in \Xi$ $I_\alpha \in BCC(S)$.

At last we obtain formula (1.44) by using Theorem 1.7 with $S = [s,t]$.

The Theorem is proved. □

Let us consider the space

$$W^*_{loc} = \{y \in X^*_{loc} \mid y' \in X^{loc}\},$$

which topology is compatible with the set of seminorms $\{\|\cdot\|_{W^*(I_\alpha)}\}_{\alpha\in\Theta}$.

In virtue of $W^*_{loc} \subset W^*_{0\,loc}$ with continuous embedding and due to the latter Theorem the next statement is true.

Corollary 1.5. $W^*_{loc} \subset C^{loc}(S;H)$ *with continuous embedding. Moreover, for every* $y, \xi \in W^*_0$ *and* $s, t \in S$*:* $s < t$ *and* $(s,t) \in BC(S)$*, formula (1.44) takes place.*

For every $n \geq 1$ and $I \in BC(S)$ let us introduce Banach space

$$W^*_n(I) = \big\{y \in X^*_n(I) \mid\ y' \in X_n(I)\big\}$$

with the norm

$$\|y\|_{W^*_n(I)} = \|y\|_{X^*_n(I)} + \|y'\|_{X_n(I)},$$

where the derivative y' is considered in sense of scalar distributions space $\mathscr{D}^*(I;H_n)$ and the space

$$W^*_{n\,loc} = \{y \in X^*_{n\,loc} \mid y' \in X^{loc}_n\},$$

which topology is compatible with the set of seminorms $\{\|\cdot\|_{W^*_n(I_\alpha)}\}_{\alpha\in\Theta}$.

As far as

$$\mathscr{D}^*(S;H_n)=\mathscr{L}(\mathscr{D}(S);H_n)\subset\mathscr{L}(\mathscr{D}(S);V_\omega^*)=\mathscr{D}^*(S;V^*)$$

it is possible to consider the derivative of an element $y\in X_n^*(S)$ in the sense of $\mathscr{D}^*(S;V^*)$. Notice that for every $n\geq 1$ $W_{n\,loc}^*\subset W_{n+1\,loc}^*\subset W_{loc}^*$.

Proposition 1.15. *For every $y\in X_{loc}^*$ and $n\geq 1$ it results in $P_ny'=(P_ny)'$, where we mean the derivative of an element of X_{loc}^* in the sense of the scalar distributions space $\mathscr{D}^*(S;V^*)$.*

Remark 1.16. We point out that in virtue of the previous assumptions the derivatives of an element of $X_{n\,loc}^*$ in the sense of $\mathscr{D}(S;V^*)$ and in the sense of $\mathscr{D}(S;H_n)$ coincide.

Proof. It is sufficient to show that for every $\varphi\in\mathscr{D}(S)$ $P_ny'(\varphi)=(P_ny)'(\varphi)$. In virtue of definition of derivative in sense of $\mathscr{D}^*(S;V^*)$ we have

$$\forall\varphi\in D(S)\quad P_ny'(\varphi)=-P_ny(\varphi')=-P_n\int_S y(\tau)\varphi'(\tau)d\tau$$
$$=-\int_S P_ny(\tau)\varphi'(\tau)d\tau=-(P_ny)(\varphi')=(P_ny)'(\varphi).$$

The Proposition is proved. □

From Propositions 1.13, 1.12, 1.15 it follows the next

Proposition 1.16. *For every $n\geq 1$ $W_{n\,loc}^*=P_nW_{loc}^*$, i.e.*

$$W_{n\,loc}^*=\{P_ny(\cdot)\mid y(\cdot)\in W_{loc}^*\}.$$

Moreover, if the triple $(\{H_i\}_{i\geq 1};V_j;H)$, $j=1,2$ satisfies Condition (γ) with $C=C_j$, then for every $y\in W_{loc}^*$, $n\geq 1$ and $\alpha\in\Theta$ it results in

$$\|P_ny(\cdot)\|_{W^*(I_\alpha)}\leq\max\{C_1,C_2\}\cdot\|y(\cdot)\|_{W^*(I_\alpha)}.$$

Theorem 1.17. *Let the triple $(\{H_i\}_{i\geq 1};V_j;H)$, $j=1,2$ satisfy Condition (γ) with $C=C_j$. Moreover, let $D\subset X_{loc}^*$ be bounded in X_{loc}^* set and $E\subset X_{loc}$ bounded in X_{loc}. For every $n\geq 1$ let us consider*

$$D_n:=\{y_n\in X_{n\,loc}^*\mid y_n\in D\text{ and }y_n'\in P_nE\}\subset W_{n\,loc}^*.$$

Then for each $\alpha\in\Theta$, $n\geq 1$ and $y_n\in D_n$

$$\|y_n\|_{W^*(I_\alpha)}\leq\|D\|_+^\alpha+C\cdot\|E\|_+^\alpha,\tag{1.45}$$

where $C = \max\{C_1, C_2\}$, $\|D\|_+^\alpha = \sup_{y\in D} \|y\|_{X^*(I_\alpha)}$ *and* $\|E\|_+^\alpha = \sup_{f\in E} \|f\|_{X(I_\alpha)}$, *i.e. the set*

$$\bigcup_{n\geq 1} D_n \quad \text{is bounded in} \quad W^*_{loc}$$

and, consequently, bounded in $C^{loc}(S; H)$.

Remark 1.17. Due to Proposition 1.12 D_n is well-defined and $D_n \subset W^*_{n\,loc}$.

Remark 1.18. A priori estimates (like (1.45)) appear at studying of global solvability of differential-operator equations, inclusions and evolution variation inequalities in nonreflexive Banach and Frechet spaces with maps of w_λ-pseudomonotone type by using the Faedo–Galerkin method at boundary transition, when it is necessary to obtain a priori estimates of approximate solutions y_n in X^*_{loc} and its derivatives y_n' in X^{loc}.

Proof. The assertion of the Theorem is immediate consequence of the inequality $\forall n \geq 1$, $\alpha \in \Theta$, $y_n \in D_n$

$$\begin{aligned}\|y_n\|_{W^*(I_\alpha)} &= \|y_n\|_{X^*(I_\alpha)} + \|y_n'\|_{X(I_\alpha)} \leq \|D\|_+^\alpha + \|P_n E\|_+^\alpha \\ &\leq \|D\|_+^\alpha + \max\{C_1, C_2\} \cdot \|E\|_+^\alpha,\end{aligned}$$

that is valid in virtue of Theorem 1.8. □

Further, let B_0, B_1, B_2 be some Banach spaces that satisfy (1.28) and (1.29), $p_0, p_1 \in [1; +\infty)$ be arbitrary numbers. Let us denote by S a set which can be presented as no more than numerable join of convex sets in $\mathbb{R}$. We denote by $BC(S) = \{I_\alpha\}_{\alpha\in\Theta}$ the family of all convex bounded sets from S, distinct from a point. Furthermore for $p > 1$ we set

$$L_p^{loc}(S; V) = \left\{ y : S \to V \;\middle|\; \forall I \in BC(S) \quad y|_I \in L_p(I; B) \right\}.$$

In this space the local base of topology is the following:

$$\mathscr{B} := \left\{ \bigcap_{k=1}^{n} V(\alpha_k, \varepsilon_k) \;\middle|\; \varepsilon_k > 0,\ \alpha_k \in \Theta,\ k = \overline{1,n}, \quad n \geq 1 \right\},$$

where for every $\alpha \in \Theta$ and $\varepsilon > 0$:

$$V(\alpha, \varepsilon) = \left\{ u \in L_p^{loc}(S; V) \;\middle|\; ||u||_{L_p(I_\alpha;V)} < \varepsilon \right\}.$$

Lemma 1.8. $\mathscr{B}$ *is local base of some topology* τ *in* $L_p^{loc}(S; V)$, *which converts the given space in a separable locally convex linear topological space and, moreover,*

(a) τ is compatible with the set of seminorms

$$\big\{\rho_\alpha(\cdot) = \|\cdot\|_{L_p(I_\alpha;V)}\big\}_{\alpha\in\Theta} \quad on \quad L_p^{loc}(S;V); \tag{1.46}$$

(b) A set $E \subset L_p^{loc}(S;V)$ is bounded only when $\forall\alpha \in \Theta$ ρ_α is bounded on E.

Proof. We prove the system of seminorms $\{\rho_\alpha\}_{\alpha\in\Theta}$ divides points on $L_p^{loc}(S;V)$. Let $u \in L_p^{loc}(S;V) \setminus \{\bar{0}\}$, then $\lambda(t \in S \mid u(t) \neq \bar{0}) > 0$, where λ is Lebesgue measure on $\mathbb{R}$. Because of $\bigcup\limits_{\alpha\in\Theta} I_\alpha = S$, we have $\exists\alpha_0 \in \Theta : ||u||_{L_p(I_{\alpha_0};V)} > 0$. From here it follows $\rho_{\alpha_0}(u) > 0$.

From [GGZ74, Theorem 1.37] it follows, that the given system of seminorms $\{\rho_\alpha\}_{\alpha\in\Theta}$ generates some locally convex topology τ on $L_p^{loc}(S;V)$, which converts the given space in locally convex linear topological space, whose local base we obtain by find intersections of such sets:

$$\left\{ V(I_\alpha,\varepsilon) = \left\{u \in L_p^{loc}(S;V) \mid \rho_\alpha(u) < \varepsilon\right\} \,\middle|\, \alpha \in \Theta,\ \varepsilon > 0 \right\}.$$

The statement (b) follows from the same Theorem.

The Lemma is proved. □

Let $W^{loc} = \{y \in L_{p_0}^{loc}(S;B_0) \mid y' \in L_{p_1}^{loc}(S;B_2)\}$, where the derivative y' of an element $y \in L_{p_0}^{loc}(S;B_0)$ is regarded in the sense of space of distributions $\mathscr{D}^*(S;B_2)$; $1 < p_i < +\infty$, $i = 0, 1$. In the given space a subbase of topology σ is assigned through the following sets:

$$\mathscr{C} = \left\{ U(\alpha,\varepsilon) = \left\{ u \in W^{loc} \,\middle|\, ||u||_{L_p(I_\alpha;B_0)} + ||u'||_{L_{p_1}(I_\alpha;B_2)} < \varepsilon \right\} \,\middle|\, \alpha \in \Theta,\ \varepsilon > 0 \right\}.$$

Lemma 1.9. *$\mathscr{C}$ is a subbase of some topology σ in W^{loc}, which turns the given space into separable locally convex linear topological space and, moreover:*

(a) σ is compatible with the set of seminorms $\{\rho_\alpha(u) = ||u||_{L_p(I_\alpha;B_0)} + ||u'||_{L_{p_1}(I_\alpha;B_2)}\}_{\alpha\in\Theta}$, divides points on W^{loc}.

(b) A set $E \subset W^{loc}$ is bounded only when for every $\alpha \in \Theta$ ρ_α is bounded on E.

Proof. As well as in Lemma 1.8 it is enough to show, that the system of seminorms $\{\rho_\alpha\}_{\alpha\in\Theta}$ divides points on W^{loc}. Let $u \in W^{loc} \setminus \{\bar{0}\}$. Then $\lambda(t \in S \mid u(t) \neq \bar{0}) > 0$. Because of $\bigcup\limits_{\alpha\in\Theta} I_\alpha = S$, we have $\exists\alpha_0 \in \Theta : ||u||_{L_p(I_{\alpha_0};V)} + ||u'||_{L_{p_1}(I_\alpha;B_2)} \geq ||u||_{L_p(I_{\alpha_0};V)} > 0$. From here $\rho_{\alpha_0}(u) > 0$, as it was to be shown.

The Lemma is proved. □

For every $\alpha \in BC(S)$ we consider the set with the natural operations

$$W(I_\alpha) = \{v \in L_{p_0}(I_\alpha; B_0) \mid v' \in L_{p_1}(I_\alpha; B_2)\},$$

where the derivative v' of an element $v \in L_{p_0}(I_\alpha; B_0)$ is considered in the sense of the scalar distribution space $\mathscr{D}(I_\alpha; B_2)$. It is obvious that

$$W(I_\alpha) \subset L_{p_0}(I_\alpha; B_0).$$

Let us also consider the set

$$W^{loc} = \{y \in L_{p_0}^{loc}(S; B_0) \mid y' \in L_{p_1}^{loc}(S; B_2)\},$$

It is clear, that

$$W^{loc} \subset L_{p_0}^{loc}(S; B_0).$$

Theorem 1.18. *W^{loc} with the natural operations, which is topologically compatible with set of seminorms $\{\rho_\alpha(\cdot) = \|\cdot\|_{W(I_\alpha)}\}_{\alpha\in\Theta}$ is a Frechet space.*

Proof. Since the set S can be presented as no more than numerable join of convex sets in $\mathbb{R}$ there exists $\Xi \subset \Theta$ ($\mathrm{card}\,\Xi \leq \aleph_0$) such that

$$S = \bigcup_{\alpha\in\Xi} I_\alpha.$$

Thus, as well as in Lemma 1.7, we can prove that no more than numerable system of seminorms $\{\rho_\alpha(\cdot)\}_{\alpha\in\Xi}$ divides points on W^{loc}. Thus, the families of seminorms $\{\rho_\alpha(\cdot)\}_{\alpha\in\Xi}$ and $\{\rho_\alpha(\cdot)\}_{\alpha\in\Theta}$ are equivalent and the locally convex linear topological space $\left(W^{loc}, \{\rho_\alpha(\cdot)\}_{\alpha\in\Xi}\right)$ is metrizable.

Now let us prove that the metrizable space W^{loc} is complete. Let us consider a Cauchy sequence $\{y_n\}_{n\geq 1} \subset W^{loc}$; without loss of generality we can assume that for every $\alpha, \beta \in \Xi$: $\alpha \neq \beta$ it follows that $I_\alpha \cap I_\beta = \emptyset$. We also consider

$$\Xi = \{\alpha_1 < \alpha_2 < \ldots < \alpha_n < \alpha_{n+1} < \ldots\}.$$

(i_1) Because of $\{y_n\}_{n\geq 1} \subset W^{loc}$ is a Cauchy sequence also $\{y_n|_{I_{\alpha_1}}\}_{n\geq 1}$ is a Cauchy sequence in $W(I_{\alpha_1})$. Thus in virtue of Theorem 1.12 there is a subsequence $\{v_{1,n}\}_{n\geq 1}$ of $\{y_n\}_{n\geq 1}$ that converges in $W(I_{\alpha_1})$ to some $x_1 \in W(I_{\alpha_1})$;

(i_2) Analogously to (i_1), due to $\{v_{1,n}\}_{n\geq 1} \subset W^{loc}$ is a Cauchy sequence the same follows for $\{v_{1,n}|_{I_{\alpha_2}}\}_{n\geq 1}$ in $W(I_{\alpha_2})$. Thus there is a subsequence $\{v_{2,n}\}_{n\geq 1}$ of $\{v_{1,n}\}_{n\geq 1}$ that converges in $W(I_{\alpha_2})$ to some $x_2 \in W(I_{\alpha_2})$;

. . . .

(i_m) Due to $\{v_{m,n}\}_{n\geq 1} \subset W^{loc}$ is a Cauchy sequence the same follows for $\{v_{m,n}|_{I_{\alpha_m}}\}_{n\geq 1}$ in $W(I_{\alpha_m})$. Thus there is a subsequence $\{v_{m+1,n}\}_{n\geq 1}$ of $\{v_{m,n}\}_{n\geq 1}$ that converges in $W(I_{\alpha_m})$ to some $x_m \in W(I_{\alpha_m})$;

. . . .

Thanks to $(i_1), (i_2), \dots,$ using the diagonal Cantor method, we can choose a subsequence $\{y_{n_k}\}_{k\geq 1} = \{v_{n,n}\}_{n\geq 1}$ from $\{y_n\}_{n\geq 1}$ that converges in $W(I_{\alpha_m})$ to $x_m \in W(I_{\alpha_m})$ for every $m \geq 1$.

By setting $y(t) = x_m(t)$, $t \in I_{\alpha_m}$, $m \geq 1$ we obtain that for every $\alpha \in \Xi$ $\rho_\alpha(y_{n_k} - y) \to 0$ as $k \to \infty$.

To conclude the proof we remark that $y \in W^{loc}$ in virtue of the definition W^{loc} and the condition: $\forall \alpha, \beta \in \Xi \colon \alpha \neq \beta$ it follows that $I_\alpha \cap I_\beta = \emptyset$.

The Theorem is proved. □

Analogously with the proof of Theorem 1.18 we can obtain the next:

Theorem 1.19. *The set W^*_{loc} (respectively $W^*_{i\,loc}$, $i = \overline{0,2}$) with the natural operations, which topology is compatible with the set of seminorms $\{\|\cdot\|_{W^*(I_\alpha)}\}_{\alpha\in\Theta}$ (respectively $\{\|\cdot\|_{W_i^*(I_\alpha)}\}_{\alpha\in\Theta}$, $i = \overline{0,2}$) is a Frechet space.*

Theorem 1.20. *Under conditions (1.28)–(1.29), we have $W^{loc} \subset C^{loc}(S; B_2)$ with the continuous embedding.*

Proof. At first let us prove the embedding $W^{loc} \subset C^{loc}(S; B_2)$ in the sense of the set theory. Let $y \in W^{loc}$ be fixed. Then for every $t \in S$, since the set S can be presented as no more than numerable join of convex compact sets in $\mathbb{R}$, distinct from a point, there is $I \in BCC(S)$ such that $t \in I$. Moreover, we can consider that t is an interior point of I in the space $(S, |\cdot|)$. Hence, due to the definition of W^{loc} and Theorem 1.8 it follows that $y\big|_I \in W(I) \subset C(I; B_2)$. Thus the function $y : S \to B_2$ is continuous in the point t.

Now let us prove the continuous embedding $W^{loc} \subset C^{loc}(S; B_2)$. Since the set S can be presented as no more than numerable join of convex compact sets in $\mathbb{R}$, distinct from a point, there exists $\Xi \subset \Delta$ (card $\Xi \leq \aleph_0$) such that

$$\bigcup_{\alpha\in\Xi} I_\alpha = S.$$

So, it is enough to show that for every $\alpha \in \Xi$ there is a continuous seminorm $\mu_\alpha : C^{loc}(S; B_2) \to \mathbb{R}$ and a constant $C_\alpha > 0$ such that

$$\|y\|_{W_0^*(I_\alpha)} \leq C_\alpha \mu_\alpha(u) \quad \forall u \in W^{loc}.$$

In fact for every $\alpha \in \Xi$ $I_\alpha \in BCC(S)$. Thus the above inequality is true in virtue of Theorem 1.14.

The Theorem is proved. □

Theorem 1.21. *Under above assumptions, the embedding W^{loc} in $L^{loc}_{p_0}(S; B_1)$ is compact, that is, an arbitrary bounded in W^{loc} set is precompact in $L^{loc}_{p_0}(S; B_1)$.*

Proof. Arguing by contradiction, let $\{y_n\}_{n\geq 1} \subset W^{loc}$ be bounded in W^{loc} sequence that has no any accumulation point in $L^{loc}_{p_0}(S; B_1)$. From [RUD73, Theorem 1.37] it follows, that for every convex bounded set $S_\alpha \subset S$

$$\sup_{n\geq 1} \left(\|y_n\|_{L_{p_0}(S_\alpha;B_0)} + \|y_n'\|_{L_{p_1}(S_\alpha;B_2)} \right) < +\infty. \tag{1.47}$$

As on real line the arbitrary convex set can be presented as join no more than numerable number of bounded convex sets. Without loss of generality we suppose $S = \bigcup_{\alpha\in\Xi} S_\alpha$, where S_α is bounded convex set in $\mathbb{R}$ $\forall \alpha \in \Xi$ and $\operatorname{card} \Xi \leq \aleph_0$. Further we consider only those $\alpha \in \Xi$ for which $\lambda(S_\alpha) > 0$.

Let it be $\Xi = \{\alpha_n\}_{n\geq 1}$, then:

(i_1) From (1.47) and Theorem 1.14 about compactness we obtain there is a subsequence $\{v_{1,n}\}_{n\geq 1}$ of $\{y_n\}_{n\geq 1}$, that is fundamental in the space $L_{p_0}(S_{\alpha_1}; B_1)$.
(i_2) Analogously to (i_1), from (1.47) and Theorem 1.14 about compactness it follows, there exists $\{v_{2,n}\}_{n\geq 1} \subset \{v_{1,n}\}_{n\geq 1}$ that is fundamental in $L_{p_0}(S_{\alpha_2}; B_1)$.

....

(i_m) From (1.47) and Theorem 1.14 about compactness it follows, there exists $\{v_{m,n}\}_{n\geq 1} \subset \{v_{m-1,n}\}_{n\geq 1}$, that is fundamental in $L_{p_0}(S_{\alpha_m}; B_1)$.

....

Thanks to (i_1), (i_2), ..., using the diagonal Cantor method, we can choose a subsequence $\{y_{n_k}\}_{k\geq 1} = \{v_{n,n}\}_{n\geq 1}$ from $\{y_n\}_{n\geq 1}$ that is fundamental in $L^{loc}_{p_0}(S; B_1)$. This is a contradiction.

The Theorem is proved. □

By the analogy with the last Theorem, due to Lemma 1.2, we can obtain the next:

Theorem 1.22. *Let assumptions (1.28)–(1.29) for Banach spaces B_0, B_1 and B_2 are valid, $p_1 \in [1; +\infty)$, $S = [0, T]$ and the set $K \subset L^{loc}_{p_1}(S; B_0)$ such that*

(a) K is precompact set in $L^{loc}_{p_1}(S; B_2)$.
(b) K is bounded set in $L^{loc}_{p_1}(S; B_0)$.

Then K is precompact set in $L^{loc}_{p_1}(S; B_1)$.

Now we combine all results to obtain the necessary a priori estimates.

Theorem 1.23. *Let all conditions of Theorem 1.17 are satisfied and $V \subset H$ with compact embedding. Then estimate (1.45) is true and the set*

$$\bigcup_{n\geq 1} D_n \text{ is bounded in } C^{loc}(S; H) \text{ and precompact in } L^{loc}_p(S; H)$$

for every $p \geq 1$.

Proof. Estimate (1.45) follows from Theorem 1.17. Now we apply the compactness Theorem 1.21 with

$$B_0 = V,\ B_1 = H,\ B_2 = V^*,\ p_0 = 1,\ p_1 = 1.$$

Notice that $X^*_{loc} \subset L^{loc}_1(S; V)$ and $X^{loc} \subset L_1(S; V^*)$ with continuous embedding. Hence, the set

$$\bigcup_{n\geq 1} D_n \text{ is precompact in } L^{loc}_1(S; H).$$

In virtue of (1.45) and of Theorem 1.16 on continuous embedding W^*_{loc} in C^{loc} $(S; H)$ it follows that the set

$$\bigcup_{n\geq 1} D_n \text{ is bounded in } C^{loc}(S; H).$$

Further, we complete the proof by using standard conclusions, Lebesgue Theorem and the diagonal Cantor method.

The Theorem is proved. □

Let B, B_1 be Banach spaces, $\mathscr{D}$ some set supplied with function $M : \mathscr{D} \to \mathbb{R}$, and $\mathscr{D} \subset B \subset B_1$ where the embedding B in B_1 is continuous; $M(\lambda v) = |\lambda| M(v)$ $\forall v \in \mathscr{D}\ \forall \lambda \in \mathbb{R}$ and the set

$$\{v \in \mathscr{D} \mid M(v) \leq 1\} \quad \text{is relative compact in } B.$$

Let S be a set which can be presented as no more than numerable join of convex sets in $\mathbb{R}$;

$$\mathscr{F}^{loc} = \left\{ v : S \to \mathscr{D} \,\middle|\, \begin{array}{l} v \in L^{loc}_1(S; B_1); \\ \int_{S_\alpha} M\big(v(\tau)\big)^{p_0} d\tau \leq c_\alpha < +\infty\ \forall S_\alpha \subset S \ - \text{ convex,} \\ \qquad \text{bounded; } v' \in E \subset L^{loc}_{p_1}(S; B_1), \\ \qquad\qquad E \text{ is bounded in } L^{loc}_{p_1}(S; B_1), \end{array} \right\} \tag{1.48}$$

where $1 < p_i < +\infty$, $i = 0, 1$. A derivative v' of an element $v \in L^{loc}_1(S; B_1)$ is regarded in the sense of space of distributions $\mathscr{D}^*(S; B_1)$.

Theorem 1.24. *Under above assumptions, $\mathscr{F}^{loc} \subset L^{loc}_{p_0}(S; B)$ and $\mathscr{F}^{loc}$ is relative compact in $L^{loc}_{p_0}(S; B)$.*

Remark 1.19. This Theorem referred to Banach spaces $L_p(S; B)$ is proved by Yu.A. Dubinsky in [DU65].

Proof. Analogously to the proof of Theorem 1.21, the set S can be presented as no more than numerable join of bounded convex sets, i.e. $S = \bigcup_{\alpha\in I} S_\alpha$, where S_α

is the bounded convex set in $\mathbb{R}$ $\forall \alpha \in I$ and card $I \leq \aleph_0$. Further we consider only those $\alpha \in I$ for which $\lambda(S_\alpha) > 0$; λ is Lebesgue measure on $\mathbb{R}$. Let $\{v_n\}_{n\geq 1}$ be a sequence in $\mathscr{F}^{loc}$. We want to show, that from this sequence it is possible to eliminate a Cauchy subsequence in $L^{loc}_{p_0}(S; B)$.

Let it be $I = \{\alpha_n\}_{n\geq 1}$, then:

(i_1) From (1.48), (1.46) and [LIO69, Theorem 12.1 Chap. I] it follows that there is a subsequence $\{v_{1,n}\}_{n\geq 1}$ of sequence $\{v_n\}_{n\geq 1}$, that is fundamental in space $L_{p_0}(S_{\alpha_1}; B)$.

(i_2) Analogously to (i_1), from (1.48), (1.46) and [LIO69, Theorem 12.1 Chap. I] it follows that there exists $\{v_{2,n}\}_{n\geq 1} \subset \{v_{1,n}\}_{n\geq 1}$, that is fundamental in $L_{p_0}(S_{\alpha_2}; B)$.

....

(i_m) From (1.48), (1.46) and [LIO69, Theorem 12.1 Chap. I] it follows that there exists $\{v_{m,n}\}_{n\geq 1} \subset \{v_{m-1,n}\}_{n\geq 1}$, that is fundamental in $L_{p_0}(S_{\alpha_m}; B)$.

....

Thanks to $(i_1), (i_2), \ldots$, using the diagonal Cantor method we can choose a subsequence $\{v_{n_k}\}_{k\geq 1} = \{v_{n,n}\}_{n\geq 1}$ from sequence $\{v_n\}_{n\geq 1}$, that is fundamental in $L^{loc}_{p_0}(S; B)$.

The Theorem is proved. □

1.1.1 On Schauder Basis in Some Banach Spaces

In order to study of solvability of differential-operator equations and inclusions, evolutionary inequalities in functional Banach spaces with maps of w_λ-pseudomonotone type by using Faedo–Galerkin method there is a problem of choosing a basis in "evolution triple" $V \subset H \subset V^*$. In this paper, using the theory of interpolation of Banach spaces [TRI78], we partially solve this problem, covering wide enough class of spaces.

Let $(V, \|\cdot\|_V)$ be a separable Banach space continuously and densely embedded in the Hilbert space $(H, (\cdot,\cdot))$. As V is separable, there exists countable vector system $\{h_i\}_{i\geq 1} \subset V$ complete in V, and consequently in H. For arbitrary $n \geq 1$ let H_n be linear capsule, stretched on $\{h_i\}_{i=1}^n$. On H_n we consider inner product induced from H, that we denote again as $(\cdot,\cdot)$. Moreover let $P_n : H \to H_n \subset H$ be an operator of orthogonal projection from H on H_n, i.e. $\forall h \in H \quad P_n h = \arg\min_{h_n \in H_n} \|h - h_n\|_H$.

Definition 1.6. The vector system $\{h_i\}_{i\geq 1}$ from separable Hilbert space $(V; (\cdot,\cdot)_V)$, continuously and densely embedded in a Hilbert space $(H; (\cdot,\cdot)_H)$, is called *special basis* for the pair of spaces $(V; H)$, if it satisfies the following conditions:

(i) $\{h_i\}_{i\geq 1}$ is orthonormal in $(H, (\cdot,\cdot)_H)$.
(ii) $\{h_i\}_{i\geq 1}$ is orthogonal in $(V, (\cdot,\cdot)_V)$.
(iii) $\forall i \geq 1$ $(h_i, v)_V = \lambda_i (h_i, v)_H$ $\forall v \in V$, where $0 \leq \lambda_1 \leq \lambda_2, \ldots, \lambda_j \longrightarrow \infty$ as $j \longrightarrow \infty$.

Lemma 1.10. *If V is a Hilbert space, compactly and densely embedded in a Hilbert space H, then there exists a special basis $\{h_i\}_{i\geq 1}$ for $(V; H)$. Moreover, for an arbitrary such system, the triple $(\{h_i\}_{i\geq 1}; V; H)$ satisfies Condition (γ) with constant $C = 1$.*

Proof. From [TEM75, pages 54–58] under these assumptions it is well-known, that there exists a special basis $\{h_i\}_{i\geq 1}$ for the pair $(V; H)$. So, in order to complete the proof it is enough to show that the triple $(\{h_i\}_{i\geq 1}; V; H)$ satisfies Condition (γ) with constant $C = 1$ for an arbitrary special basis $\{h_i\}_{i\geq 1}$ for $(V; H)$. Therefore, let us take as H_n a linear span, stretched on $\{h_i\}_{i=1}^n$. We point out H_n is a finite-dimensional space. Thus, the norms $\|\cdot\|_H$ and $\|\cdot\|_V$ are equivalent on H_n. From here it follows $\forall n \geq 1\ \exists c_n > 0,\ \exists C > 0 :\ \forall h \in H_n\ \ \|P_n h\|_V \leq c_n \|P_n h\|_H \leq c_n \|h\|_H \leq c_n C \|h\|_V$. It also means that $P_n \in \mathscr{L}(V, V)$.

Further let us prove that $\forall n \geq 1$

$$\|P_n h\|_V \leq \|h\|_V \qquad \forall h \in \bigcup_{m\geq 1} H_m. \tag{1.49}$$

Let $n \geq 1$ be fixed , then $\forall h \in \bigcup_{m\geq 1} H_m \Rightarrow \exists m_0 \geq n+1 : h \in H_{m_0}$. From here, taking into account (i) and (ii), we have $h = \sum_{i=1}^{m_0} (h, h_i)_H h_i$, $P_n h = \sum_{i=1}^{n} (h, h_i)_H h_i$. In order to obtain (1.49) it is necessary to show that $P_n h$ is orthogonal to $(h - P_n h)$ in V. Because of $(P_n h, h - P_n h)_V = (\sum_{i=1}^{n} (h, h_i)_H h_i, \sum_{j=n+1}^{m_0} (h, h_j)_H h_j)_V = \sum_{i=1}^{n} \sum_{j=n+1}^{m} (h, h_i)_H (h, h_j)_H (h_i, h_j)_V = 0$, $\{h_i\}_{i\geq 1}$ is orthogonal in V. So, in virtue of continuity of $\|\cdot\|_V$ and P_n on V $\forall n \geq 1$ we have that (1.19) is true. □

For interpolating pair A_0, A_1 (i.e. for Banach spaces A_0 and A_1, that are linearly and continuously embedded in some linear topological space) on a set $A_0 + A_1$ let us consider the functional

$$K(t, x) = \inf_{x=x_0+x_1:\ x_0\in A_0,\ x_1\in A_1} \Big(\|x_0\|_{A_0} + t\|x_1\|_{A_1}\Big), \qquad t \geq 0,\ x \in A_0 + A_1.$$

For fixed $x \in A_0 + A_1$, this map is monotone increasing, continuous and convex upwards function of the variable $t \geq 0$ (see [TRI78, Lemma 1.3.1]).

For $\theta \in (0, 1)$ and $1 < p < +\infty$ let us consider the following space:

$$(A_0, A_1)_{\theta,p} = \left\{x \in A_0 + A_1 \ \Big|\ \int_0^{+\infty} \Big[t^{-\theta} K(t, x)\Big]^p \frac{dt}{t} < +\infty\right\}. \tag{1.50}$$

$(A_0, A_1)_{\theta,p}$ with $\|x\|_{\theta,p} = (\int\limits_0^{+\infty} [t^{-\theta} K(t,x)]^p \frac{dt}{t})^{\frac{1}{p}}$ is a Banach space (for more details see [TRI78, Section 1.3]) and it results in (see [TRI78, Theorem 1.3.3]):

$$A_0 \cap A_1 \subset (A_0, A_1)_{\theta,p} \subset A_0 + A_1 \qquad \forall \theta \in (0,1),\ \forall 1 < p < +\infty \tag{1.51}$$

with dense and continuous embedding.

Definition 1.7. Let it be $1 \le r < 2$. We say that the filter of Banach spaces $\{Z_p\}_{p \ge r}$ and Hilbert space H satisfy *main conditions*, if

(a) $\forall p_2 > p_1 > r$ $Z_{p_2} \subset Z_{p_1} \subset H$ with continuous and dense embedding;
(b) $\forall p_2 > p > p_1 > r$ $(Z_{p_1}, Z_{p_2})_{\theta,p} = Z_p$, where $\theta = \theta(p) \in (0,1)$: $\frac{1}{p} = \frac{1-\theta}{p_1} + \frac{\theta}{p_2}$;
(c) Z_2 is a Hilbert space.
(d) For every complete system $\{h_j\}_{j \ge 1} \subset Z_p$, $p \ge r$, $\forall M > 2$, $\forall N \in (r,2)$ the set $I_+(I_-)$ contains its minimal (maximal) element (when $I_+(I_-) \neq \emptyset$), where $I_+ = I \cap [2,M)$, $I_- = I \cap (N,2]$, $I = \{p \ge r \,|\, (\{h_i\}_{i \ge 1}; Z_p; H)$ is not satisfies Condition (γ) with constant $C\}$, $C \ge 1$, $\|\cdot\|_{Z_p} = \|\cdot\|_{(Z_N,Z_M)_{\theta(p),p}}$ $\forall p \in (N;M)$.

Theorem 1.25. *Let us assume:* $1 \le r < 2$, *filter of Banach spaces* $\{Z_p\}_{p \ge r}$ *and Hilbert space* H *satisfy main conditions, vector system* $\{h_i\}_{i \ge 1} \subset Z_2$ *such that the triple* $\left(\{h_i\}_{i \ge 1}; Z_2; H\right)$ *satisfies Condition* (γ) *with constant* $C \ge 1$ *and* $\{h_i\}_{i \ge 1} \subset Z_p$ *for all* $p > r$. *Then, for all* $p > r$ *the triple* $\left(\{h_i\}_{i \ge 1}; Z_p; H\right)$ *satisfies Condition* (γ).

Remark 1.20. In the case $Z_2 \subset H$ with compact embedding, thanks to Lemma 1.10, as a vector system $\{h_i\}_{i \ge 1}$ we can choose a special basis for the pair $(Z_2; H)$. In particular, the above result means that the special basis for $(Z_2; H)$ is a Schauder basis for an arbitrary space Z_p at $r < p \le 2$.

Proof. For $1 \le r < 2$ let $\{h_i\}_{i \ge 1} \subset Z_r$ be a vector system such that the triple $\left(\{h_i\}_{i \ge 1}; Z_2; H\right)$ satisfies Condition (γ).

Let us prove that $\forall p > r$ the triple $(\{h_i\}_{i \ge 1}; Z_p; H)$ satisfies Condition (γ) with constant C.

At first we consider the case $p \ge 2$. Let $N > 2$ be an arbitrary fixed number. We check, that $\forall p \in [2,N)$ the triple $(\{h_i\}_{i \ge 1}; Z_p; H)$ satisfies Condition (γ) with constant C. For the proof of this fact we benefit from modified transfinitary induction method. The set $W = [2,N)$ is ordered by order "$\prec$" := "$\le$".

For an arbitrary $p \in W$ the statement $G(p)$ consists of the triple $(\{h_i\}_{i \ge 1}; Z_p; H)$ satisfies Condition (γ) with constant C. So,

(1) As $p = 2$ (for the first element of W) the statement $G(p)$ holds, thanks to conditions of this Theorem;

(2) Let p be an arbitrary element in W. Assuming $G(q)$ is true for all $q \in I(p) = [2, p)$, we prove that from here the statement $G(p)$ follows. Let x be a fixed element in the space Z_N, dense in Z_p, $a \in (r, 2)$ is arbitrary. Then $\forall q \in [2, p]$, in virtue of (1.50) and the main condition (b) for $\{Z_p\}_{p \geq r}$ and H with $p = q$, $p_1 = a$, $p_2 = N$, it results in:

$$\|x\|_{Z_q} = \|x\|_{(Z_a, Z_N)_{\theta,q}} = \left(\int_0^{+\infty} \left[t^{-\theta} K(t, x)\right]^q \frac{dt}{t} \right)^{\frac{1}{q}}, \tag{1.52}$$

where $\theta = \theta(q) = \frac{\frac{1}{a} - \frac{1}{q}}{\frac{1}{a} - \frac{1}{N}} \in [\theta(2), \theta(p)] = \left[\frac{\frac{1}{a} - \frac{1}{2}}{\frac{1}{a} - \frac{1}{N}}, \frac{\frac{1}{a} - \frac{1}{p}}{\frac{1}{a} - \frac{1}{N}} \right] \subset (0, 1)$, i.e. $\frac{1}{q} = \frac{1-\theta}{a} + \frac{\theta}{N}$.

In the following we prove

$$\|x\|_{Z_q} \to \|x\|_{Z_p} \quad \text{as} \quad q \to p \ (q \in [2, p)). \tag{1.53}$$

Denoted by

$$f(t, q) = \left[t^{-\theta(q)} K(t, x)\right]^q \frac{1}{t}, \qquad \forall (t, q) \in [0, +\infty) \times [2, p],$$

from (1.50) and (1.52) it obviously follows that for every $\forall q \in [2, p]$ $f(\cdot, q) \in L_1[0, +\infty)$; moreover for almost every $t \in [0, +\infty)$ $f(t, \cdot) \in C[2, p]$. Furthermore, pointing out that for every $t > 0$ and $q \in [2, p]$

$$\left[t^{-\theta(q)} K(t, x)\right]^q \frac{1}{t} \leq \max \left\{ \left[t^{-\theta(2)} K(t, x)\right]^2; \left[t^{-\theta(2)} K(t, x)\right]^p; \right.$$
$$\left. \left[t^{-\theta(p)} K(t, x)\right]^2; \left[t^{-\theta(p)} K(t, x)\right]^p \right\} \frac{1}{t} =: g(t);$$

having in mind (1.51) and $x \in Z_N = Z_a \cap Z_N$, we have:

$$\int_0^{+\infty} |g(t)| dt = \int_0^{+\infty} g(t) dt \leq \max \left\{ \int_0^{+\infty} \left[t^{-\theta(2)} K(t, x)\right]^2 \frac{dt}{t}; \int_0^{+\infty} \left[t^{-\theta(2)} K(t, x)\right]^p \frac{dt}{t}; \right.$$
$$\left. \int_0^{+\infty} \left[t^{-\theta(p)} K(t, x)\right]^2 \frac{dt}{t}; \int_0^{+\infty} \left[t^{-\theta(p)} K(t, x)\right]^p \frac{dt}{t} \right\}$$
$$= \max \left\{ \|x\|^2_{(Z_a, Z_N)_{\theta(2),2}}; \|x\|^p_{(Z_a, Z_N)_{\theta(2),p}}; \|x\|^2_{(Z_a, Z_N)_{\theta(p),2}}; \right.$$
$$\left. \|x\|^p_{(Z_a, Z_N)_{\theta(p),p}} \right\} < +\infty.$$

Thus, the Theorem of continuous dependence of Lebesgue integral on parameter all conditions of the Theorem on continuous association of an integral of Lebesgue on parameter assures the convergence (1.53).

By using the induction assumption

$$\forall q \in [2, p) \quad \forall x \in Z_N \quad \forall n \geq 1 \qquad \|P_n x\|_{Z_q} \leq C\,\|x\|_{Z_q}.$$

and passing to the limit as $q \nearrow p$ in the last inequality, we obtain

$$\|P_n x\|_{Z_p} \leq C\,\|x\|_{Z_p} \qquad \forall x \in Z_N \quad \forall n \geq 1.$$

Then from density Z_N in Z_p and continuity P_n on Z_p $\forall n \geq 1$ the statement $G(p)$ follows. So, for all $p \in [2, N)$ the statement $G(p)$ is true. Because of N is arbitrary greater than 2, the triple $\left(\{h_i\}_{i\geq 1}; Z_p; H\right)$ satisfies Condition (γ) for every $p \geq 2$.

In order to conclude the proof of the Theorem it is necessary to remark that the case $p \in (r, 2]$ can be proved similarly to the case $p \geq 2$, by replacing "$\prec$" with "$>$" and setting $W = (N, 2]$, where $N \in (r, 2)$ is arbitrary.

The Theorem is proved. □

Corollary 1.6. *Let V_1, V_2 be Banach spaces, continuously embedded in the Hilbert space H. Let us assume that for some filters of Banach spaces $\{Z^i_p\}_{p\geq r_i}$ $(r_i \in [1; 2)$, $i = 1, 2)$, that together with H satisfy main conditions, there exist $p_i > r_i$ such that $V_i = Z^i_{p_i}$ $(i = 1, 2)$, within to equivalence of norms. Moreover, there exists Hilbert space $Z \subset V_1 \cap V_2$, compactly embedded in H, such that for special basis $\{h_i\}_{i\geq 1}$ for pair $(Z; H)$, for some $0 \leq \mu_1 \leq \mu_2, \ldots, \mu_j \longrightarrow \infty$ at $j \longrightarrow \infty$ and $s_i > 0$ $(i = 1, 2)$*

$$Z^i_2 = \left\{u \in H \;\middle|\; \sum_{j=1}^{\infty} \mu_j^{s_i} (u, h_j)^2 < +\infty\right\}$$

be a Hilbert space with inner product

$$(u, v)_{Z^i_2} = \sum_{j=1}^{\infty} \mu_j^{2s_i} (u, h_j)(v, h_j). \tag{1.54}$$

Then triple $\left(\{h_i\}_{i\geq 1}; V_i; H\right)$ satisfies Condition (γ) $(i = 1, 2)$.

Proof. Having in mind Lemma and Theorem, it is enough to show that $\{h_i\}_{i\geq 1}$ is a special basis for $(Z^i_2; H)$ $(i = 1, 2)$. Condition (i) of Definition 1.6 is obviously satisfied. Using (1.54) and condition (i) we have

$$\forall i, j \geq 1 \quad (h_i, h_j)_{Z^i_2} = \sum_{k=1}^{\infty} \mu_k^{2s} (h_i, h_k)(h_j, h_k) = \mu_i^{2s} \delta_{ij} = \mu_i^{2s} \begin{cases} 1, & i = j, \\ 0, & i \neq j, \end{cases}$$

so the condition (ii) holds. Finally condition (iii) follows from the last equality.

The Lemma is proved. □

Example 1.1. Let Ω be a bounded domain in $\mathbb{R}^n$, that satisfies the cone condition (see [TRI78, Definition 4.2.3]). Let us note that cubes and bounded domains with rather smooth interface satisfy such condition (see [TRI78, Remark 4.2.3.5]). We denote by $W^{k,p}(\Omega)$ the Sobolev space with $p > 1$, $k = 0, 1, 2 \ldots$. Let $\{h_n\}_{n\geq 1}$ be a special basis for $(H_{k+1}(\Omega); L_2(\Omega))$, where $H_{k+1}(\Omega) = W^{k+1,2}(\Omega)$.

Next result is the direct consequence of corollary and interpolation Theorem of Sobolev spaces [TRI78, Theorem 4.3.1.2].

Corollary 1.7. *Hence we obtain that for every $p > 1$ the triple $(\{h_n\}_{n\geq 1}; W^{k,p}(\Omega); L_2(\Omega))$ satisfies Condition (γ), under the main condition (d).*

1.2 The Classes of W_{λ_0}-Pseudomonotone Maps

In this section we consider the main properties of multivalued w_{λ_0}-pseudomonotone maps.

Let $(X, \|\cdot\|_X)$ be some Banach space, $A : X \to 2^{X^*}$ be a strict multivalued map, i.e.

$$A(y) \neq \emptyset \quad \forall y \in X.$$

We consider the associated maps

$$\text{co}A : X \to 2^{X^*} \quad \text{and} \quad \overset{*}{\overline{\text{co}}}\, A : X \to 2^{X^*},$$

defined by relations

$$(\text{co}A)(y) = \text{co}(A(y)) \quad \text{and} \quad (\overset{*}{\overline{\text{co}}}\, A(y)) = \overset{*}{\overline{\text{co}}}\, (A(y))$$

respectively, where $\underline{*}$ is weakly star closure in the space X^*.

For the multivalued map $A : X \rightrightarrows X^*$ we define the *upper*

$$[A(y), \omega]_+ = \sup_{d \in A(y)} \langle d, w \rangle_X$$

and *lower*

$$[A(y), \omega]_- = \inf_{d \in A(y)} \langle d, w \rangle_X$$

support functions, where $y, \omega \in X$ and also the *upper*

$$\|A(y)\|_+ = \sup_{d \in \mathscr{A}(y)} \|d\|_{X^*}$$

and *lower*

$$\|A(y)\|_- = \inf_{d \in \mathscr{A}(y)} \|d\|_{X^*}$$

norms.

Definition 1.8. Later on $y_n \rightharpoonup y$ in X means that y_n converges weakly to y in the space X. If the space X is reflexive, then $y_n \rightharpoonup y$ in X^* will mean that y_n converges weakly to y in the space X^*. If not, then y_n converges to y weakly star in the space X^*.

Denote by $C_v(X^*)$ the set of all nonempty, convex, weakly star closed subsets of X^*.

Let W be also a normalized space with the norm $\|\cdot\|_W$. We consider $W \subset X$ with continuous embedding, $C \in \Phi$, i.e. $C(r_1;\cdot) : \mathbb{R}_+ \to \mathbb{R}$ is a continuous function for every $r_1 \geq 0$ and such that $\tau^{-1}C(r_1; \tau r_2) \to 0$ for $\tau \to 0+ \ \forall\, r_1, r_2 \geq 0$ and $\|\cdot\|'_W$ is the (semi)norm on X, that is relatively compact on W and relatively continuous on X.

Definition 1.9. The multivalued map $A : D(A) \subset X \to 2^{X^*}$ with the convex definitional domain $D(A)$ is called:

- *Monotone* on $D(A)$, if

$$[A(y_1), y_1 - y_2]_- \geq [A(y_2), y_1 - y_2]_+ \qquad \forall\, y_1, y_2 \in D(A).$$

- *Maximal monotone* on $D(A)$, if it is monotone and from

$$\langle w - d(u), v - u\rangle_X \geq 0$$

for each $u \in D(L)$, $d(u) \in A(u)$ it follows that $v \in D(A)$ and $\omega \in A(v)$.
- *An operator with semibounded variation on W (with (X, W)-semibounded variation)* if $\forall\, y_1, y_2 \in D(A)$, $\|y_1\|_X \leq R$, $\|y_2\|_X \leq R$

$$[A(y_1), y_1 - y_2]_- \geq [A(y_2), y_1 - y_2]_+ - C(R; \|y_1 - y_2\|'_W);$$

- *An operator with N-semibounded variation on W* if

$$[A(y_1), y_1 - y_2]_- \geq [A(y_2), y_1 - y_2]_- - C(R; \|y_1 - y_2\|'_W);$$

- *An operator with V-semibounded variation on W* if

$$[A(y_1), y_1 - y_2]_+ \geq [A(y_2), y_1 - y_2]_+ - C(R; \|y_1 - y_2\|'_W).$$

- *λ-pseudomonotone on W (w_λ-pseudomonotone),* if for every sequence $\{y_n\}_{n\geq 0} \subset W \cap D(A)$ such that $y_n \rightharpoonup y_0$ in W, from the inequality

$$\overline{\lim_{n\to\infty}} \langle d_n, y_n - y_0\rangle_X \leq 0, \tag{1.55}$$

where $d_n \in A(y_n)$ $\forall n \geq 1$, it follows the existence of $\{y_{n_k}\}_{k\geq 1}$ from $\{y_n\}_{n\geq 1}$ and $\{d_{n_k}\}_{k\geq 1}$ from $\{d_n\}_{n\geq 1}$ such that

$$\underline{\lim_{k\to\infty}} \langle d_{n_k}, y_{n_k} - w\rangle_X \geq [A(y_0), y_0 - w]_- \quad \forall w \in D(A); \tag{1.56}$$

- *λ_0-pseudomonotone on W ($w\lambda_0$-pseudomonotone),* if for every sequence $\{y_n\}_{n\geq 0} \subset W \cap D(A)$ such that $y_n \rightharpoonup y_0$ in W, $d_n \rightharpoonup d_0$ in X^*, where $d_n \in A(y_n)$ $\forall n \geq 1$, from the inequality (1.55), it follows the existence of $\{y_{n_k}\}_{k\geq 1}$ from $\{y_n\}_{n\geq 1}$ and $\{d_{n_k}\}_{k\geq 1}$ from $\{d_n\}_{n\geq 1}$ such that (1.56) is true.

Remark 1.21. The idea of using subsequences in Definition 1.9 for single-valued pseudomonotone operators was introduced by I.V.Skrypnik [SKR94].

Let X be again some Banach space. Further $A : X \rightrightarrows X^*$ will mean that A maps X into $2^{X^*} \setminus \emptyset$, i.e. A is a multivalued map with nonempty bounded values.

Proposition 1.17. *Let Y_1, Y_2 be reflexive spaces,*

$$L : D(L) \subset Y_1 \to Y_2$$

is closed linear operator i.e.

if $D(L) \ni y_n \to y$ in Y_1 and $Ly_n \to \chi \in Y_2$ in Y_2 then $y \in D(L)$ and $\chi = Ly$. Then the normalized space $D(L)$ with the graph norm

$$\|y\|_{D(L)} = \|y\|_{Y_1} + \|Ly\|_{Y_2} \quad \forall y \in D(L) \tag{1.57}$$

is a reflexive Banach space.

Proof. Let us prove the reflexivity of the normalized space $D(L)$ with the norm (1.57). To this aim, taking into account that under this condition the given space is a Banach space, let us consider the reflexive Banach space $Z = Y_2 \times Y_1$ with the norm

$$\|z\|_Z = \|z_1\|_{Y_2} + \|z_2\|_{Y_1} \quad \forall z = (z_1, z_2) \in Z$$

and the linear variety

$$\mathscr{L} = \{(Ly, y) \mid y \in D(L)\} \subset Z.$$

Let us prove that $\mathscr{L}$ is closed in Z. Under the following assumption

$$\mathscr{L} \ni z_n = (Lx_n, x_n) \to (\chi, x) = z \in Z \quad \text{in} \quad Z \quad \text{as} \quad n \to +\infty,$$

we have

$$x_n \to x \quad \text{in} \quad Y_1 \quad \text{and} \quad Lx_n \to \chi \quad \text{in} \quad Y_2 \quad \text{as} \quad n \to +\infty.$$

Whence, due to the closureness of the operator L on $D(L)$ it follows that $x \in D(L)$ and $Lx = \chi$. So, $z = (\chi, x) \in \mathscr{L}$.

Further, since an arbitrary closed linear subspace of a reflexive Banach space is reflexive [HP00, Corollary 1, p.51], then the space $(\mathscr{L}, \|\cdot\|_Z)$ is a reflexive Banach space too.

Now we consider the isometric isomorphism I between Banach spaces $D(L)$ and $\mathscr{L}$ defined in the following way:

$$D(L) \ni y \to Iy = (Ly, y) \in \mathscr{L}.$$

Let us prove the reflexivity of $D(L)$ by using the reflexivity criterium [PET67, p.57]. Let us consider an arbitrary bounded sequence $\{y_n\}_{n\geq 1} \subset D(L)$, so $\{Iy_n\}_{n\geq 1}$ is bounded in $\mathscr{L}$. Hence, due to the reflexivity criterium, there exists a subsequence $\{y_{n_k}\}_{k\geq 1} \subset \{y_n\}_{n\geq 1}$ such that

$$Iy_{n_k} \rightharpoonup y \quad \text{in} \quad \mathscr{L} \quad \text{as} \quad k \to +\infty$$

for some $y \in \mathscr{L}$. From here, for an arbitrary $f \in D(L)^*$ (we observe that $f \circ I^{-1} \in \mathscr{L}^*$) if follows that

$$f(y_{n_k}) = f(I^{-1}(Iy_{n_k})) \to f(I^{-1}y) \quad \text{as} \quad k \to \infty.$$

Thus

$$y_{n_k} \rightharpoonup z = I^{-1}y \in D(L) \quad \text{in} \quad D(L) \quad \text{as} \quad k \to \infty.$$

Therefore, due to the reflexivity criterium, the space $D(L)$ is reflexive with respect to the norm (1.57). □

Corollary 1.8. *Let Y be reflexive Banach space, $L : D(L) \subset Y \to Y^*$ – be linear maximally monotone on $D(L)$ operator. Then for each bounded in $D(L)$ sequence there is weakly converged in $D(L)$ subsequence.*

Proof. At first let us check the conditions of Proposition 1.17. It is enough to check the closureness of L on $D(L)$. Let $\{y_n\}_{n\geq 1} \subset D(L)$ be such sequence that for some $y \in Y$ and $\chi \in Y^*$

$$y_n \rightharpoonup y \quad \text{in} \quad Y \quad \text{and} \quad Ly_n \rightharpoonup \chi \quad \text{in} \quad Y^* \quad \text{as} \quad n \to \infty. \tag{1.58}$$

Let us show, that

$$\langle \chi - Lu, y - u\rangle_Y \geq 0 \qquad \text{for any } u \in D(L).$$

Really, due to (1.58) and monotone of L on $D(L)$ it follows that for each $n \geq 1$ and $u \in D(L)$

$$0 \leq \langle Ly_n - Lu, y_n - u\rangle_Y \to \langle \chi - Lu, y - u\rangle_Y \quad \text{as } n \to \infty.$$

From the maximal monotony of L follows that $y \in D(L)$ and $\chi = Ly$. It means that L is closed on $D(L)$. Thus from Proposition 1.17 and from the Banach–Alaoglu Theorem it follows the necessary statement.

The Corollary is proved. □

Let W be again a normalized space with the norm $\|\cdot\|_W$. We consider $W \subset X$ with continuous embedding, $C \in \Phi$, i.e. $C(r_1; \cdot) : \mathbb{R}_+ \to \mathbb{R}$ is a continuous function for every $r_1 \geq 0$ and such that $\tau^{-1}C(r_1; \tau r_2) \to 0$ for $\tau \to 0+ \; \forall \, r_1, r_2 \geq 0$ and $\|\cdot\|'_W$ is the (semi)norm on X, that is relatively compact on W and relatively continuous on X.

The first three lemmas order classes of multi-valued maps with semi-bounded variation (s.b.v) and classes of locally bounded maps with bounded values which satisfy Property (Π) and Property $(\kappa)_+$. The obtained statements allow us to retract additional hypotheses for a supplementary boundedness of the generated multi-valued maps with s.b.v. in applications when investigating differential-operators inclusions and multi-variational inequalities which describe mathematical models for non-linear geophysical processes and fields which contain partial differential equations with discontinuous or multi-valued relationship between determinative parameters of the problem. These results are the fundamental generalization of well-known corresponding results for single-valued maps of monotone type (see for example [GGZ74, IM88, ZM99, ZM04] and references there).

Lemma 1.11. *Every strict multivalued operator $A : X \to 2^{X^*}$ with $(X; W)$-semibounded variation is bounded-valued, i.e. $A : X \rightrightarrows X^*$.*

Proof. We remark that for every $y \in X$

$$X \ni \omega \to [A(y), \omega]_+ \in \mathbb{R} \cup \{+\infty\}, \quad X \ni \omega \to [A(y), \omega]_- \in \mathbb{R} \cup \{-\infty\}.$$

So, due to the definition of the semibounded variation on (X, W) we obtain that for all $\omega \in X$, for some $R = R(\omega, y) > 0$

$$[A(y), \omega]_+ \leq [A(y + \omega), \omega]_- + C_A(R; \|\omega\|'_W) < +\infty.$$

From last, in virtue of Banach–Steinhaus Theorem, it follows that $\|A(y)\|_+ < +\infty$ for every $y \in X$. □

Lemma 1.12. *The multivalued operator $A : X \rightrightarrows X^*$ with $(X; W)$-semibounded variation is locally bounded.*

Proof. We obtain this statement arguing by contradiction. If A is not locally bounded then for some $y \in X$ there exists a sequence $\{y_n\}_{n \geq 1} \subset X$ such that $y_n \to y$ in X and $\|A(y_n)\|_+ \to +\infty$ as $n \to +\infty$. We suppose that

$$\alpha_n = 1 + \|A(y_n)\|_+ \|y_n - y\|_X$$

for every $n \geq 1$. Then, due to Proposition 1, $\forall \, \omega \in X$ and some $R > 0$ we have

$$\alpha_n^{-1}[A(y_n),\omega]_+ \le \alpha_n^{-1}\Big\{[A(y_n), y_n - y]_+ + [A(y_n),\omega + y - y_n]_+\Big\}$$
$$\le \alpha_n^{-1}\Big\{[A(y_n), y_n - y]_+ + [A(y+\omega), y+\omega - y_n]_+$$
$$+ C_A(R; \|y_n - y - \omega\|'_W)\Big\}.$$

Since the sequence $\{\alpha_n^{-1}\}$ is bounded and $\|y_n - y - \omega\|'_W \to \|\omega\|'_W$ (according to the assumption $\|\xi\|'_W \le k\|\xi\|_X$ for all $y \in X$), due to Proposition 1, we have

$$\forall n \ge 1 \quad \alpha_n^{-1}[A(y_n),\omega]_+ \le \alpha_n^{-1}\Big\{C_A(R; \|y_n - y - \omega\|'_W)$$
$$+ \|A(y+\omega)\|_+ \cdot \|y + \omega - y_n\|_X\Big\} + 1 \le N_1,$$

where N_1 does not depend on $n \ge 1$. Thus,

$$\sup_{n\ge 1} \left|\alpha_n^{-1}[A(y_n),\omega]_+\right| < \infty \;\; \forall\, \omega \in X.$$

Therefore, since the Banach–Steinhaus Theorem, there exists $N > 0$ such that

$$\|A(y_n)\|_+ \le N\alpha_n = N\left(1 + \|A(y_n)\|_+ \cdot \|y_n - y\|_X\right) \quad \forall n \ge 1.$$

By choosing $n_0 \ge 1$ from the condition $N\|y_n - y\| \le 1/2 \; \forall n \ge n_0$ we obtain that for every $n \ge n_0$ $\|A(y_n)\|_+ \le 2N$, which contradicts the assumption. So, the local boundness is proved. □

Lemma 1.13. *The multivalued operator $A : X \rightrightarrows X^*$ with $(X;W)$-semibounded variation has Property (Π).*

Proof. In virtue of the locally boundness of A there exist $\varepsilon > 0$ and $M_\varepsilon > 0$ such that $\|A(\xi)\|_+ \le M_\varepsilon \;\; \forall\; \|\xi\|_X \le \varepsilon$. It means that for some $R \ge \varepsilon$

$$\|d(y)\|_{X^*} = \sup_{\|\xi\|_X\le\varepsilon} \frac{1}{\varepsilon}\langle d(y),\xi\rangle_X \le \sup_{\|\xi\|_X\le\varepsilon} \frac{1}{\varepsilon}\{[A(y),\xi - y]_+ + \langle d(y), y\rangle_X\}$$
$$\le \sup_{\|\xi\|_X\le\varepsilon} \frac{1}{\varepsilon}\Big\{[A(\xi),\xi - y]_- + \langle d(y), y\rangle_X + C_A(R; \|y - \xi\|'_W)\Big\}$$
$$\le \sup_{\|\xi\|_X\le\varepsilon} \frac{1}{\varepsilon}\Big\{\|A(\xi)\|_+ \cdot \|\xi - y\|_X + \langle d(y), y\rangle_X + C_A(R; \|y - \xi\|'_W)\Big\}$$
$$\le \frac{1}{\varepsilon}(\varepsilon M_\varepsilon + k_2 M_\varepsilon + k_1 + l) = C,$$

where $l = \sup_{\|y\|_X\le k_2} \sup_{\|\xi\|_X\le\varepsilon} C_A(R; \|y - \xi\|'_W) < +\infty$, since $C(R;\cdot) : \mathbb{R}_+ \to \mathbb{R}$ is a continuous function and $\|\cdot\|'_W$ is relatively continuous $\|\cdot\|_X$ on X. □

Lemma 1.14. *Let X be a reflexive Banach space. Then every λ-pseudomonotone on W map is λ_0-pseudomonotone on W. For bounded maps the converse implication is true.*

Proof. The direct implication is obvious. Let us prove the converse implication. We consider the λ_0-pseudomonotone on W map $A : X \rightrightarrows X^*$, $y_n \to y$ weakly in W, the (1.55) holds, where $d_n \in \overset{*}{\overline{\text{co}}}\, A(y_n)$. From the boundness of the operator A it immediately follows the boundness of $\overset{*}{\overline{\text{co}}}\, A$ and so the boundness of the sequence $\{d_n\}$ in X^*. Consequently, there exists a subsequence $\{d_m\} \subset \{d_n\}$ and, respectively, $\{y_m\} \subset \{y_n\}$, such that $d_m \to d$ weakly in X^* and at the same time

$$\overline{\lim_{m\to\infty}}\, \langle d_m, y_m - v\rangle_X \le \overline{\lim_{n\to\infty}}\, \langle d_n, y_n - v\rangle_X \le 0.$$

However the operator A is λ_0-pseudomonotone on W, therefore there exist the subsequences $\{y_{n_k}\}_{k\ge1} \subset \{y_m\}$ and $\{d_{n_k}\}_{k\ge1} \subset \{d_m\}$ for what (1.56) is true. This proves our statement. □

Remark 1.22. Let us pay our attention to the fact that for the classical definitions (not passing to the subsequences) this statement is problematic!

In F. Browder and P. Hess work [BP72] the class of generalized pseudomonotone operators has been introduced.

Definition 1.10. The operator $A : X \to C_v(X^*)$ is called generalized pseudomonotone on W, if for each pair of sequences $\{y_n\}_{n\ge1} \subset W$ and $\{d_n\}_{n\ge1} \subset X^*$ such that $d_n \in A(y_n)$, $y_n \to y$ weakly in W, $d_n \to d$ *-weakly in X^*, from the inequality

$$\overline{\lim_{n\to\infty}}\, \langle d_n, y_n\rangle_X \le \langle d, y\rangle_X \tag{1.59}$$

we have $d \in A(y)$ and $\langle d_n, y_n\rangle_X \to \langle d, y\rangle_X$.

Proposition 1.18. *Every generalized pseudomonotone on W operator is λ_0-pseudomonotone on W.*

Proof. Let $y_n \to y$ weakly in W, $A(y_n) \ni d_n \to d$ *-weakly in X^* and (1.59) holds (we remark that in this case the inequality (1.55) is also true). Then, in view of the generalized pseudomonotony, $\langle d_n, y_n\rangle_X \to \langle d, y\rangle_X$, $d \in A(y)$, consequently, in virtue of Proposition 1,

$$\lim_{n\to\infty}\, \langle d_n, y_n - v\rangle_X = \langle d, y - v\rangle_X \ge [A(y), y - v]_- \quad \forall v \in X.$$

The Proposition is proved. □

The converse statement in Proposition 1.18 is not true, but

Proposition 1.19. *Let $A : X \rightrightarrows X^*$ be a λ_0-pseudomonotone operator. Then the next property takes place:*

from $y_n \to y$ weakly in W, $\overset{*}{\overline{\text{co}}}\, A(y_n) \ni d_n \to d$ $*$-weakly in X^* and from the inequality (1.55) the existence of the subsequences $\{y_m\} \subset \{y_n\}$ and $\{d_m\} \subset \{d_n\}$ such that $\langle d_m, y_m\rangle_X \to \langle d, y\rangle_X$, with $d \in \overset{*}{\overline{\text{co}}}\, A(y)$, follows.

Proof. Let $\{y_n\}$, $\{d_n\}$ be required sequences, consequently, one can choose such subsequences $\{y_m\}$, $\{d_m\}$, that the inequality (1.56) is true. By fixing in the last relation $\omega = y$, we get

$$\langle d_m, y_m - y\rangle_X \to 0$$

or

$$\langle d_m, y_m\rangle_X \to \langle d, y\rangle_X,$$
$$\langle d, y - v\rangle_X = \lim_{m\to\infty} \langle d_m, y_m - v\rangle_X \geq [A(y), y - v]_- \quad \forall\, v \in X.$$

From here, in virtue of Proposition 1 we obtain that $d \in \overset{*}{\overline{\text{co}}}\, A(y)$. □

Proposition 1.20. *Let $A = A_0 + A_1 : X \rightrightarrows X^*$, where $A_0 : X \rightrightarrows X^*$ is a monotone map, and the operator $A_1 : X \rightrightarrows X^*$ has the following properties:*

(1) There exists a linear normalized space Z in which W is compactly and densely enclosed and $X \subset Z$ with continuous and dense embedding.
(2) The operator $A_1 : Z \rightrightarrows Z^*$ is locally polynomial, i.e. $\forall R > 0$ there exists $n = n(R)$ and a polynomial function $P_R(t) = \sum_{0<\alpha\leq n} \lambda_\alpha(R)t^\alpha$ with continuous factors $\lambda_\alpha(R) \geq 0$ such that the estimation is valid

$$\|A_1(y_1) - A_1(y_2)\|_+^{(Z^*)} \leq P_R\left(\|y_1 - y_2\|_Z\right) \quad \forall\, \|y_i\|_Z \leq R, \quad i = 1, 2.$$

Then A is the operator with semibounded variation on W.

Proposition 1.21. *Let in the previous Proposition the operator $A_0 : X \rightrightarrows X^*$ be N-monotone, and instead of the condition (2) we make the following one:*

(2') a map (multivalued) $A_1 : Z \rightrightarrows Z^*$ is locally polynomial in the sense that $\forall R > 0$ there exists $n = n(R)$ and a polynomial $P_R(t)$ for which

$$\text{dist}\left(A_1(y_1), A_1(y_2)\right) \leq P_R\left(\|y_1 - y_2\|_Z\right) \quad \forall\, \|y_i\|_Z \leq R, \quad i = 1, 2. \tag{1.60}$$

Then $A = A_0 + A_1$ is the operator with N-semibounded variation on W.

Proof. We give the proof in Proposition 1.21. In the case of Proposition 1.20 the reasonings are similar. Since for each $y_1, y_2 \in X$

$$[A_0(y_1), y_1 - y_2]_- \geq [A_0(y_2), y_1 - y_2]_+,$$

we must estimate $[A_1(y_1), y_1 - y_2]_- - [A_1(y_2), y_1 - y_2]_-$.

For any $d_1 \in A_1(y_1)$, $d_2 \in A_1(y_2)$ we find

$$\begin{aligned}\langle d_2, y_1 - y_2\rangle_X - \langle d_1, y_1 - y_2\rangle_X &= \langle d_2, y_1 - y_2\rangle_Z - \langle d_1, y_1 - y_2\rangle_Z \\ &\leq \|d_1 - d_2\|_{Z^*} \|y_1 - y_2\|_Z,\end{aligned}$$

hence

$$\begin{aligned}&[A_1(y_2), y_1 - y_2]_- - [A_1(y_1), y_1 - y_2]_- \\ &\leq \operatorname{dist}(A_1(y_1), A_1(y_2)) \|y_1 - y_2\|_Z.\end{aligned}$$

From here and from estimation (1.60) as $\|y_i\|_Z \leq R$ $(i = 1, 2)$ (respectively $\|y_i\|_X \leq \hat{R}$, $R = R(\hat{R})$) we obtain

$$[A_1(y_1), y_1 - y_2]_- \geq [A_1(y_2), y_1 - y_2]_- - C\left(\hat{R}; \|y_1 - y_2\|'_W\right),$$

where $\|\cdot\|'_W = \|\cdot\|_Z$, $C(R, t) = P_R(t)\, t$.

It is easy to check that $C \in \Phi$. □

New classes of λ_0-pseudomonotone maps are defined in the next statement. It is fundamentally used when validating an existence of a solution for the second order differential-operator inclusion by the singular perturbations method [PKZ08,ZK07].

Proposition 1.22. *Let one of two conditions hold:*

(1) $A : X \rightrightarrows X^*$ is radially lower semicontinuous operator with semibounded variation on W.

(2) $A : X \rightrightarrows X^*$ is radially continuous from above operator with N-semibounded variation on W with compact values in X^*.

Then A is λ_0-pseudomonotone on W map.

Proof. Let $y_n \to y$ weakly in W, $\overline{\text{co}}^{*} A(y_n) \ni d_n \to d$ weakly star in X^* and (1.55) is true. By using the property of semibounded variation on W of the operator A, we conclude that for every $v \in X$

$$\langle d_n, y_n - v\rangle_X \geq [A(y_n), y_n - v]_- \geq [A(v), y_n - v]_+ - C\left(R; \|y_n - v\|'_W\right).$$

The function $X \ni w \mapsto [A(v), w]_+$ is convex and semicontinuous from below, and so it is weakly semicontinuous from below, therefore by substituting in the last inequality $v = y$ and passing to the limit, in view of the properties of the function C, we obtain $\varliminf_{n\to\infty} \langle d_n, y_n - y\rangle_X \geq 0$, i.e $\langle d_n, y_n - y\rangle_X \to 0$.

For any $h \in X$ and $\tau \in [0, 1]$ we shall put $\omega_\tau = \tau h + (1 - \tau)\, y$, then

$$\langle d_n, y_n - \omega_\tau\rangle_X \geq [A(\omega_\tau), y_n - \omega_\tau]_+ - C\left(R; \|y_n - \omega_\tau\|'_W\right)$$

or by passing to the limit

$$\tau \varliminf_{n\to\infty} \langle d_n, y - h\rangle_X \geq \tau\,[A(w_\tau), y - h]_+ - C\left(R; \tau\,\|y - h\|'_W\right).$$

By dividing the last inequality by τ and by passing to the limit as $\tau \to 0+$, in view of the radial lower semicontinuity of A and of the properties of the function C, we obtain that for each $h \in X$ $\varliminf_{n\to\infty} \langle d_n, y - h\rangle_X$

$$\geq \varliminf_{\tau\to 0+} [A(\omega_\tau), y - h]_+ + \lim_{\tau\to 0+} \frac{1}{\tau} C\left(R; \tau\,\|y - h\|'_W\right) \geq [A(y), y - h]_-.$$

Moreover as $\langle d_n, y_n - y\rangle_X \to 0$ we get

$$\varliminf_{n\to\infty} \langle d_n, y_n - h\rangle_X = \varliminf_{n\to\infty} \langle d_n, y - h\rangle_X \geq [A(y), y - h]_- \quad \forall h \in X,$$

and this proves the first statement of Proposition 1.22.

Now we stop on the basic distinctive moments of the second statement. Because of the N-semiboundness of the variation for the operator A we conclude that

$$\begin{aligned}\varliminf_{n\to\infty} \langle d_n, y_n - v\rangle_X &\geq \varliminf_{n\to\infty} [A(y_n), y_n - v]_- \\ &\geq \varliminf_{n\to\infty} [A(v), y_n - v]_- - C\left(R; \|y - v\|'_W\right) \end{aligned} \tag{1.61}$$

Let us estimate the first member in the right part of (1.61). Let us prove that the function $X \ni h \mapsto [A(v), h]_-$ is weakly lower semicontinuous $\forall v \in X$. Let $z_n \to z$ weakly in X, then for each $n = 1, 2, \ldots \exists \xi_n \in \overset{*}{\overline{\mathrm{co}}}\, A(v)$ such that

$$[A(v), z_n]_- = \langle \xi_n, z_n\rangle_X.$$

From the sequence $\{\xi_n; z_n\}$ we take a subsequence $\{\xi_m; z_m\}$ such that

$$\varliminf_{n\to\infty} [A(v), z_n]_- = \varliminf_{n\to\infty} \langle \xi_n, z_n\rangle_X = \lim_{m\to\infty} \langle \xi_m, z_m\rangle_X$$

and by virtue of the compactness of the set $\overset{*}{\overline{\mathrm{co}}}\, A(v)$ we find that $\xi_m \to \xi$ strongly in X^* with $\xi \in \overset{*}{\overline{\mathrm{co}}}\, A(v)$. Hence

$$\underline{\lim}_{n\to\infty} [A(v), z_n]_- = \lim_{n\to\infty} \langle \xi_m, z_m\rangle_X = \langle \xi, z\rangle_X = [A(v), z]_-,$$

and this proves the weak lower semicontinuity of the function $h \mapsto [A(v), h]_-$.

So from (1.61) we get

$$\begin{aligned}\underline{\lim}_{n\to\infty} \langle d_n, y_n - v\rangle_X &\geq \underline{\lim}_{n\to\infty} [A(y_n), y_n - v]_- \\ &\geq [A(v), y - v]_- - C\left(R; \|y - v\|'_W\right).\end{aligned}$$

Then by substituting v with y in the last inequality we have $\langle d_n, y_n - y\rangle_X \to 0$, therefore

$$\underline{\lim}_{n\to\infty} \langle d_n, y_n - v\rangle_X \geq [A(v), v - w]_- - C\left(R; \|y - v\|'_X\right) \quad \forall v \in X.$$

By substituting in the last inequality v for $tw + (1-t)y$, where $w \in X$, $t \in [0, 1]$, then by dividing the result on t and by passing to the limit as $t \to 0+$, because of the radial semicontinuity from above we find

$$\underline{\lim}_{n\to\infty} \langle d_n, y_n - w\rangle_X \geq [A(y), y - w]_- \quad \forall w \in X.$$

The Proposition is proved. □

Definition 1.11. The multivalued map $A : X \rightrightarrows X^*$ is called:

- *s-radial lower semicontinuous on* W, if for fixed $y, \xi \in W$ and any selector $d \in \overline{\text{co}}^* A$
$$\underline{\lim}_{t\to 0+} \langle d(y - t\xi), \xi\rangle_X \geq [A(y), \xi]_-;$$
- s_{λ_0}*-pseudomonotone on* W, if for every $\{y_n\}_{n\geq 0} \subset W$ such that $y_n \rightharpoonup y_0$ in W, $\overline{\text{co}}^* A(y_n) \ni d_n \rightharpoonup d_0$ in X^*, from the inequality (1.55) it follows the existence of $\{y_{n_k}\}_{k\geq 1}$ from $\{y_n\}_{n\geq 1}$ and $\{d_{n_k}\}_{k\geq 1}$ from $\{d_n\}_{n\geq 1}$ such that
$$\langle d_{n_k}, y_{n_k} - y_0\rangle_X \to 0 \text{ as } k \to \infty \quad \text{and} \quad d \in \overline{\text{co}}^* A(y).$$

The next Proposition is some generalization of the Minty Lemma on W.

Proposition 1.23. *Let $A : X \rightrightarrows X^*$ be radial lower semicontinuous on W with $(X; W)$-semibounded variation, the space W is dense in X, $y \in W$. Then the inequality*

$$\langle f, y - v\rangle_X \geq [A(v), y - v]_+ - C(R; \|y - v\|'_W) \quad \forall v \in W \tag{1.62}$$

is equivalent to the inclusion

$$f \in \overline{\text{co}}^* A(y). \tag{1.63}$$

Proof. The implication (1.63) $\Rightarrow$ (1.62) is obvious. Let us consider (1.62) $\Rightarrow$ (1.63). For each $h \in W$ let us set in (1.62) $v = y$-th, $t > 0$, divide all parts on t and pass to the limit as $t \to 0+$. Since the radial lower semicontinuity of A we obtain, that

$$\langle f, h\rangle_X \geq \varliminf_{t\to 0+} [A(y - tv), h]_+ \geq [A(y), h]_- \quad \forall h \in W.$$

For each $h \in X$ let $W \ni h_n \to h$ in X. Thus for each $n \geq 1$ there is $d_n \in \overline{\text{co}}^* A(y)$ such that $[A(y), h_n]_- = \langle d_n, h_n\rangle_X$. In particular, we may assume, that $d_n \to d \in \overline{\text{co}}^* A(y)$ weakly star in X^*. Therefore,

$$\langle f, h\rangle_X = \lim_{n\to\infty} \langle f, h_n\rangle_X \geq \lim_{n\to\infty} [A(y), h_n]_- = \lim_{n\to\infty} \langle d_n, h_n\rangle_X = \langle d, h\rangle_X \quad \forall h \in X,$$

i.e. $\langle f, h\rangle_X \geq \langle d, h\rangle_X \quad \forall h \in X$. Thus $f = d \in \overline{\text{co}}^* A(y)$. □

Corollary 1.9. *Let $A : X \rightrightarrows X^*$ be s-radial lower semicontinuous on W with $(X; W)$-semibounded variation, the space W is dense in X. Then the inequality (1.62) is equivalent to the inclusion $f \in \overline{\text{co}}^* A(y)$.*

Definition 1.12. The multivalued map $A : X \rightrightarrows X^*$ is called *the variation calculus operator on* W, if it can be represented in the form $A(y) = \widehat{A}(y, y)$ where the mapping $\widehat{A} : X \times X \rightrightarrows X^*$ possesses the following properties:

(a) $\forall\, \xi \in W \;\; \widehat{A}(\xi, \cdot) : X \rightrightarrows X^*$ is s-radially lower semicontinuous operator with $(X; W)$-semibounded variation ($(X; W)$-s.b.v.).
(b) $\forall\, \xi \in W$ the mapping $W \ni y \to \widehat{A}(y, \xi) \subset X^*$ is weakly precompact, i.e. for each set B bounded in W the set $\overline{\text{co}}^* \widehat{A}(B, \omega)$ is compact in $\sigma(X^*; X)$-topology of the space X (thus, if $\{y_n\}$ is any bounded in W sequence, $d \in \overline{\text{co}}^* \widehat{A}(\cdot, \omega)$ is some selector ($d_n = d(y_n, \omega) \in \overline{\text{co}}^* \widehat{A}(y_n, \omega)$), then it is possible to extract such a subsequences $\{y_m\} \subset \{y_n\}$ and $\{d_m\} \subset \{d_n\}$ that $d_m = d(y_m, \omega) \to$ æ(ω) weakly star in X^*.
(c) Let $y_n \to y$ weakly in W and for some selector $d \in \overline{\text{co}}^* \widehat{A}$

$$\langle d(y_n, y_n) - d(y_n, y), y_n - y\rangle_X \to 0$$

($d(y_n, y_n) \in \overline{\text{co}}^* \widehat{A}(y_n, y_n), d(y_n, y) \in \overline{\text{co}}^* \widehat{A}(y_n, y)$). Then there exists $\{y_m\} \subset \{y_n\}$ such that $\forall\, \xi \in W \; d(y_m, \xi) \to \chi$ weakly star in X^*, at that $\chi \in \overline{\text{co}}^* \widehat{A}(y, \omega)$.
(d) If $y_n \to y$ weakly in W and for some $\xi \in W \; d(y_n, \xi) \to \chi$ weakly star in X^*, where $d(y_n, \xi) \in \overline{\text{co}}^* \widehat{A}(y_n, \xi)$, then $\langle d(y_n, \xi), y_n\rangle_X \to \langle \chi, y\rangle_X$.

The next statement orders classes of w_{λ_0}-pseudo-monotone maps. Particularly, these results show that energetic extensions of differential operators of Leray–Lions type, Dubinskii type, operators of hydrodynamic type etc are the maps of w_{λ_0}-pseudo-monotone type [BR68, ET99, KOG01, PAP87]. Using this statement and results about an existence of the generalized solution and functional-topological properties of the resolving operator of differential-operator inclusions and multi-variational inequalities we obtained new results concerning properties of set of solutions for partial differential equations with non-linear, discontinuous or multi-valued relationship between determinative parameters of non-linear geophysical processes and fields.

Proposition 1.24. *Let the multivalued map $A : X \rightrightarrows X^*$ has the bounded values. Then the next implications are true: "A is s-radially lower semicontinuous operator with $(X; W)$-s.b.v." $\overset{1}{\Rightarrow}$ "A is a variation calculus operator on W" $\overset{2}{\Rightarrow}$ "A is λ-pseudomonotone on W" $\overset{3}{\Rightarrow}$ "A is λ_0-pseudomonotone on W" $\overset{4}{\Rightarrow}$ "A is $s\lambda_0$-pseudomonotone on W" $\overset{5}{\Rightarrow}$ "A satisfies the (M) condition on W [ZMN04]".*

Proof. Let us consider the first implication. Let $\widehat{A} : X \times X \rightrightarrows X^*$, defined by

$$\widehat{A}(v, y) = \widehat{A}(\omega, y), \ \widehat{A}(y, y) = A(y) \qquad \forall v, \omega, y \in X,$$

i.e. the constant map by the first argument, when the second fixed. Let us check the conditions of Definition 1.12.

The condition (a) is clear. The condition (b) follows from the next conclusions: $\widehat{A}(\cdot, \omega) : W \rightrightarrows X^*$ is constant map for each $\omega \in W$. Hence, $\widehat{A}(B, \omega) = A(\omega)$ is bounded set in X^* for each $B \subset W$. Thus, in virtue of the Banach–Alaoglu Theorem [RUD73], $\overline{\text{co}}^*\widehat{A}(B, \omega)$ is a compact set in weakly star topology o the space X^*. Let us check the condition (c). Let $y_n \rightharpoonup y$ in W and $d(y_n, y_n) - d(y_n, y), y_n - y\rangle_X \to 0$ with some selector $d \in \overline{\text{co}}^*\widehat{A}$, at that for each $\omega \in W$ $æ(v) = d(y_n, v) \in \overline{\text{co}}^*\widehat{A}(y_n, v) = \overline{\text{co}}^* A(v)$. Hence, there is a subsequence $\{y_m\} \subset \{y_n\}$ such that $æ(v) = d(y_n, v) \to æ(v)$ weakly star in X^*. Moreover, $æ(v) \overset{\Delta}{=} d(y, v) \in \overline{\text{co}}^*\widehat{A}(y, v) = \overline{\text{co}}^* A(v)$.

Condition (d). Let $y_n \rightharpoonup y$ in W and for each $v \in W$, for each selector $d \in \overline{\text{co}}^* A$ $æ(v) = d(y_n, v) \in \overline{\text{co}}^*\widehat{A}(y_n, v)$, i.e. $d(y_n, v) \to æ(v)$ weakly star in X^*. Thus, $\langle d(y_n, v), y_n\rangle_X = \langle æ(v), y_n\rangle_X = \langle æ(v), y\rangle_X$, that proves the first implication.

Now we consider the second implication. Let $y_n \rightharpoonup y$ in W and

$$\varlimsup_{n\to\infty} \langle d_n, y_n - y\rangle_X \le 0, \tag{1.64}$$

where $d_n = g(y_n)$, $g \in \overline{\text{co}}^* A$ is some selector, at that

$$A(y) = \widehat{A}(y, y), \quad \widehat{A} : X \times X \rightrightarrows X^*.$$

Due to the condition (b) there is a subsequence $\{y_m\} \subset \{y_n\}$ such, that $g(y_m, y) \to f(y)$ weakly star in X^*. Since the condition (d)

$$\langle g(y_m, y), y_m\rangle_X \to \langle f(y), y\rangle_X.$$

Hence,

$$\langle g(y_m, y), y_m - y\rangle_X \to 0. \tag{1.65}$$

From (1.64) we find

$$\varlimsup_{m\to\infty} \langle d_m, y_m - y\rangle_X \le \varlimsup_{n\to\infty} \langle d_n, y_n - y\rangle_X \le 0,$$

i.e.

$$\varlimsup_{m\to\infty} \langle d_m - g(y_m, y), y_m - y\rangle_X = \varlimsup_{m\to\infty} \langle g(y_m, y_m) - g(y_m, y), y_m - y\rangle_X. \tag{1.66}$$

In virtue of the condition (a)

$$\langle g(y_m, y_m), y_m - y\rangle_X \ge \langle g(y_m, y), y_m - y\rangle_X - C(R; \|y_m - y\|'_W).$$

Thus from (1.65) and from the properties for real function C we have the estimation

$$\varliminf_{m\to\infty} \langle g(y_m, y_m) - g(y_m, y), y_m - y\rangle_X \ge 0.$$

The inequality (1.66) gives

$$\langle g(y_m, y_m) - g(y_m, y), y_m - y\rangle_X \to 0. \tag{1.67}$$

In virtue of the condition (c) for each $v \in W$ $g(y_m, v) \to æ(y, v) \in \overline{\mathrm{co}}^* \widehat{A}(y, v)$ weakly star in X^*. Due to the condition (d) we obtain

$$\langle g(y_m, v), y_m - y\rangle_X \to 0, \tag{1.68}$$

because $\langle g(y_m, v), y_m\rangle_X \to \langle æ(y, v), y\rangle_X$.

Further for each $\omega \in W$ and $\tau \in [0, 1]$ let $v(\tau) = y + \tau(\omega - y)$. Then from the condition (a)

$$\begin{aligned}\tau\langle g(y_m, y_m), y - v\rangle_X &\ge [\widehat{A}(y_m, v(\tau)), y_m - v(\tau)]_+ \\ &\quad - \langle g(y_m, y_m), y_m - y\rangle_X - C(R; \|y_m - v(\tau)\|'_W) \\ &\ge [\widehat{A}(y_m, v(\tau)), y_m - y]_- + \tau[\widehat{A}(y_m, v(\tau)), y - \omega]_+ \\ &\quad - \langle g(y_m, y_m), y_m - y\rangle_X - C(R; \|y_m - v(\tau)\|'_W).\end{aligned} \tag{1.69}$$

For each m there is $\xi_m \in \overline{\mathrm{co}}^* \widehat{A}(y_m, v(\tau))$ such that

$$[\widehat{A}(y_m, v(\tau)), y_m - y]_- = \langle \xi_m, y_m - y\rangle_X,$$

i.e. there exists a selector $r \in \overline{\mathrm{co}}^* \widehat{A}$ such that $\xi_m = r(y_m, v(\tau))$. Furthermore, due to the condition (b) we may conclude that up to a subsequence $r(y_m, v(\tau)) \to \xi(y, v(\tau))$ weakly star in X^*. In virtue of the condition (d)

$$\langle r(y_m, v(\tau)), y_m - y\rangle_X \to 0$$

or

$$[\widehat{A}(y_m, v(\tau)), y_m - y]_- \to 0.$$

Hence, up to a limit in (1.69) as $m \to \infty$, we get

$$\tau \varliminf_{m\to\infty} \langle g(y_m, y_m), y - v\rangle_X \geq \tau \varliminf_{m\to\infty} [\widehat{A}(y_m, v(\tau)), y - \omega]_+ - C(R; \tau\|y - \omega\|'_W),$$

respectively since (1.67)

$$\begin{aligned}\tau \varliminf_{m\to\infty} \langle g(y_m, y_m), y_m - v\rangle_X &\geq \tau \varliminf_{m\to\infty} [\widehat{A}(y_m, v(\tau)), y - \omega]_+ - C(R; \tau\|y-\omega\|'_W)\\ &\geq \tau \lim_{m\to\infty} \langle g(y_m, v(\tau)), y - v\rangle_X - C(R; \tau\|y-\omega\|'_W).\end{aligned}$$

Due to the conditions (b) and (c) we may consider that $g(y_m, v(\tau)) \to æ(y, v(t)) \in \overline{\mathrm{co}}^* \widehat{A}(y, v(\tau))$ weakly star in X^*. Thus

$$\tau \varliminf_{m\to\infty} \langle g(y_m, y_m), y_m - v\rangle_X \geq \tau\langle æ(y, v(t)), y - v\rangle_X - C(R; \tau\|y - \omega\|'_W).$$

If we divide the last inequality on τ and pass to the limit as $\tau \to 0+$, since the s-radial lower semicontinuous of the operator $\widehat{A}(y, \cdot) : W \rightrightarrows W^*$ we finally obtain

$$\begin{aligned}\varliminf_{m\to\infty} \langle d_m, y_m - \omega\rangle_X &= \varliminf_{m\to\infty} \langle g(y_m, y_m), y_m - \omega\rangle_X\\ &\geq [\widehat{A}(y, y), y - \omega]_- = [A(y), y - \omega]_- \quad \forall \omega \in W,\end{aligned}$$

that proves the second implication.

The third implication is obvious. Let us consider the fourth implication. Let $y_n \rightharpoonup y$ in W, $\overline{\mathrm{co}}^* A(y_n) \ni d_n \to d$ weakly star in X^* and $\varlimsup_{n\to\infty} \langle d_n, y_n - y\rangle_X \leq 0$. Then due to the λ_0-pseudomonotony of A on W there are such subsequences $\{y_m\} \subset \{y_n\}$ and $\{d_m\} \subset \{d_n\}$ such that inequality (1.56) is true. Hence, $\lim_{m\to\infty} \langle d_m, y_m - y\rangle_X = 0$, or $\lim_{m\to\infty} \langle d_m, y_m\rangle_X = \langle d, y\rangle_X$. Therefore, from (1.56) we find

$$\langle d, y - v\rangle_X = \varliminf_{n\to\infty} \langle d_n, y_n - v\rangle_X \geq [A(y), y - v]_- \quad \forall v \in X,$$

i.e. $\langle d, \omega\rangle_X \geq [A(y), \omega]_-$ $\forall \omega \in X$. Since the set $A(y)$ is bounded in X^* we obtain that $d \in \overline{\mathrm{co}}^* A(y)$.

The implication 5 checks directly. The Proposition is proved. □

The next propositions concern $w\lambda_0$-pseudo-monotone demi-closed maps disturbed by $w\lambda_0$-pseudo-monotone maps. Such results allows us (by the help of corresponding theorems about properties of the resolving operator) to consider new classes of problems. So, we can state the existence of solutions and continuous (in some sense) dependence of solutions on functional parameters of the given problem.

Proposition 1.25. *Let $A : X \rightrightarrows X^*$ be a λ_0-pseudomonotone on W operator, and let the map $B : X \rightrightarrows X^*$ possess the following properties:*

1. *The map $\overset{*}{\overline{\text{co}}}\, B : W \rightrightarrows X^*$ is compact, i.e. the image of a bounded set in W is precompact in X^*.*
2. *The graph of $\overset{*}{\overline{\text{co}}}\, B$ is closed in $W_w \times X^*$ (i.e. with respect to the weak topology in W and the strong one in X^*).*

Then the map $C = A + B$ is λ_0-pseudomonotone on W.

Proof. Let $y_n \to y$ weakly in W, $d_n \in \overset{*}{\overline{\text{co}}}\, C(y_n)$, $d_n \to d$ weakly star in X^* and

$$\varlimsup_{n\to\infty} \langle d_n, y_n - y\rangle_X \le 0.$$

As the operator $B : W \rightrightarrows X^*$ is bounded, $\overset{*}{\overline{\text{co}}}\, C = \overset{*}{\overline{\text{co}}}\, A + \overset{*}{\overline{\text{co}}}\, B$. Hence $d_n = d_n' + d_n''$, $d_n' \in \overline{\text{co}}^* A(y_n)$, $d_n'' \in \overset{*}{\overline{\text{co}}}\, B(y_n)$. By virtue of the boundness of B, we obtain that $d_n'' \to d''$ weakly star in X^*, so that $d_n' \to d' = d - d''$ weakly star in X^*.

From inequality (1.55), passing to the subsequence $\{y_m\} \subset \{y_n\}$, we find

$$\begin{aligned} 0 &\ge \varlimsup_{n\to\infty} \langle d_n, y_n - y\rangle_X \\ &\ge \varlimsup_{n\to\infty} \langle d_n', y_n - y\rangle_X + \varliminf_{n\to\infty} \langle d_n'', y_n - y\rangle_X \\ &\ge \varlimsup_{m\to\infty} \langle d_m', y_m - y\rangle_X + \lim_{m\to\infty} \langle d_m'', y_m - y\rangle_X. \end{aligned} \tag{1.70}$$

Since $\overset{*}{\overline{\text{co}}}\, B$ is compact and the graph is closed in $W_w \times X^*$, we can consider that $d_m'' \to d''$ strongly in X^* and, moreover, $d'' \in \overset{*}{\overline{\text{co}}}\, B(y)$. Then

$$\varlimsup_{m\to\infty} \langle d_m', y_m - y\rangle_X \le 0.$$

Again, passing to subsequences, as A is λ_0-pseudomonotone, we get

$$\varliminf_{m\to\infty} \langle d_m', y_m - v\rangle_X \ge [A(y), y - v]_-, \quad \forall v \in X,$$

and then

$$\underline{\lim}_{m\to\infty} \langle d_m, y_m - v\rangle_X = \underline{\lim}_{m\to\infty} \langle d'_m, y_m - v\rangle_X + \lim_{m\to\infty} \langle d''_m, y_m - v\rangle_X$$
$$\geq [\overset{*}{\overline{\mathrm{co}}}\, A(y), y - v]_- + \langle d'', y - v\rangle_X$$
$$\geq [\overset{*}{\overline{\mathrm{co}}}\, C(y), y - v]_-, \ \forall v \in X. \qquad \square$$

Proposition 1.26. *Let $A : X \rightrightarrows X^*$ be a λ_0-pseudomonotone operator on W, the embedding of W in Banach space Y be compact and dense, the embedding of X in Y be continuous and dense, and let $\overset{*}{\overline{\mathrm{co}}}\, B : Y \rightrightarrows Y^*$ be a locally bounded map such that the graph of $\overset{*}{\overline{\mathrm{co}}}\, B$ is closed in $Y \times Y^*_w$ (i.e. with respect to the strong topology of Y and the weakly star one in Y^*). Then $C = A + B$ is a λ_0-pseudomonotone on W map.*

Proof. Let (1.55) be fulfilled. The operator $\overset{*}{\overline{\mathrm{co}}}\, B$ is locally bounded, i.e. for all $y \in Y$ there exist $N > 0$ and $\varepsilon > 0$ such that

$$\| \overset{*}{\overline{\mathrm{co}}}\, B(\xi)\|_+ \leq N, \text{ if } \|\xi - y\|_Y \leq \varepsilon.$$

Obviously, a locally bounded operator is bounded-valued. Therefore, $\overset{*}{\overline{\mathrm{co}}}\, C(y) = \overset{*}{\overline{\mathrm{co}}}\, A(y) + \overset{*}{\overline{\mathrm{co}}}\, B(y)$ and $d_n = d'_n + d''_n$, $d'_n \in \overset{*}{\overline{\mathrm{co}}}\, A(y_n)$, $d''_n \in \overset{*}{\overline{\mathrm{co}}}\, B(y_n)$. Since the imbedding $W \subset Y$ is compact, we have that $y_n \to y$ strongly in Y and by virtue of local the boundness of $\overset{*}{\overline{\mathrm{co}}}\, B$ the sequence $\{d''_n\}$ is bounded in Y^* (and then also in X^*), which means that there will be a subsequence $\{d''_m\} \subset \{d''_n\}$ such that $d''_m \to d''$ weakly star in Y^*. The embedding operator $I^* : Y^* \to X^*$ is continuous, so that I^* remains continuous also in the weakly star topologies [RS80]. Hence, $d''_m \to d''$ weakly star in X^*, so that $d'_m = d_m - d''_m \to d' = d - d''$ weakly star in X^*. Therefore

$$\langle d''_m, y_m - y\rangle_X \to 0.$$

Then from (1.70) we get $\overline{\lim}_{m\to\infty} \langle d'_m, y_m - v\rangle_X \leq 0$, whence up to a subsequence

$$\underline{\lim}_{m_k\to\infty} \left\langle d'_{m_k}, y_{m_k} - v\right\rangle_X \geq \left[\overset{*}{\overline{\mathrm{co}}}\, A(y, y - v)\right]_-, \ \forall\, v \in X.$$

Further, as the operator $\overset{*}{\overline{\mathrm{co}}}\, B$ is closed in $Y \times Y^*_w$, $d'' \in \overset{*}{\overline{\mathrm{co}}}\, B(y)$ and

$$\underline{\lim}_{m_k\to\infty} \left\langle d_{m_k}, y_{m_k} - v\right\rangle_X = \underline{\lim}_{m_k\to\infty} \left\langle d'_{m_k}, y_{m_k} - v\right\rangle_X + \lim_{m_k\to\infty} \left\langle d^{**}_{m_k}, y_{m_k} - v\right\rangle_X$$
$$\geq \left[\overset{*}{\overline{\mathrm{co}}}\, A(y), y - v\right]_- + [\overline{\mathrm{co}}^*\, B(y), y - v]_-$$
$$= \left[\overset{*}{\overline{\mathrm{co}}}\, C(y), y - v\right]_-, \forall\, v \in X.$$

The Proposition is proved. $\square$

Proposition 1.27. *Let the functional $\varphi : X \to \mathbb{R}$ be convex, lower semicontinuous on X. Then the multivalued map $B = \partial\varphi : X \to C_v(X^*)$ is λ_0-pseudomonotone on X and it satisfies Condition (Π).*

Proof. (a) Property (Π). Let $k > 0$ and the bounded set $B \subset X$ be arbitrary fixed. Then $\forall y \in B$ and $\forall d(y) \in \partial\varphi(y)$ $\langle d(y), y - y_0\rangle_X \le k$ is fulfilled. Let $u \in X$ be arbitrary fixed, so

$$\begin{aligned}\langle d(y), u\rangle_X &= \langle d(y), u - y\rangle_X + \langle d(y), y\rangle_X \le \varphi(u) - \varphi(y) + k\\ &\le \varphi(u) - \inf_{y\in B}\varphi(y) + k \equiv const < +\infty,\end{aligned}$$

since every convex lower semicontinuous functional is lower bounded by every bounded set. Hence, thanks to the Banach–Steinhaus Theorem, there exists $N = N(y_0, k, B)$ such that $\|d(y)\|_{X^*} \le N$ for each $y \in B$;

(b) λ_0-pseudomonotony on X. Let $y_n \rightharpoonup y_0$ in X, $\partial\varphi(y_n) \ni d_n \rightharpoonup d$ in X^* and the inequality (1.55) true. Then, due to the monotony of $\partial\varphi$, for each $d_0 \in \partial\varphi(y_0)$ and for each $n \ge 1$

$$\langle d_n, y_n - y_0\rangle_X = \langle d_n - d_0, y_n - y_0\rangle_X + \langle d_0, y_n - y_0\rangle_X \ge \langle d_0, y_n - y_0\rangle_X.$$

Hence

$$\varliminf_{n\to+\infty} \langle d_n, y_n - y_0\rangle_X \ge \varliminf_{n\to+\infty} \langle d_0, y_n - y_0\rangle_X = 0.$$

Because of the last inequality and of inequality (1.55) it results in

$$\lim_{n\to+\infty} \langle d_n, y_n - y_0\rangle_X = 0.$$

Thus for each $w \in X$

$$\begin{aligned}\varliminf_{n\to+\infty} \langle d_n, y_n - w\rangle_X \ge{}& \lim_{n\to+\infty} \langle d_n, y_n - y_0\rangle_X\\ &+ \varliminf_{n\to+\infty} \langle d_n, y_0 - w\rangle_X = \langle d_0, y_0 - w\rangle_X. \quad (1.71)\end{aligned}$$

From another side we have

$$\begin{aligned}\langle d_0, w - y_0\rangle_X \le{}& \varlimsup_{n\to+\infty} \langle d_n, w - y_n\rangle_X \le \varphi(w)\\ &- \varliminf_{n\to+\infty} \varphi(y_n) \le \varphi(w) - \varphi(y_0), \qquad (1.72)\end{aligned}$$

since every convex lower semicontinuous functional is weakly lower semicontinuous. From (1.72) it follows that $d_0 \in \partial\varphi(y_0)$. From here, due to the inequality (1.71), we obtain inequality (1.56) as $B = \partial\varphi$ on X. □

Now we consider a functional $\varphi : X \mapsto \mathbb{R}$.

Definition 1.13. The functional φ refers to the locally Lipschitz, if for any $x_0 \in X$ there are $r, c > 0$ such, that

$$|\varphi(x) - \varphi(y)| \leq c\|x - y\|_X \qquad \forall x, y \in B_r(x_0) = \{x \in X \mid \|x - x_0\|_X < r\}.$$

For a locally Lipschitz functional φ, defined on a Banach space X, we consider the *upper Clarke derivative* [Cl90]

$$\varphi_{Cl}^{\uparrow}(x, h) = \overline{\lim_{v \to x, \alpha \searrow 0+}} \frac{1}{\alpha}(\varphi(v + \alpha h) - \varphi(v)) \in \mathbb{R}, \quad x, h \in X$$

and *Clarke generalized gradient*

$$\partial_{Cl}\varphi(x) = \{p \in X^* \mid \langle p, v - x\rangle_X \leq \varphi_{Cl}^{\uparrow}(x, v - x) \quad \forall v \in X\}, \quad x \in X.$$

Proposition 1.28. *Let W be a Banach space compactly embedded in some Banach space Y, $\varphi : Y \mapsto \mathbb{R}$ be a locally Lipschitz functional. Then the Clarkes generalized gradient $\partial_{Cl}\varphi : Y \rightrightarrows Y^*$ is λ_0-pseudomonotone on W.*

Proof. From the calculus of Clarkes generalized gradient (see [Cl90, Chap. 2]) we know that $\partial_{Cl}\varphi(x)$ is nonempty, closed, bounded and convex. Hence, for each $x \in Y$ $\partial_{Cl}\varphi(x) \in C_v(Y^*)$.

Now let $\{y_n\}_{n \geq 0} \subset W$ be a sequence such, that $y_n \rightharpoonup y_0$ in W, $d_n \rightharpoonup d_0$ in Y^*, where $d_n \in \partial_{Cl}\varphi(y_n)$ $\forall n \geq 1$ and the inequality (1.55) is true. Due to the compact embedding $W \subset Y$ we conclude that $y_n \to y_0$ in Y. Since the mapping $\partial_{Cl}\varphi : Y \rightrightarrows Y^*$ is weak-closed (cf. [Cl90, p. 29]), then $d_0 \in \partial_{Cl}\varphi(y_0)$. Thus,

$$\begin{aligned}\underline{\lim}_{n\to\infty} \langle d_n, y_n - \omega\rangle_Y &\geq \underline{\lim}_{n\to\infty} \langle d_n, y_n - y_0\rangle_Y + \underline{\lim}_{n\to\infty} \langle d_n, y_0 - \omega\rangle_Y \\ &= 0 + \langle d_0, y_0 - \omega\rangle_Y \geq [\partial_{Cl}\varphi(y_0), y_0 - \omega]_- \quad \forall \omega \in Y,\end{aligned}$$

which completes the proof. □

Now let $W = W_1 \cap W_2$, where $(W_1, \|\cdot\|_{W_1})$ and $(W_2, \|\cdot\|_{W_2})$ are Banach spaces such that $W_i \subset X_i$ with continuous embedding.

Lemma 1.15. *Let X_1, X_2 be reflexive Banach spaces, $A : X_1 \to C_v(X_1^*)$ and $B : X_2 \to C_v(X_2^*)$ be s-mutually bounded λ_0-pseudomonotone respectively on W_1 and on W_2 multivalued maps. Then $C := A + B : X \to C_v(X^*)$ is λ_0-pseudomonotone on W map.*

Remark 1.23. If the pair $(A; B)$ is not s-mutually bounded, then the last Proposition holds only for λ-pseudomonotone (respectively on W_1 and on W_2) maps.

Proof. At first we check that $\forall y \in X$ $C(y) \in C_v(X^*)$. The convexity of $C(y)$ follows from the same property for $A(y)$ and $B(y)$. By virtue of the Mazur Theorem, it is enough to prove that the set $C(y)$ is weakly closed. Let c be a frontier point of $C(y)$ with respect to the topology $\sigma(X^*; X^{**}) = \sigma(X^*; X)$ (the space X is reflexive). Then

$$\exists \{c_m\}_{m\geq 1} \subset C(y): \quad c_m \to c \quad \text{weakly in} \quad X^* \quad \text{as} \quad m \to +\infty.$$

From here, since the maps A and B have bounded values, due to the Banach–Alaoglu Theorem, we can assume that for each $m \geq 1$ there exist $v_m \in A(y)$ and $w_m \in B(y)$ such that $v_m + w_m = c_m$ and by passing (if it is necessary) to the subsequences we obtain:

$$v_m \rightharpoonup v \quad \text{in} \quad X_1^* \quad \text{and} \quad w_m \rightharpoonup w \quad \text{in} \quad X_2^*$$

for some $v \in A(y)$ and $w \in B(y)$. Hence $c = v + w \in C(y)$. So it is proved that the set $C(y)$ is weakly closed in X^*.

Now let $y_n \rightharpoonup y_0$ in W (from here it follows that $y_n \rightharpoonup y_0$ in W_1 and $y_n \rightharpoonup y_0$ in W_2), $C(y_n) \ni d(y_n) \rightharpoonup d_0$ in X^* and the inequality (1.55) be true. Hence

$$d_A(y_n) \in A(y_n) \quad \text{and} \quad d_B(y_n) \in B(y_n): \quad d_A(y_n) + d_B(y_n) = d(y_n).$$

Since the pair $(A; B)$ is s-mutually bounded, from the estimation

$$\langle d(y_n), y_n\rangle_X = \langle d_A(y_n)+d_B(y_n), y_n\rangle_X = \langle d_A(y_n), y_n\rangle_{X_1} + \langle d_B(y_n), y_n\rangle_{X_2} \leq k$$

we have or $\|d_A(y_n)\|_{X_1^*} \leq C$ or $\|d_B(y_n)\|_{X_2^*} \leq C$. Then, due to the reflexivity of X_1 and X_2, by passing (if it is necessary) to a subsequence we get

$$d_A(y_n) \rightharpoonup d_0' \text{ in } X_1^* \quad \text{and} \quad d_B(y_n) \rightharpoonup d_0'' \text{ in } X_2^*. \tag{1.73}$$

From the inequality (1.55) we have

$$\varliminf_{n\to\infty} \langle d_B(y_n), y_n - y_0\rangle_{X_2} + \varlimsup_{n\to\infty} \langle d_A(y_n), y_n - y_0\rangle_{X_1} \leq \varlimsup_{n\to\infty} \langle d(y_n), y_n - y_0\rangle_X \leq 0,$$

or symmetrically

$$\varliminf_{n\to\infty} \langle d_A(y_n), y_n - y_0\rangle_{X_1} + \varlimsup_{n\to\infty} \langle d_B(y_n), y_n - y_0\rangle_{X_2} \leq \varlimsup_{n\to\infty} \langle d(y_n), y_n - y_0\rangle_X \leq 0.$$

Let us consider the last inequality. It is obvious that there exists a subsequence $\{y_m\}_m \subset \{y_n\}_{n\geq 1}$ such that

$$\begin{aligned} 0 &\geq \varlimsup_{n\to\infty} \langle d_B(y_n), y_n - y_0\rangle_{X_2} + \varliminf_{n\to\infty} \langle d_A(y_n), y_n - y_0\rangle_{X_1} \\ &\geq \varlimsup_{m\to\infty} \langle d_B(y_m), y_m - y_0\rangle_{X_2} + \lim_{m\to\infty} \langle d_A(y_m), y_m - y_0\rangle_{X_1}. \end{aligned} \tag{1.74}$$

From here we obtain:

$$\text{or} \quad \lim_{m\to\infty} \langle d_A(y_m), y_m - y_0\rangle_{X_1} \le 0, \quad \text{or} \quad \overline{\lim_{m\to\infty}} \langle d_B(y_m), y_m - y_0\rangle_{X_2} \le 0.$$

Without loss of generality we suppose that

$$\lim_{m\to\infty} \langle d_A(y_m), y_m - y_0\rangle_{X_1} \le 0.$$

Then because of (1.73) and of the λ_0-pseudomonotony of A on W_1 there exists a subsequence $\{y_{m_k}\}_{k\ge 1} \subset \{y_m\}_m$ such that

$$\lim_{k\to\infty} \langle d_A(y_{m_k}), y_{m_k} - v\rangle_{X_1} \ge [A(y_0), y_0 - v]_- \quad \forall v \in X_1. \tag{1.75}$$

By substituting in the last relation v for y_0 it results in

$$\langle d_A(y_{m_k}), y_{m_k} - y_0\rangle_{X_1} \to 0 \quad \text{as} \quad k \to +\infty.$$

Therefore, taking into account (1.74), we have

$$\overline{\lim_{k\to\infty}} \langle d_B(y_{m_k}), y_{m_k} - y_0\rangle_{X_2} \le 0.$$

By virtue of the λ_0-pseudomonotony of B on W_2, up to a subsequence $\{y_{m'_k}\} \subset \{y_{m_k}\}_{k\ge 1}$ we find

$$\underline{\lim_{k\to\infty}} \langle d_B(y_{m'_k}), y_{m'_k} - w\rangle_{X_2} \ge [B(y_0), y_0 - w]_- \quad \forall w \in X_2. \tag{1.76}$$

So from the relations (1.75) and (1.76) we finally obtain

$$\begin{aligned}
&\underline{\lim_{k\to\infty}} \langle d(y_{m'_k}), y_{m'_k} - x\rangle_X \\
&\quad \ge \lim_{k\to\infty} \langle d_A(y_{m'_k}), y_{m'_k} - x\rangle_{X_1} + \underline{\lim_{k\to\infty}} \langle d_B(y_{m'_k}), y_{m'_k} - x\rangle_{X_2} \\
&\quad \ge [A(y_0), y_0 - x]_- + [B(y_0), y_0 - x]_- = [C(y_0), y_0 - x]_- \quad \forall x \in X.
\end{aligned}$$

The Lemma is proved. □

Remark 1.24. Lemma 1.15 means that the family of all λ_0-pseudomonotone on W multivalued maps which satisfy Condition (Π) makes a "convex cone" in the space $S(X; X^*)$.

References

[BR68] Brezis H (1968) Equations et inequations non lineaires dans les espaces vectoriels en dualite. Ann Inst Fourier 18:115–175

[BP72] Browder FE, Hess P (1972) Nonlinear mapping of monotone type in Banach spaces. J Funct Anal. doi:10.1016/0022-1236(72)90070-5

[Cl90] Clarke FH (1990) Optimization and nonsmooth analysis. SIAM, Philadelphia

[DU65] Dubinskii YuA (1965) Weak convergence in nonlinear elliptic and parabolic equations. Mat Sb 67:609–642

[ET99] Ekeland I, Temam R (1999) Convex analysis and variational problems. SIAM, Philadelphia

[GGZ74] Gajewski H, Gröger K, Zacharias K (1974) Nichtlineare operatorgleichungen und operatordifferentialgleichungen. Akademie-Verlag, Berlin

[HP00] Hu S, Papageorgiou NS (2000) Handbook of multivalued analysis. Volume II: Applications. Kluwer, Dordrecht

[IM88] Ivanenko VI, Mel'nik VS (1988) Variational Methods in Control Problems for Systems with distributed parameters [in Russian]. Naukova dumka, Kiev

[JAM72] James RC (1972) Super-reflexive spaces with bases. Pac J Math 41:409–419

[KMP06] Kasyanov PO, Mel'nik VS, Piccirillo AM (2006) On schauder basis in some Banach spaces. Reports NAS of Ukraine. 4, 23–30

[KOG01] Kogut PI (2001) On the variational S-compactness of conditional minimization problems. J Math Sci. doi: 10.1023/A:1012302232380

[LIO69] Lions JL (1969) Quelques methodes de resolution des problemes aux limites non lineaires. Dunod Gauthier-Villars, Paris

[PAP87] Papageorgiou NS (1987) On multivalued evolution equations and differential inclusions in Banach spaces. Comment. Math Univ San Pauli 36:21–39

[PKZ08] Perestyuk MO, Kasyanov PO, Zadoyanchuk NV (2008) On solvability for the second order evolution inclusions with the volterra type operators. Miskolc Math Notes 9: 119–135

[PET67] Petunin YuI (1967) The reflexivity criterium for a Banach space. RAS USSR. 140:56–58

[RS80] Reed M, Simon B (1980) Methods of mathematical physics. Vol. 1: functional analysis. Academic, New York

[RUD73] Rudin W (1973) Functional analysis. McGraw-Hill, New York

[SIM86] Simon J (1986) Compact sets in the space $L^p(0, T; B)$. J. Annali di Matematica Pura ed Applicata. doi: 10.1007/BF01762360

[SKR94] Skrypnik IV (1994) Methods for analysis of nonlinear elliptic boundary value problems. Providence, RI

[TEM75] Temam R (1975) A non-linear eigenvalue problem: the shape at equilibrium of a confined plasma. Avch Ration Mech Anal. doi:10.1007/BF00281469

[TRI78] Triebel H (1978) Interpolation theory, function spaces, differential operators. North-Holland Publishing Company, Amsterdam

[ZK07] Zadoyanchuk NV, Kasyanov PO (2007) Faedo–Galerkin method for nonlinear second-order evolution equations with Volterra operators. Nonlinear Oscillations 10:203–228. doi:10.1007/s11072-007-0016-y

[ZM99] Zgurovsky MZ, Mel'nik VS (1999) Nonlinear analysis and control of infinite-dimentional systems [in Russian]. Naukova dumka, Kiev

[ZM04] Zgurovsky MZ, Mel'nik VS (2004) Nonlinear analysis and control of physical processes and fields. Springer, Berlin

[ZMN04] Zgurovsky MZ, Mel'nik VS, Novikov AN (2004) Applied methods of analysis and control by nonlinear processes and fields [in Russian]. Naukova Dumka, Kiev

Chapter 2
Differential-Operator Inclusions with W_{λ_0}-Pseudomonotone Maps

Abstract In this chapter differential-operator inclusions with non-coercive maps of the Volterra type are studied qualitative and constructively. Such objects describe new mathematical models of non-linear geophysical processes and fields, in particular, piezoelectric processes which require the developing of corresponding non-coercive theory and high-precision algorithms of searching of approximate solutions. In Sect. 2.1 we validate the scheme developed by Dubinskii for $+$-coercive maps with $(W; W)$-s.b.v. and consider examples of hydrodynamic problems. Section 2.2 is devoted to the singular perturbation method for evolutional inclusions with weakly coercive mappings. In Sect. 2.3 we develop the noncoercive theory for evolutional inclusions with maps of the Volterra type mappings. In Sect. 2.4 we study periodic solutions and the Cauchy problems solutions of differential-operator inclusions with $+$-coercive maps of pseudomonotone type and consider some anisotropic problems perturbed by Clarke generalized gradient. In Sect. 2.5 we validate the finite differences method for evolutional inclusions with $+$-coercive λ_0-pseudomonotone maps and consider examples of nonlinear boundary value problems with subdifferential operators. In Sect. 2.6 we study the second order evolution inclusion with noncoercive multivalued damping. Section 2.7 devoted to some geophysical applications. The obtained results are demonstrated in the example of dynamical contact problem with the non-linear friction. The results of this chapter together with the theorems of Chap. 1 sweep new classes of differential-operator inclusions with multi-valued maps of the pseudomonotone type.

2.1 Dubinskij Method for Operators with $(X; W)$-Semibounded Variation

As before let $(V_1, \|\cdot\|_{V_1})$ and $(V_2, \|\cdot\|_{V_2})$ be reflexive Banach spaces continuously embedded in Hilbert space $(H, (\cdot,\cdot))$ such that for some numerable set $\Phi \subset V = V_1 \cap V_2$

$$\Phi \text{ is dense in } V,\ V_1,\ V_2 \text{ and in } H.$$

M.Z. Zgurovsky et al., *Evolution Inclusions and Variation Inequalities for Earth Data Processing II*, Advances in Mechanics and Mathematics 25,
DOI 10.1007/978-3-642-13878-2_2,

After identification $H \equiv H^*$ we obtain

$$V_1 \subset H \subset V_1^*, \qquad V_2 \subset H \subset V_2^*,$$

with continuous and dense embedding, $(V_i^*, \|\cdot\|_{V_i^*})$, $i = 1, 2$ is topologically adjoint of V_i space with respect to the canonical bilinear form

$$\langle \cdot, \cdot \rangle_{V_i} : V_i^* \times V_i \to \mathbb{R},$$

which coincides on H with the inner product $(\cdot, \cdot)$ on H.

Let us consider the functional spaces $X_i = L_{r_i}(S; H) \cap L_{p_i}(S; V_i)$, where S is a finite time interval, $1 < p_i \leq r_i < +\infty$. The spaces X_i are reflexive Banach spaces with the norms

$$\|y\|_{X_i} = \|y\|_{L_{p_i}(S;V_i)} + \|y\|_{L_{r_i}(S;H)},$$

$X = X_1 \cap X_2$, $\|y\|_X = \|y\|_{X_1} + \|y\|_{X_2}$. Let X_i^* $(i = 1, 2)$ be topologically adjoint with X_i. Then,

$$X^* = X_1^* + X_2^* = L_{q_1}(S; V_1^*) + L_{q_2}(S; V_2^*) + L_{r_1'}(S; H) + L_{r_2'}(S; H),$$

where $r_i^{-1} + r_i'^{-1} = p_i^{-1} + q_i^{-1} = 1$ $(i = 1, 2)$. Let us define the duality form on $X^* \times X$

$$\begin{aligned}\langle f, y \rangle &= \int_S (f_{11}(\tau), y(\tau))_H \, d\tau + \int_S (f_{12}(\tau), y(\tau))_H \, d\tau \\ &\quad + \int_S \langle f_{21}(\tau), y(\tau) \rangle_{V_1} \, d\tau + \int_S \langle f_{22}(\tau), y(\tau) \rangle_{V_2} \, d\tau \\ &= \int_S (f(\tau), y(\tau)) \, d\tau,\end{aligned}$$

where $f = f_{11} + f_{12} + f_{21} + f_{22}$, $f_{1i} \in L_{r_i'}(S; H)$, $f_{2i} \in L_{q_i}(S; V_i^*)$ $(i = 1, 2)$. Note that $\langle \cdot, \cdot \rangle$ coincides with the inner product in $\mathcal{H} = L_2(S; H)$ on $\mathcal{H}$.

Let $A : X \rightrightarrows X^*$ be the multivalued map. We consider solutions in class $W = \{y \in X \mid y' \in X^*\}$ of the following problem:

$$\begin{cases} \langle y', \xi \rangle + [A(y), \xi]_+ \geq \langle f, \xi \rangle & \forall \xi \in W, \\ y(0) = \bar{0}, \end{cases}$$

where $f \in X^*$ is arbitrary, y' is the derivative of an element $y \in X$ in the sense of scalar distributions space $\mathcal{D}^*(S; V^*) = \mathcal{L}(\mathcal{D}(S); V_w^*)$, with $V = V_1 \cap V_2$.

Let us introduce on Banach space W the graph norm $\|y\|_W = \|y\|_X + \|y'\|_{X^*}$, where

$$\|f\|_{X^*} = \inf_{\substack{f = f_{11}+f_{12}+f_{21}+f_{22}:\\ f_{1i}\in L_{r_i'}(S;H),\ f_{2i}\in L_{q_i}(S;V_i^*)\,(i=1,2)}} \max\left\{\|f_{11}\|_{L_{r_1'}(S;H)};\ \|f_{12}\|_{L_{r_2'}(S;H)};\ \|f_{21}\|_{L_{q_1}(S;V_1^*)};\ \|f_{22}\|_{L_{q_2}(S;V_2^*)}\right\}.$$

Note that the embedding $W \subset C(S;H)$ is continuous.

Now we prove the Theorem about resolvability of differential operator inclusions with nonlinear coercive w_λ-pseudomonotone maps using Dubinskij method developed in [D65]. Further we will assume either $r_1 \geq 2$ or $r_2 \geq 2$.

In order to prove the solvability for the stated problem we use the particular case of the stationary approximations method introduced by Yu.A. Dubinskii [D65] for differential-operator equations. Besides the solvability this method allows us to obtain the series of a priori estimations. Thus for example we can investigate the dynamics of solutions for a wide class of applied problems. Such problems can be described by evolutional inequalities with non-coercive multi-valued maps of the Volterra type. This method based on the conversion from first order problem to the second order differential-operator inclusion

$$-\varepsilon y_\varepsilon'' + y_\varepsilon' + \overset{*}{\overline{\mathrm{co}}}\, A y_\varepsilon \ni f, \quad y_\varepsilon(0) = \bar{0}, \quad y_\varepsilon'(T) = \bar{0}. \tag{2.1}$$

Further, when we prove the solvability for the given problem at fixed $\varepsilon > 0$, obtain a priori estimations for solutions and pass $\varepsilon \to 0+$ we obtain solutions of the initial problem with series of properties.

Now let us prove the theorem on the existence of the generalized solution for a differential-operator inclusion with nonlinear weakly coercive map of w_λ-pseudomonotone type using the scheme developed by Dubinskii. The introduced theorem allows us not only to obtain a new method of searching of solutions of the given problem with new a priori estimations but also use it further for the investigation of differential-operator inclusions and evolutional multivalued inequalities with noncoercive maps. In this theorem we validate the Dubinskii method for differential-operator inclusions with $+$-coercive maps and obtain new a priori estimations for approximative solutions. This result will be applied for the proof of Theorem 2.3 and its corollaries. As the presented results have an independent value we formulate them as separate theorem.

Theorem 2.1. *Let $A : X \rightrightarrows X^*$ be $+$-coercive, radial lower semicontinuous multivalued map with $(X;W)$-semibounded variation. Then for each $f \in X^*$ there exists at least one solution $y \in W$ of the problem:*

$$\langle y', \xi\rangle + [Ay, \xi]_+ \geq \langle f, \xi\rangle \quad \forall \xi \in W, \quad y(0) = \bar{0}. \tag{2.2}$$

Remark 2.1. Due to density of W in X and from Proposition 2 it follows that problem (2.2) is equivalent to:

$$y' + \overset{*}{\overline{\text{co}}}\, A(y) \ni f, \quad y(0) = \bar{0}, \quad y \in W.$$

Remark 2.2. In Theorem 2.1 the scheme developed by Dubinskii for evolutionary equations is generalized to evolutionary inequalities with multi-valued maps with $(X; W)$-semi-bounded variation. This scheme is new for the analyzed class of problems and enables us not only to prove solvability but also to obtain a series of estimates for the solutions, which is important for the subsequent investigations of these objects. Thus, the same scheme is considered in [D65, ZMN04] but for equations. At the same time, the other schemes are used in [DMP03, KMV08], [VM98]–[VM00] for the differential-operator inclusions with multi-valued maps of the monotone type taking weakly compact convex values.

Proof. To simplify the proof we assume that $S = [0, T]$. Let us consider the space

$$W_{\bar{0}} := \{y \in W \mid y(0) = \bar{0}\} \text{ with the norm} \|\cdot\|_W.$$

To prove the given Proposition let us use the particular case of stationary approximations method. At first, from problem (2.2) we pass to the second order differential-operator inclusion (2.1).

Let us consider the linear space

$$\widetilde{W} = \{y \in L_{p_1}(S; V_1) \cap L_{p_2}(S; V_2) \mid y' \in L_2(S; H) = \mathscr{H},\ y(0) = \bar{0}\}$$

with the norm $\|y\|_{\widetilde{W}} = \|y\|_{L_{p_1}(S;V_1)} + \|y\|_{L_{p_2}(S;V_2)} + \|y'\|_{\mathscr{H}}$, $y \in \widetilde{W}$. Note that $\widetilde{W} \subset C(S; H)$ continuously and moreover, from the assumption $\max\{r_1, r_2\} \geq 2$ it follows that $\widetilde{W} \subset W_{\bar{0}} \subset X$ continuously. If we apply the formula (1.15) to (2.1) (taking into account $y_\varepsilon'(T) = \bar{0}$ and the assumption $y_\varepsilon'' \in \mathscr{H} + X^*$) we obtain

$$\varepsilon(y_\varepsilon', \xi')_{\mathscr{H}} + \langle y_\varepsilon', \xi\rangle + [Ay_\varepsilon, \xi]_+ \geq \langle f, \xi\rangle \quad \forall \xi \in \widetilde{W}. \tag{2.3}$$

As solution of problem (2.1) we are going to find an element $y_\varepsilon \in \widetilde{W}$, for which (2.3) is true.

Now we use the coercivity condition on A. For every $y \in \widetilde{W}$

$$\begin{aligned}
&\varepsilon(y', y')_{\mathscr{H}} + \langle y', y\rangle + [Ay, y]_+ - \langle f, y\rangle \\
&\geq (\|y(T)\|_H^2 - \|y(0)\|_H^2)/2 + [Ay, y]_+ - \|f\|_{X^*}\|y\|_X \\
&\geq [Ay, y]_+ - \|f\|_{X^*}\|y\|_X \geq \|y\|_X\,(\gamma_A(\|y\|_X) - \|f\|_{X^*}),
\end{aligned}$$

where $\gamma_A(r) \to +\infty$ as $r \to +\infty$. Therefore there exists $R_1 > 0$ such that $\forall \varepsilon > 0$

$$\varepsilon(y', y')_{\mathscr{H}} + \langle y', y\rangle + [Ay, y]_+ - \langle f, y\rangle \geq 0 \quad \forall y \in \widetilde{W} : \|y\|_X = R_1. \tag{2.4}$$

Proposition 2.1. *Under the conditions of Theorem 2.1, for each $\varepsilon > 0$ problem (2.1) has at least one solution $y_\varepsilon \in \widetilde{W}$ for wich the following estimate holds true*

$$\varepsilon \|y_\varepsilon'\|_{\mathscr{H}}^2 + \|y_\varepsilon\|_X \le k, \tag{2.5}$$

where $k = k(f)$ does not depend on ε.

Proof. Since the space V is separable simultaneously with $\widetilde{W}$ (it is easy to check, using the proof of Theorem 1.6), let $\{h_n\}_{n\ge 1}$ be the complete system in $\widetilde{W}$. Remark that $h_i(0) = \bar{0}$ for $i \ge 1$. The approximate solution of (2.1) has to be searched in the form

$$y_{\varepsilon n} = \sum_{j=1}^{n} \alpha_\varepsilon^{j,n} h_j,$$

where the constants $\alpha_\varepsilon^{j,n}$ can be found from the following system of inequalities

$$\varepsilon (y_{\varepsilon n}', h')_{\mathscr{H}} + \langle y_{\varepsilon n}', h\rangle + [A y_{\varepsilon n}, h]_+ \ge \langle f, h\rangle \quad \forall h \in H_n, \tag{2.6}$$

where H_n is the linear span of $\{h_i\}_{i=1}^n$. Remark that H_n is finite-dimensional separable Banach space with the norm $\|\cdot\|_X$. Hence, let $\{v_i\}_{i\ge 1} \subset H_n$ be a dense system of vectors in H_n.

Proposition 2.2. *For each $n \ge 1$ problem (2.6) has at least one solution $y_{\varepsilon n} \in H_n$ for which the estimation $\|y_{\varepsilon n}\|_X \le R_1$ holds true.*

Proof. We approximate the solutions of (2.6) by a finite systems of algebraic inequalities

$$\begin{cases} \varepsilon (y_{\varepsilon n m}', v_i')_{\mathscr{H}} + \langle y_{\varepsilon n m}', v_i\rangle + [A y_{\varepsilon n m}, v_i]_+ \ge \langle f, v_i\rangle \\ \varepsilon (y_{\varepsilon n m}', v_i')_{\mathscr{H}} + \langle y_{\varepsilon n m}', v_i\rangle + [A y_{\varepsilon n m}, v_i]_- \le \langle f, v_i\rangle, \ i = \overline{1,m}, \end{cases} \tag{2.7}$$

where $m \ge 1$ is arbitrary and $y_{\varepsilon n m} = \sum_{j=1}^{m} \alpha_{\varepsilon,m}^{j,n} v_j$.

Proposition 2.3. *For each $m \ge 1$ problem (2.7) has at least one solution $\bar{\alpha}_{\varepsilon,n}^m = (\alpha_{\varepsilon,m}^{j,n})_{j=1}^m \in \mathbb{R}^m$ such that for $y_{\varepsilon n m}(\bar{\alpha}_{\varepsilon,n}^m) = \sum_{j=1}^{m} \alpha_{\varepsilon,m}^{j,n} v_j$ the estimation $\|y_{\varepsilon n m}(\bar{\alpha}_{\varepsilon,n}^m)\|_X \le R_1$ holds true and the set*

$$G_{\varepsilon n}(m) = \Big\{ y_{\varepsilon n m}(\bar{\alpha}_{\varepsilon,n}^m) \in H_n \,\Big|\, y_{\varepsilon n m}(\bar{\alpha}_{\varepsilon,n}^m) \text{ is a solution (2.7)}, \ \|y_{\varepsilon n m}(\bar{\alpha}_{\varepsilon,n}^m)\|_X \le R_1 \Big\}$$

is compact in H_n.

Proof. For any fixed $m \geq 1$ let us consider the multivalued map $B : \mathbb{R}^m \to C_v(\mathbb{R}^m)$ defined in the following way:

$$\forall \bar{\alpha} \in \mathbb{R}^m \quad B(\bar{\alpha}) = (B_i(\bar{\alpha}))_{i=1}^m,$$

where for each $i = \overline{1, m}$

$$B_i(\bar{\alpha}) = \Big[\, \varepsilon(y'(\bar{\alpha}), v_i')_{\mathscr{H}} + \langle y'(\bar{\alpha}), v_i\rangle + [Ay(\bar{\alpha}), v_i]_- - \langle f, v_i\rangle,$$
$$\varepsilon(y'(\bar{\alpha}), v_i')_{\mathscr{H}} + \langle y'(\bar{\alpha}), v_i\rangle + [Ay(\bar{\alpha}), v_i]_+ - \langle f, v_i\rangle \,\Big] \in C_v(\mathbb{R}),$$
$$\bar{\alpha} = (\alpha_i)_{i=1}^m, \quad y(\bar{\alpha}) = \sum_{i=1}^m \alpha_i v_i, \quad y'(\bar{\alpha}) = \sum_{i=1}^m \alpha_i v_i'.$$

We consider the norm in $\mathbb{R}^m$

$$\|\bar{\alpha}\|_{\mathbb{R}^m} = \left\| \sum_{i=1}^m \alpha_i v_i \right\|_X \qquad \forall \bar{\alpha} = (\alpha_i)_{i=1}^m \in \mathbb{R}^m$$

and paring

$$\langle \bar{\alpha}, \bar{\beta}\rangle = \sum_{i=1}^m \alpha_i \beta_i \quad \forall \bar{\alpha} = (\alpha_i)_{i=1}^m \in \mathbb{R}^m,\ \bar{\beta} = (\beta_i)_{i=1}^m \in \mathbb{R}^m.$$

Due to Proposition 1 for each $\bar{\alpha} = (\alpha_i)_{i=1}^m \in \mathbb{R}^m$

$$\begin{aligned}
[B(\bar{\alpha}), \bar{\alpha}]_+ &= \sup\left\{ \sum_{i=1}^m b_i \alpha_i \,\middle|\, b_i \in B_i(\bar{\alpha}),\ i = \overline{1, n} \right\} \\
&\geq \sum_{i=1}^m \Big(\varepsilon(y'(\bar{\alpha}), v_i')_{\mathscr{H}} + \langle y'(\bar{\alpha}), v_i\rangle + [Ay(\bar{\alpha}), v_i]_+ - \langle f, v_i\rangle \Big)\alpha_i \\
&= \sum_{i=1}^m \varepsilon(y'(\bar{\alpha}), \alpha_i v_i')_{\mathscr{H}} + \langle y'(\bar{\alpha}), \alpha_i v_i\rangle + [Ay(\bar{\alpha}), \alpha_i v_i]_+ - \langle f, \alpha_i v_i\rangle \\
&\geq \varepsilon(y'(\bar{\alpha}), y'(\bar{\alpha}))_{\mathscr{H}} + \langle y'(\bar{\alpha}), y(\bar{\alpha})\rangle + [Ay(\bar{\alpha}), y(\bar{\alpha})]_+ - \langle f, y(\bar{\alpha})\rangle \geq 0,
\end{aligned}$$

as $\|\bar{\alpha}\|_{\mathbb{R}^m} = \|y(\bar{\alpha})\|_X = R_1$. Therefore,

$$[B(\bar{\alpha}), \bar{\alpha}]_+ \geq 0 \quad \text{for all} \quad \bar{\alpha} \in \mathbb{R}^m : \|\bar{\alpha}\|_{\mathbb{R}^m} = R_1. \tag{2.8}$$

Similarly for each $\bar{\alpha} = (\alpha_i)_{i=1}^m \in \mathbb{R}^m$ and $\bar{\beta} = (\beta_i)_{i=1}^m \in \mathbb{R}^m$

$$[B(\bar{\alpha}), \bar{\beta}]_+ = \sup \left\{ \sum_{i=1}^m b_i \beta_i \;\middle|\; b_i \in B_i(\bar{\alpha}),\ i = \overline{1,n} \right\}$$
$$= \sum_{i=1}^m \Big(\varepsilon (y'(\bar{\alpha}), v_i')_{\mathscr{H}} + \langle y'(\bar{\alpha}), v_i \rangle - \langle f, v_i \rangle \Big) \beta_i$$
$$+ \sum_{i=1}^m \Big(\max\{[Ay(\bar{\alpha}), v_i]_+,\ [Ay(\bar{\alpha}), -v_i]_+\} \Big) \cdot |\beta_i|.$$

The map

$$\mathbb{R}^m \ni \bar{\alpha} \to \sum_{i=1}^m \Big(\varepsilon (y'(\bar{\alpha}), v_i')_{\mathscr{H}} + \langle y'(\bar{\alpha}), v_i \rangle - \langle f, v_i \rangle \Big) \beta_i$$

is affine and hence continuous. The upper semicontinuity of the map

$$\mathbb{R}^m \ni \bar{\alpha} \to \max\{[Ay(\bar{\alpha}), v_i]_+,\ [Ay(\bar{\alpha}), -v_i]_+\} \quad \forall i = \overline{1,m}$$

follows from the same statement for

$$\mathbb{R}^m \ni \bar{\alpha} \to [Ay(\bar{\alpha}), v_i]_+ \quad \text{and} \quad \mathbb{R}^m \ni \bar{\alpha} \to [Ay(\bar{\alpha}), -v_i]_+ \quad i = \overline{1,m}.$$

The latter follows from (finite-dimensional) locally boundness and from $(X; W)$-semibounded variation of A (Lemma 1.13). Hence for each $\bar{\beta} \in \mathbb{R}^m$ the map $\mathbb{R}^m \ni \bar{\alpha} \to [B(\bar{\alpha}), \bar{\beta}]_+$ is upper semicontinuous. So, in virtue of Castaing Theorem [AE84], B is upper semicontinuous map on $\mathbb{R}^m$.

Now, due to the relation (2.8) we can apply the analogue of "acute angle" Lemma (see Corollary 2) for the map B. We obtain that for each $m \geq 1$ there exists at least one solution (2.7) $\bar{\alpha}_{\varepsilon n}^m = (\alpha_{\varepsilon,m}^{j,n})_{j=1}^m \in \mathbb{R}^m$ such that $\|\bar{\alpha}_{\varepsilon n}^m\|_{\mathbb{R}^m} \leq R_1$. Hence, for $y_{\varepsilon nm} = \sum_{j=1}^m \alpha_\varepsilon^{j,n} v_j$ the estimation $\|y_{\varepsilon nm}\|_X \leq R_1$ holds true.

The compactness of $G_{\varepsilon n}(m)$ obviously follows from the boundness of $G_{\varepsilon n}(m)$ and upper semicontinuity of B on $\mathbb{R}^m$. □

Let us consider the set

$$G_{\varepsilon n} = \bigcap_{m \geq 1} G_{\varepsilon n}(m).$$

It is not empty, since for each $m \geq 1$ $G_{\varepsilon n}(m+1) \subset G_{\varepsilon n}(m)$ and $G_{\varepsilon n}(m)$ is a compact set. So, there exists $y_{\varepsilon n} \in G_{\varepsilon n}$ such that $\|y_{\varepsilon n}\|_X \leq R_1$ and

$$\varepsilon (y_{\varepsilon n}', v_i')_{\mathscr{H}} + \langle y_{\varepsilon n}', v_i \rangle + [Ay_{\varepsilon n}, v_i]_+ \geq \langle f, v_i \rangle \quad \forall i \geq 1.$$

Since $\{v_i\}_{i \geq 1}$ is dense in H_n the validity of (2.6) follows. □

From Propositions 2, 3 and (2.6) we obtain that for any $\varepsilon > 0$ and $n \geq 1$ there exists $d_{\varepsilon n} \in \overset{*}{\overline{\text{co}}}\, A y_{\varepsilon n}$ such that

$$\varepsilon(y'_{\varepsilon n}, \xi')_{\mathscr{H}} + \langle y'_{\varepsilon n}, \xi\rangle + \langle d_{\varepsilon n}, \xi\rangle = \langle f, \xi\rangle \quad \forall \xi \in H_n. \tag{2.9}$$

Substituting in the last relation $\xi = y_{\varepsilon n} \in H_n$ we obtain

$$\varepsilon\|y'_{\varepsilon n}\|^2_{\mathscr{H}} + \frac{1}{2}\|y_{\varepsilon n}(T)\|^2_H + \langle d_{\varepsilon n}, y_{\varepsilon n}\rangle = \langle f, y_{\varepsilon n}\rangle. \tag{2.10}$$

Hence

$$\langle d_{\varepsilon n}, y_{\varepsilon n}\rangle \leq \|f\|_{X^*} R_1 \quad \forall \varepsilon > 0,\ n \geq 1.$$

So, due to Proposition 2.2 and Property (Π) for A (see. Lemma 1.13) there exists $C > 0$ such that

$$\|d_{\varepsilon n}\|_{X^*} \leq C \quad \forall \varepsilon > 0,\ n \geq 1. \tag{2.11}$$

From the estimate (2.10) it follows that

$$\varepsilon\|y'_{\varepsilon n}\|^2_{\mathscr{H}} \leq (C + \|f\|_{X^*})R_1 \quad \forall \varepsilon > 0,\ n \geq 1.$$

Due to this and to Proposition 2.2 the estimation:

$$\varepsilon\|y'_{\varepsilon n}\|^2_{\mathscr{H}} + \|y_{\varepsilon n}\|_X \leq k(f) \quad \forall \varepsilon > 0,\ n \geq 1 \tag{2.12}$$

follows. Therefore, we may assume that for the arbitrary $\varepsilon > 0$ the sequence $\{y_{\varepsilon n}\}_{n\geq 1}$ (more exactly, some of its subsequences) is weakly convergent in reflexive Banach space $\widetilde{W}$ to the function y_ε and consequently $y'_{\varepsilon n} \rightharpoonup y'_\varepsilon$ in $\mathscr{H}$ and $y_\varepsilon(0) = \bar{0}$.

In virtue of (2.11) for all $\varepsilon > 0$ we may consider that up to a subsequence $d_{\varepsilon n} \rightharpoonup \varkappa_\varepsilon$ in X^*. By passing to the limit in (2.9) we obtain

$$\varepsilon(y'_\varepsilon, \xi')_{\mathscr{H}} + \langle y'_\varepsilon, \xi\rangle + \langle \varkappa_\varepsilon, \xi\rangle = \langle f, \xi\rangle \quad \forall \xi \in \widetilde{W}. \tag{2.13}$$

Now, we show that $\varkappa_\varepsilon \in \overset{*}{\overline{\text{co}}}\, A y_\varepsilon$. Because of $(X; W)$-semibounded variation of A we have that

$$\begin{aligned}&\varepsilon(y'_{\varepsilon n} - \xi', y'_{\varepsilon n} - \xi')_{\mathscr{H}} + \langle y'_{\varepsilon n} - \xi', y_{\varepsilon n} - \xi\rangle + [A y_{\varepsilon n}, y_{\varepsilon n} - \xi]_- \\ &\geq [A\xi, y_{\varepsilon n} - \xi]_+ - C_A(R; \|y_{\varepsilon n} - \xi\|'_W) \quad \forall \xi \in \widetilde{W},\end{aligned} \tag{2.14}$$

where $\|\cdot\|'_W$ is a seminorm compact with respect to the norm in W and to the norm in $\widetilde{W}$. Here $R > 0$ such that $\|y_{\varepsilon n}\|_X \leq R$, $\|\xi\|_X \leq R$, $k(f) \leq R$.

Since $\cup_{n\geq 1} H_n$ is dense in $\widetilde{W}$ there exists a sequence $v_{\varepsilon n} \in H_n$ such that for fixed $\varepsilon > 0$

$$v_{\varepsilon n} \to y_\varepsilon \quad \text{strongly in} \quad \widetilde{W} \quad \text{as} \quad n \to \infty.$$

Hence in virtue of (2.9) and (2.14) we obtain

$$\begin{aligned}
&\varepsilon(y'_{\varepsilon n}, y'_{\varepsilon n} - \xi')_{\mathscr{H}} + \langle y'_{\varepsilon n}, y_{\varepsilon n} - \xi\rangle + \langle d_{\varepsilon n}, y_{\varepsilon n} - \xi\rangle - \varepsilon(\xi', y'_{\varepsilon n} - \xi')_{\mathscr{H}} - \langle \xi', y_{\varepsilon n} - \xi\rangle \\
&\quad = \langle f, y_{\varepsilon n} - v_{\varepsilon n}\rangle + \varepsilon(y'_{\varepsilon n}, v'_{\varepsilon n} - \xi')_{\mathscr{H}} + \langle y'_{\varepsilon n}, v_{\varepsilon n} - \xi\rangle + \langle d_{\varepsilon n}, v_{\varepsilon n} - \xi\rangle \\
&\qquad - \varepsilon(\xi', y'_{\varepsilon n} - \xi')_{\mathscr{H}} - \langle \xi', y_{\varepsilon n} - \xi\rangle \\
&\quad \geq \varepsilon(y'_{\varepsilon n} - \xi', y'_{\varepsilon n} - \xi')_{\mathscr{H}} + \langle y'_{\varepsilon n} - \xi', y_{\varepsilon n} - \xi\rangle + [Ay_{\varepsilon n}, y_{\varepsilon n} - \xi]_- \\
&\quad \geq [A\xi, y_{\varepsilon n} - \xi]_+ - C_A(R; \|y_{\varepsilon n} - \xi\|'_W).
\end{aligned} \tag{2.15}$$

Following relations hold true:

$$\langle f, y_{\varepsilon n} - v_{\varepsilon n}\rangle \to 0, \quad (y'_{\varepsilon n}, v'_{\varepsilon n} - \xi')_{\mathscr{H}} \to (y'_\varepsilon, y'_\varepsilon - \xi')_{\mathscr{H}},$$

$$\langle y'_{\varepsilon n}, v_{\varepsilon n} - \xi\rangle \to \langle y'_\varepsilon, y_\varepsilon - \xi\rangle, \quad \langle d_{\varepsilon n}, v_{\varepsilon n} - \xi\rangle \to \langle \varkappa_\varepsilon, y_\varepsilon - \xi\rangle,$$
$$(\xi', y'_{\varepsilon n} - \xi')_{\mathscr{H}} \to (\xi', y'_\varepsilon - \xi')_{\mathscr{H}}, \quad \langle \xi', y_{\varepsilon n} - \xi\rangle \to \langle \xi', y_\varepsilon - \xi\rangle,$$
$$\varliminf_{n\to\infty} [A\xi, y_{\varepsilon n} - \xi]_+ \geq [A\xi, y_\varepsilon - \xi]_+,$$
$$C_A(R; \|y_{\varepsilon n} - \xi\|'_W) \to C_A(R; \|y_\varepsilon - \xi\|'_W),$$

as $n \to \infty$. Then by passing to the limit in (2.15) as $n \to \infty$, we have

$$\begin{aligned}
&\varepsilon(y'_\varepsilon, y'_\varepsilon - \xi')_{\mathscr{H}} + \langle y'_\varepsilon, y_\varepsilon - \xi\rangle + \langle \varkappa_\varepsilon, y_\varepsilon - \xi\rangle - \varepsilon(\xi', y'_\varepsilon - \xi')_{\mathscr{H}} - \langle \xi', y_\varepsilon - \xi\rangle \\
&\quad = \varepsilon(y'_\varepsilon - \xi', y'_\varepsilon - \xi')_{\mathscr{H}} + \langle y'_\varepsilon - \xi', y_\varepsilon - \xi\rangle + \langle \varkappa_\varepsilon, y_\varepsilon - \xi\rangle \\
&\quad \geq [A\xi, y_\varepsilon - \xi]_+ - C_A(R; \|y_\varepsilon - \xi\|'_W).
\end{aligned} \tag{2.16}$$

In last inequality we set $\xi = y_\varepsilon - \tau\omega$ where $\omega \in \widetilde{W}$. Hence

$$\tau\varepsilon(\omega', \omega')_{\mathscr{H}} + \tau\langle \omega', \omega\rangle + \langle \varkappa_\varepsilon, \omega\rangle \geq [A(y_\varepsilon - \tau\omega), \omega]_+ - \frac{1}{\tau} C_A(R; \|\tau\omega\|'_W).$$

Due to radial lower semicontinuity of A we can pass to the limit as $\tau \to 0+$, we have

$$\langle \varkappa_\varepsilon, \omega\rangle \geq [Ay_\varepsilon, \omega]_- \quad \forall \omega \in \widetilde{W}.$$

So $\varkappa_\varepsilon \in \overset{*}{\overline{\mathrm{co}}}\, Ay_\varepsilon$ namely y_ε satisfies inequality (2.3).

Remark also, that from (2.13) and from the definition of the derivative in the sense of $\mathscr{D}(S; V^*)$ it follows that

$$\varepsilon y''_\varepsilon = d_{1\varepsilon} + d_{2\varepsilon} \quad y'_\varepsilon(T) = \bar{0} \tag{2.17}$$

where $d_{1\varepsilon} = \varkappa_\varepsilon - f \in \overset{*}{\overline{\mathrm{co}}}\, A(y_\varepsilon) - f \subset X^*$ and $d_{2\varepsilon} = y'_\varepsilon \in \mathscr{H}$. Hence $y''_\varepsilon \in X^* + \mathscr{H}$. □

Let $\varepsilon \to 0$. Then up to a subsequence $y_\varepsilon \rightharpoonup y$ in X. So, in virtue of (2.17) and due to $y_\varepsilon'(T) = \bar{0}$ we have:

$$y_\varepsilon'(t) = -\varepsilon^{-1} \int_0^{T-t} d_{1\varepsilon}(T-\tau)e^{-(T-t-\tau)/\varepsilon} d\tau, \quad t \in S,$$

where $\{d_{1\varepsilon}\}$ is a bounded set in X^*. Since $\varepsilon^{-1} \int_0^\infty e^{-(\tau/\varepsilon)} d\tau = 1$ then from the convolution product inequality it follows that $\{y_\varepsilon'\}$ is a bounded set in X^*. Therefore we may assume that $y_\varepsilon \rightharpoonup y$ in W up to a subsequence.

Using (2.5) we obtain

$$\varepsilon|(y_\varepsilon', \xi')_{\mathscr{H}}| \le \varepsilon\|y_\varepsilon'\|_{\mathscr{H}}\|\xi'\|_{\mathscr{H}} \le \sqrt{k}\sqrt{\varepsilon}\|\xi'\|_{\mathscr{H}} \to 0 \text{ as } \varepsilon \to 0+.$$

Then due to (2.13) (remark that $\varkappa_\varepsilon \in \overset{*}{\overline{\text{co}}}\, Ay_\varepsilon$),

$$\langle y', \xi\rangle + \langle \varkappa, \xi\rangle = \langle f, \xi\rangle \quad \forall\, \xi \in \widetilde{W}, \tag{2.18}$$

where $\varkappa$ is a weak limit of the sequence $\varkappa_\varepsilon$ in X^*. Note, though, that the equality (2.18) holds true $\forall \xi \in W$. Due to Proposition 2 to prove the Theorem it is sufficient to show that $\langle \varkappa, \xi\rangle \le [Ay, \xi]_+ \ \forall \xi \in W$. From the inequality (2.16), which is valid for y_ε, and in virtue of (2.13) the relation

$$\langle f, y_\varepsilon - \xi\rangle - \varepsilon(\xi', y_\varepsilon' - \xi')_{\mathscr{H}} - \langle \xi', y_\varepsilon - \xi\rangle \ge [A\xi, y_\varepsilon - \xi]_+ - C_A(R; \|y_\varepsilon - \xi\|_W'),$$

follows and after passing to the limit as $\varepsilon \to 0+$ (due to (2.18)), up to a subsequence we have that

$$\langle y' - \xi', y - \xi\rangle + \langle \varkappa, y - \xi\rangle \ge [A\xi, y - \xi]_+ - C_A(R; \|y - \xi\|_W')$$

is valid $\forall\, \xi \in W$. Setting in it $\xi = y - \tau\omega$, $\omega \in W$ we find

$$\tau\langle \omega', \omega\rangle + \langle \varkappa, \omega\rangle \ge [A(y - \tau\omega), \omega]_+ - \frac{1}{\tau}C_A(R; \tau\|\omega\|_W').$$

Hence passing to the limit as $\tau \to 0+$ due to radial lower semicontinuity of the operator A and properties of the function C_A we obtain the required relation:

$$\langle \varkappa, \omega\rangle \ge [A(y), \omega]_- \quad \forall \omega \in W.$$

The Theorem is proved. □

Remark 2.3. Let in the last Theorem A be *Volterra type operator*, namely for any $u, v \in X$, $t \in S$ from the equality $u(s) = v(s)$ for a.e. $s \in [0, t]$ it follows that

$[A(u),\xi_t]_+ = [A(v),\xi_t]_+ \ \forall \xi_t \in X\colon \xi_t(s) = 0$ for a.e. $s \in S\backslash[0,t]$; $[\cdot]_{V_i}$ is some seminorm on V_i $(i = 1,2)$, assume that $[y]_{X_i} = \left(\int\limits_S [y(\tau)]_{V_i}^{p_i} d\tau\right)^{1/p_i}$. Obviously that $[\cdot]_{X_i}$ is a seminorm on X. We can obtain the uniform boundness of solutions of problem (2.6), if instead of +-coercivity of the operator $A : X \rightrightarrows X^*$ the following condition hold true:

$$\exists\, \lambda_0 > 0,\ \beta > 0,\ \gamma_1 > 0,\ \gamma_2 > 0,\ \alpha \in \mathbb{R} \text{ and } f \in X^* \text{ such that } \forall y \in W$$

$$\langle y', y\rangle + \langle d(y), y\rangle \le \langle f, y\rangle \text{ for the given } d \in \overset{*}{\overline{\text{co}}}\, A, \tag{2.19}$$

$$[y]_{X_1} + [y]_{X_2} + \lambda_0 \|y\|_{L_{p_0}(S;H)} \ge \beta \|y\|_X, \tag{2.20}$$

$$\langle d(y), y\rangle \ge \gamma_1 [y]_{X_1}^{p_1} + \gamma_2 [y]_{X_2}^{p_2} + \alpha, \tag{2.21}$$

where $p_0 = \max\{r_1, r_2\}$.

Remark that sufficient conditions for (2.19)–(2.21) are following: for each $y \in X$

$$[y]_{X_1} + [y]_{X_2} + \lambda_0 \|y\|_{L_{p_0}(S;H)} \ge \beta \|y\|_X, \tag{2.22}$$

$$[Ay, y]_- \ge \gamma_1 [y]_{X_1}^{p_1} + \gamma_2 [y]_{X_2}^{p_2} + \alpha, \tag{2.23}$$

but the conditions (2.19)–(2.21) are weaker then (2.20), (2.21).

Proof. All the difference lies in the estimation obtaining:

$$\|y_{\varepsilon n}\|_X \le R_1 \quad \text{for each } \varepsilon > 0,\ n \ge 1,$$

since the resolvability of problem (2.6) often follows from the resolvability of algebraic inequalities and inclusions theory (in Theorem 2.1 then, firstly, we choose a bounded set in which solutions will lie and here we show that solutions of problem (2.6) (if they do exist) are uniformly bounded). Since A is Volterra operator with semibounded variation on W, then from (2.19) to (2.21) it follows that for each $t, \in S$

$$\frac{\|y(t)\|_H^2 - \|y(0)\|_H^2}{2} + \langle d(y), y\rangle_{X_t} \le \langle f, y\rangle_{X_t} \text{ for given } d \in A, \tag{2.24}$$

$$[y]_{X_{1t}} + [y]_{X_{2t}} + \lambda_0 \|y\|_{L_{p_0}(0,t;H)} \ge \beta \|y\|_{X_t}, \tag{2.25}$$

and

$$\langle d(y), y\rangle_{X_t} \ge \gamma_1 [y]_{X_{1t}}^{p_1} + \gamma_2 [y]_{X_{2t}}^{p_2} + \tilde{\alpha} \tag{2.26}$$

where for $i = 1,2$ $[y]_{X_{it}} = (\int\limits_0^t [y(\tau)]_V\, p_i d\tau)1/p_i$ is a seminorm on $X_t = X([0,t])$, $\langle\cdot,\cdot\rangle_{X_t}$ is paring on $X_t^* \times X_t$. Then, using Cauchi and Young inequalities, and also (2.9), (2.10), (2.24), (2.25), (2.26) and $y_{\varepsilon n}(0) = \bar{0}$ we obtain

$$
\begin{aligned}
&\frac{1}{2}\|y_{\varepsilon n}(t)\|_H^2 + \gamma_1 \int_0^t [y_{\varepsilon n}(\tau)]_{V_1}^{p_1} d\tau + \gamma_2 \int_0^t [y_{\varepsilon n}(\tau)]_{V_2}^{p_2} d\tau \\
&\quad \le \frac{1}{2}\|y_{\varepsilon n}(t)\|_H^2 + \langle d_{\varepsilon n}, y_{\varepsilon n}\rangle_{X_t} - \tilde{\alpha} \\
&\quad \le \langle f, y_{\varepsilon n}\rangle_{X_t} - \tilde{\alpha} \le \|f\|_{X^*{}_t}\|y_{\varepsilon n}\|_{X_t} - \tilde{\alpha} \\
&\quad \le \frac{1}{\beta}\|f\|_{X^*{}_t}\Big([y_{\varepsilon n}]_{X_{1t}} + [y_{\varepsilon n}]_{X_{2t}} + \lambda_0\|y_{\varepsilon n}\|_{L_{p_0}(0,t;H)}\Big) - \tilde{\alpha} \\
&\quad \le C_1 + \frac{\gamma_1}{2}\int_0^t [y_{\varepsilon n}(\tau)]_{V_1}^{p_1} d\tau + \frac{\gamma_2}{2}\int_0^t [y_{\varepsilon n}(\tau)]_{V_2}^{p_2} d\tau + C_2\Big(\int_0^t \|y_{\varepsilon n}(\tau)\|_H^{p_0} d\tau\Big)^{2/p_0} \\
&\quad \le C_1 + \frac{\gamma_1}{2}[y_{\varepsilon n}]_{X_{1t}}^{p_1} + \frac{\gamma_2}{2}[y_{\varepsilon n}]_{X_{2t}}^{p_2} + C_2\|y_{\varepsilon n}\|_{L_{p_0}([0,t];H)}^2.
\end{aligned}
$$

Therefore

$$
\|y_{\varepsilon n}(t)\|_H^2 + \gamma_1[y_{\varepsilon n}]_{X_{1t}}^{p_1} + \gamma_2[y_{\varepsilon n}]_{X_{2t}}^{p_2} \le 2C_1 + 2C_2\|y_{\varepsilon n}\|_{L_{p_0}([0,t];H)}^2,
$$

hence

$$
\|y_{\varepsilon n}(t)\|_H^{p_0} \le C_3 + C_3\int_0^t \|y_{\varepsilon n}(\tau)\|_H^{p_0} d\tau \quad \text{for each } t \in S,\ \varepsilon > 0 \text{ and } n \ge 1.
$$

So, $\|y_{\varepsilon n}(t)\|_H^{p_0} \le C_4 e^{C_4 t}$ namely $\|y_{\varepsilon n}(t)\|_H \le C_5$. Then $[y_{\varepsilon n}]_{X_{it}} \le C_6, i = 1, 2$ and $\|y_{\varepsilon n}\|_{X_t} \le C_7$. The arbitrariness of $t \in S$ gives us the required estimation. □

2.2 Brezis Method for W_{λ_0}-Pseudomonotone Maps

2.2.1 *Problem Definition*

Let X be the reflexive Banach space, X^* be its topologically adjoint,

$$
\langle \cdot, \cdot \rangle_X : X^* \times X \to \mathbb{R} \text{ is canonical paring.}
$$

We assume that for some interpolation pair of reflexive Banach spaces $X_1, X_2, X = X_1 \cap X_2$. Then due to Theorem 1.3 $X^* = X_1^* + X_2^*$. Remark also that

$$
\langle f, y\rangle_X = \langle f_1, y\rangle_{X_1} + \langle f_2, y\rangle_{X_2} \quad \forall f \in X^*\ \forall y \in X,
$$

where $f = f_1 + f_2$, $f_i \in X_i^*, i = 1, 2$.

Let $L : D(L) \subset X \to X^*$ be linear operator, with dense definitional domain, $A : X_1 \to C_v(X_1^*)$, $B : X_2 \to C_v(X_2^*)$ are the multivalued maps. We consider the following problem:

$$Ly + A(y) + B(y) \ni f, \quad y \in D(L), \tag{2.27}$$

where $f \in X^*$ is arbitrary fixed.

Remark 2.4. Further we will assume that $D(L)$ is reflexive Banach space with the norm

$$\|y\|_{D(L)} = \|y\|_X + \|Ly\|_{X^*} \quad \forall y \in D(L).$$

The given condition is true through the maximal monotony of L on $D(L)$ (Corollary 1.8).

2.2.2 *Singular Perturbations Method*

Let us consider generally speaking the multivalued duality map

$$J(y) = \{\xi \in X^* \mid \langle \xi, y\rangle_X = \|\xi\|_{X^*}^2 = \|y\|_X^2\} \in C_v(X^*) \quad \forall y \in X,$$

namely

$$J(y) = \partial(\|\cdot\|_X^2/2)(y) \quad \forall y \in X,$$

which is the corollary of Proposition 8. From Theorem 2 it follows that this map is defined on the whole space X, and from [AE84] its maximal monotony follows. Besides, due to [AE84, Theorem 4, p. 202 and Proposition 8, p. 203] for each $f \in X^*$ the map

$$\left.\begin{aligned} J^{-1}(f) &= \{y \in X \mid f \in J(y)\} \\ &= \{y \in X \mid \langle f, y\rangle_X = \|f\|_{X^*}^2 = \|y\|_X^2\} \in C_v(X). \end{aligned}\right\} \tag{2.28}$$

is also defined on the whole space X and it is the maximal monotone multivalued map.

We will approximate the inclusion from (2.27) by the following:

$$\varepsilon L^* J^{-1}(Ly_\varepsilon) + Ly_\varepsilon + A(y_\varepsilon) + B(y_\varepsilon) \ni f. \tag{2.29}$$

Definition 2.1. We will say that a solution $y \in D(L)$ of problem (2.27) turns out by *Singular perturbations method*, if y is a weak limit of a subsequence $\{y_{\varepsilon_{n_k}}\}_{k\ge1}$ of the sequence $\{y_{\varepsilon_n}\}_{n\ge1}$ ($\varepsilon_n \searrow 0+$ as $n \to \infty$) in the space $D(L)$ that for every $n \ge 1$ $D(L) \ni y_{\varepsilon_n}$ is a solution of problem (2.29).

2.2.3 The Main Result

Theorem 2.2. *Let X be a reflexive Banach space,*

$$L : D(L) \subset X \to X^*$$

be a linear, densely defined, maximal monotone on $D(L)$ operator, $A : X_1 \to C_v(X_1^*)$ and $B : X_2 \to C_v(X_2^*)$ be finite-dimensionally locally bounded, s-mutually bounded, λ_0-pseudomonotone on $D(L)$ multivalued maps, for which Condition (Π) is valid. Also let for some $f \in X^*$ there exists $R > 0$ such that

$$[A(y), y]_+ + [B(y), y]_+ - \langle f, y\rangle_X \geq 0 \quad \forall y \in X : \|y\|_X = R. \tag{2.30}$$

Then there exists at least one solution $y \in D(L)$ of problem (2.27).

Proof. Firstly let us show that for each $\varepsilon > 0$ problem (2.29) has a solution in the space $D(L)$.

For each $y, \omega \in D(L)$ we set

$$\left.\begin{aligned} \Pi_\varepsilon(y,\omega) = \varepsilon\left[L\omega, J^{-1}(Ly)\right]_+ + \langle Ly, \omega\rangle_X \\ + [A(y),\omega]_+ + [B(y),\omega]_+ - \langle f,\omega\rangle_X. \end{aligned}\right\} \tag{2.31}$$

Remark that for any $y \in D(L)$ the form

$$D(L) \ni \omega \to \Pi_\varepsilon(y,\omega)$$

is positively homogeneous, convex and lower semicontinuous in the graph norm L

$$\|\omega\|_{D(L)} = \|\omega\|_X + \|L\omega\|_{X^*} \quad \forall \omega \in D(L).$$

Therefore due to Proposition 3 it follows that for each $\varepsilon > 0$ there exists a multivalued map

$$\mathscr{B}_\varepsilon(y) : D(L) \to C_v\big((D(L))^*\big),$$

such that

$$\Pi_\varepsilon(y,\omega) = [\mathscr{B}_\varepsilon(y), \omega]_+ \quad \text{for any} \quad y, \omega \in D(L). \tag{2.32}$$

Proposition 2.4. *For every $\varepsilon > 0$ $\mathscr{B}_\varepsilon : D(L) \to C_v\big((D(L))^*\big)$ is finitedimensionally locally bounded, λ_0-pseudomonotone on $D(L)$ multivalued map.*

Proof. Since for each $y \in D(L)$ the form

$$D(L) \ni \omega \to \varepsilon[L\omega, J^{-1}(Ly)]_+ + \langle Ly, \omega\rangle_X$$

is positively homogeneous, convex and lower semicontinuous on $D(L)$ then due to Proposition 3 we define new multivalued map

$$\mathscr{M}_\varepsilon : D(L) \to C_v\big((D(L))^*\big)$$

in the following way:

$$[\mathscr{M}_\varepsilon(y), v]_+ = \varepsilon[Lv, J^{-1}(Ly)]_+ + \langle Ly, v\rangle_X \quad \forall y, v \in D(L). \tag{2.33}$$

Remark that $\mathscr{M}_\varepsilon$ has bounded values since for each $y, v \in D(L)$ the right side of the last equality is bounded above.

Lemma 2.1. *For every $\varepsilon > 0$ the multivalued map $\mathscr{M}_\varepsilon$ is monotone, bounded and upper hemicontinuous on $D(L)$.*

Proof. Firstly let us prove monotony of $\mathscr{M}_\varepsilon$. From (2.33) and from Proposition 1 for all $y_1, y_2 \in D(L)$ we obtain:

$$[\mathscr{M}_\varepsilon(y_1), y_1 - y_2]_- = \varepsilon\left[Ly_1 - Ly_2, J^{-1}(Ly_1)\right]_- + \langle Ly_1, y_1 - y_2\rangle_X,$$

$$[\mathscr{M}_\varepsilon(y_2), y_1 - y_2]_+ = \varepsilon\left[Ly_1 - Ly_2, J^{-1}(Ly_2)\right]_+ + \langle Ly_2, y_1 - y_2\rangle_X.$$

Since J^{-1} is a monotone map, and the operator L is positive, then comparing the last two equalities we have

$$[\mathscr{M}_\varepsilon(y_1), y_1 - y_2]_- \geq [\mathscr{M}_\varepsilon(y_2), y_1 - y_2]_+ \quad \text{for any } y_1, y_2 \in D(L).$$

So monotony of $\mathscr{M}_\varepsilon$ on $D(L)$ is proved.

Now we prove hemicontinuity of $\mathscr{M}_\varepsilon$ on $D(L)$. We will prove the statement by contradiction. Let $\omega \in D(L)$, $y_n \to y$ in $D(L)$ and $\beta > 0$ such that

$$\forall n \geq 1 \quad [\mathscr{M}_\varepsilon(y_n), \omega]_+ \geq \beta + [\mathscr{M}_\varepsilon(y), \omega]_+ . \tag{2.34}$$

Remark that

$$[\mathscr{M}_\varepsilon(y_n), \omega]_+ = \langle Ly_n, \omega\rangle_X + \varepsilon\left[L\omega, J^{-1}(Ly_n)\right]_+ \quad \forall n \geq 1.$$

In virtue of Proposition 3 and the definition of the upper support function, for each $n \geq 1$ the existence of $l_n = l_n(\omega) \in J^{-1}(Ly_n)$ such that

$$\left[L\omega, J^{-1}(Ly_n)\right]_+ = \langle L\omega, l_n\rangle_X$$

follows. Hence similarly for any $n \geq 1$ the existence of $m_{\varepsilon n} = m_{\varepsilon n}(\omega) \in \mathscr{M}_\varepsilon(y_n)$ such that

$$\langle m_{\varepsilon n}, \omega\rangle_{D(L)} = \langle Ly_n, \omega\rangle_X + \varepsilon\langle L\omega, l_n\rangle_X = [\mathscr{M}_\varepsilon(y_n), \omega]_+$$

follows. Due to boundness of J^{-1} (see (2.28) and Proposition 1.17) we have:

$$Ly_n \to Ly \quad \text{in} \quad X^* \quad \text{as} \quad m \to \infty, \tag{2.35}$$

and therefore, in virtue of the Banach–Alaoglu Theorem the existence of a subsequence $\{l_{n_m}\}_{m\ge 1} \subset \{l_n\}_{n\ge 1}$ such that

$$l_{n_m} \rightharpoonup l \quad \text{in} \quad X \quad \text{as} \quad n \to \infty \quad \text{for some} \quad l \in X \tag{2.36}$$

follows. This implies that

$$\overline{\lim_{m\to\infty}} \left[L\omega, J^{-1}(Ly_{n_m})\right]_+ = \langle L\omega, l\rangle_X.$$

Now we prove that $l \in J^{-1}(Ly)$. From $l_{n_m} \in J^{-1}(y_{n_m})$ and from (2.28) it follows that for all $m \ge 1$

$$\langle Ly_{n_m}, l_{n_m}\rangle_X = \|Ly_{n_m}\|_{X^*}^2 = \|l_{n_m}\|_X^2.$$

Hence by passing to the limit as $m \to \infty$ in virtue of (2.35) and (2.36) we have:

$$\langle Ly, l\rangle_X = \|Ly\|_{X^*}^2 \ge \|l\|_X^2,$$

but

$$\langle Ly, l\rangle_X \le \|Ly\|_{X^*}\|l\|_X.$$

So,

$$\|Ly\|_{X^*}^2 = \|l\|_X^2 = \langle Ly, l\rangle_X$$

and due to (2.28) it implies that $l \in J^{-1}(Ly)$.

Finally we have

$$\overline{\lim_{m\to\infty}} \left[L\omega, J^{-1}\left(Ly_{n_m}\right)\right]_+ = \langle L\omega, l\rangle_X \le \left[L\omega, J^{-1}\left(Ly\right)\right]_+$$

and

$$\begin{aligned}\overline{\lim_{m\to\infty}} \left[M_\varepsilon\left(y_{n_m}\right), \omega\right]_+ &= \langle Ly, \omega\rangle_X + \varepsilon\langle L\omega, l\rangle_X \\ &\le \langle Ly, \omega\rangle_X + \varepsilon\left[L\omega, J^{-1}\left(Ly\right)\right]_+ = \left[M_\varepsilon\left(y\right), \omega\right]_+.\end{aligned}$$

We obtained the contradiction with (2.34).

All we have to do is to prove boundness of M_ε on $D(L)$. In virtue of (2.33) and (2.28) it implies that for every bounded in $D(L)$ set $D \subset D(L)$ for all $y \in D$ and $v \in D(L)$

$$\begin{aligned}[M_\varepsilon(y), v]_+ &\le \varepsilon\|Lv\|_{X^*}\|J^{-1}(Ly)\|_+ + \|Ly\|_{X^*}\|v\|_X \\ &\le \varepsilon\|v\|_{D(L)}\|Ly\|_{X^*} + \|y\|_{D(L)}\|v\|_{D(L)} \\ &\le (\varepsilon + 1)\|y\|_{D(L)}\|v\|_{D(L)} \le (\varepsilon + 1)\|D\|_+\|v\|_{D(L)} < +\infty.\end{aligned}$$

Therefore in virtue of Banach–Steinhaus Theorem

$$\|M_\varepsilon(D)\|_+ \le (\varepsilon + 1)\|D\|_+ < +\infty.$$

Boundedness above is proved. Lemma 2.1 is proved. □

Since every hemicontinuous monotone on $D(L)$ multivalued map is λ_0-pseudomonotone on $D(L)$ (see [ZM00, Proposition 6 and Remark 3, c.63]) then the multivalued map M_ε also is λ_0-pseudomonotone on $D(L)$. Hence, due to Lemma 1.15 and s-mutual bounededness of (A, B) on X the same is valid for $\mathscr{B}_\varepsilon$. Finite-dimensional local boundness of $\mathscr{B}_\varepsilon$ follows from the same property for A, B and M_ε and from (2.31), (2.32). The Proposition is proved. □

From Lemma 1.15 and from Remark 1 it follows that the map $A + B : X \rightrightarrows X^*$ is λ_0-pseudomonotone on $D(L)$ and satisfies Condition (Π) on X.

The relations (2.33) and (2.28) imply for arbitrary $\varepsilon > 0$ and $y \in D(L)$

$$[M_\varepsilon(y), y]_+ = \varepsilon[Ly, J^{-1}(Ly)]_+ + \langle Ly, y\rangle_X \ge \varepsilon\|Ly\|^2_{X^*} \ge 0.$$

Hence, due to Proposition 1 and (2.30) for every $\varepsilon > 0$

$$\begin{cases} [\mathscr{B}_\varepsilon(y), y]_+ = \varepsilon\left[Ly, J^{-1}(Ly)\right]_+ \\ \qquad\qquad +\langle Ly, y\rangle_X + [A(y), y]_+ + [B(y), y]_+ \\ \qquad\qquad -\langle f, y\rangle_X \ge 0 \qquad \forall y \in D(L) : \|y\|_X = R. \end{cases} \tag{2.37}$$

Proposition 2.5. *For every $\varepsilon > 0$ there exists at least one solution of the problem*

$$\bar{0} \in \mathscr{B}_\varepsilon(y_\varepsilon), \quad y_\varepsilon \in D(L), \quad \|y_\varepsilon\|_X \le R. \tag{2.38}$$

Remark 2.5. Remark that $\|\cdot\|_X$ is the norm on $D(L)$.

Proof. Let $\mathscr{F}$ be a filter of all finite-dimensional subspaces of the space $D(L)$. In an arbitrary subspace $F \in \mathscr{F}$ let us consider the norm $\|\cdot\|_F = \|\cdot\|_X\big|_F$ induced from the space X to the space $F \subset X$. Now we fix an arbitrary $\varepsilon > 0$ and prove the validity of the following Proposition.

Proposition 2.6. *For any $F \in \mathscr{F}$ there exists at least one solution $y_{\varepsilon F} \in F$ of the problem*

$$[\mathscr{B}_\varepsilon(y_{\varepsilon F}), h]_+ \ge 0 \quad \forall h \in F \tag{2.39}$$

for which the estimation $\|y_{\varepsilon F}\|_X \le R$ is true.

Proof. Let a space $F \in \mathscr{F}$ be arbitrary fixed. Since F is finite-dimensional separable Banach space with respect to the norm $\|\cdot\|_F$ on F then there exists a vector system $\{v_i\}_{i\ge1} \subset F$ which is dense in F. We will approximate (2.39) by the finite system of algebraic inequalities

$$[\mathscr{B}_\varepsilon(y_{\varepsilon F m}), v_i]_- \leq 0 \leq [\mathscr{B}_\varepsilon(y_{\varepsilon F m}), v_i]_+ \quad i = \overline{1,m}, \tag{2.40}$$

where $m \geq 1$ and the solution $y_{\varepsilon F m} \in F$ are represented in the form of $\sum_{j=1}^{m} \alpha_{\varepsilon,m}^{j,F} v_j$.

Proposition 2.7. *For $m \geq 1$ problem (2.40) has at least one solution $\bar{\alpha}^m_{\varepsilon,F} = (\alpha_{\varepsilon,m}^{j,F})_{j=1}^m \in \mathbb{R}^m$ such that for*

$$y_{\varepsilon F m}(\bar{\alpha}^m_{\varepsilon,F}) = \sum_{j=1}^{m} \alpha_{\varepsilon,m}^{j,F} v_j$$

the estimation

$$\|y_{\varepsilon F m}(\bar{\alpha}^m_{\varepsilon,F})\|_X \leq R$$

holds true. Moreover, the set

$$G_{\varepsilon F}(m) = \Big\{ y_{\varepsilon F m}(\bar{\alpha}^m_{\varepsilon,F}) \in F \mid y_{\varepsilon F m}(\bar{\alpha}^m_{\varepsilon,F}) \text{ satisfies (2.40)} \\ \text{and } \|y_{\varepsilon F m}(\bar{\alpha}^m_{\varepsilon,F})\|_X \leq R \Big\}$$

is compact in F.

Proof. For any fixed $m \geq 1$ let us consider the multivalued map $B : \mathbb{R}^m \to C_v(\mathbb{R}^m)$ which is defined in the following way:

$$\forall \bar{\alpha} \in \mathbb{R}^m \quad B(\bar{\alpha}) = (B_i(\bar{\alpha}))_{i=1}^m,$$

where for every $i = \overline{1,m}$

$$B_i(\bar{\alpha}) = \Big[[\mathscr{B}_\varepsilon(y(\bar{\alpha})), v_i]_-, [\mathscr{B}_\varepsilon(y(\bar{\alpha})), v_i]_+ \Big] \in C_v(\mathbb{R}), \quad \bar{\alpha} = (\alpha_i)_{i=1}^m.$$

In $\mathbb{R}^m$ we consider the norm

$$\|\bar{\alpha}\|_{\mathbb{R}^m} = \|y(\bar{\alpha})\|_X = \left\| \sum_{i=1}^{m} \alpha_i v_i \right\|_X \quad \forall \bar{\alpha} = (\alpha_i)_{i=1}^m \in \mathbb{R}^m$$

and the paring

$$\langle \bar{\alpha}, \bar{\beta} \rangle = \sum_{i=1}^{m} \alpha_i \beta_i \quad \forall \bar{\alpha} = (\alpha_i)_{i=1}^m \in \mathbb{R}^m,\ \bar{\beta} = (\beta_i)_{i=1}^m \in \mathbb{R}^m.$$

Due to Proposition 1 and (2.37) it implies that for each $\bar{\alpha} = (\alpha_i)_{i=1}^m \in \mathbb{R}^m$

$$
\begin{aligned}
[B(\bar{\alpha}), \bar{\alpha}]_+ &= \sup\left\{ \sum_{i=1}^m b_i \alpha_i \,\middle|\, b_i \in B_i(\bar{\alpha}),\ i = \overline{1,n} \right\} \\
&\geq \sum_{i=1}^m [\mathscr{B}_\varepsilon(y(\bar{\alpha})), v_i]_+ \alpha_i = \sum_{i=1}^m [\mathscr{B}_\varepsilon(y(\bar{\alpha})), \alpha_i v_i]_+ \\
&\geq \left[\mathscr{B}_\varepsilon(y(\bar{\alpha})), \sum_{i=1}^m \alpha_i v_i \right]_+ = [\mathscr{B}_\varepsilon(y(\bar{\alpha})), y(\bar{\alpha})]_+ \geq 0,
\end{aligned}
$$

as $\|\bar{\alpha}\|_{\mathbb{R}^m} = \|y(\bar{\alpha})\|_X = R$. Hence,

$$
[B(\bar{\alpha}), \bar{\alpha}]_+ \geq 0 \quad \text{for all} \quad \bar{\alpha} \in \mathbb{R}^m : \|\bar{\alpha}\|_{\mathbb{R}^m} = R. \tag{2.41}
$$

Similarly for all $\bar{\alpha} = (\alpha_i)_{i=1}^m \in \mathbb{R}^m$ and $\bar{\beta} = (\beta_i)_{i=1}^m \in \mathbb{R}^m$

$$
\begin{aligned}
[B(\bar{\alpha}), \bar{\beta}]_+ &= \sup\left\{ \sum_{i=1}^m b_i \beta_i \,\middle|\, b_i \in B_i(\bar{\alpha}),\ i = \overline{1,n} \right\} \\
&= \sum_{i=1}^m \Big(\max\{[\mathscr{B}_\varepsilon(y(\bar{\alpha})), v_i]_+,\ [\mathscr{B}_\varepsilon(y(\bar{\alpha})), -v_i]_+\} \Big) \cdot |\beta_i|.
\end{aligned}
$$

Upper semicontinuity of the map

$$
\mathbb{R}^m \ni \bar{\alpha} \to \max\{[\mathscr{B}_\varepsilon(y(\bar{\alpha})), v_i]_+,\ [\mathscr{B}_\varepsilon(y(\bar{\alpha})), -v_i]_+\} \quad \forall i = \overline{1,m}
$$

follows from the same Proposition for the maps

$$
\mathbb{R}^m \ni \bar{\alpha} \to [\mathscr{B}_\varepsilon(y(\bar{\alpha})), v_i]_+ \quad \text{and} \quad \mathbb{R}^m \ni \bar{\alpha} \to [\mathscr{B}_\varepsilon(y(\bar{\alpha})), -v_i]_+ \quad i = \overline{1,m}. \tag{2.42}
$$

The latter follows from finite-dimensional local boundness and λ_0-pseudomonotony of $\mathscr{B}_\varepsilon$ on $D(L)$. Indeed due to Lemma 1.14 and Proposition 2.4 we have that $\mathscr{B}_\varepsilon$ is λ-pseudomonotone and locally bounded on F,and hence, due to [M00, Lemma 1, p. 1516] and [AE84, Proposition 2, p. 127] for each $\bar{\beta} \in \mathbb{R}^m$ the map (2.42) is upper semicontinuous. Therefore,

$$
\mathbb{R}^m \ni \bar{\alpha} \to [B(\bar{\alpha}), \bar{\beta}]_+
$$

is upper semicontinuous. So, in virtue of Castaing Theorem [AE84, p. 132], B is upper semicontinuous on $\mathbb{R}^m$. Now, due to (2.41) we can apply the "acute angle" Lemma for multivalued maps (see Lemma 2) to the map B. The given Lemma implies that for every $m \geq 1$ there exists at least one solution (2.40)

$$
\bar{\alpha}_{\varepsilon F}^m = (\alpha_{\varepsilon,m}^{j,F})_{j=1}^m \in \mathbb{R}^m \quad \text{such that} \quad \|\bar{\alpha}_{\varepsilon F}^m\|_{\mathbb{R}^m} \leq R.
$$

Hence, for

$$y_{\varepsilon F m} = \sum_{j=1}^{m} \alpha_{\varepsilon}^{j,F} v_j$$

the estimation $\|y_{\varepsilon F m}\|_X \le R$ is valid.

The compactness of $G_{\varepsilon F}(m)$ directly follows from boundness of $G_{\varepsilon F}(m)$ and hemicontinuity of B on $\mathbb{R}^m$. Proposition 2.7 is proved. □

We continue the proof of Proposition 2.6. Let us consider the set

$$G_{\varepsilon F} = \bigcap_{m\ge 1} G_{\varepsilon F}(m).$$

It is not empty, since for any $m \ge 1$

$$G_{\varepsilon F}(m+1) \subset G_{\varepsilon F}(m),$$

and $G_{\varepsilon F}(m)$ is the compact (namely, the family $\{G_{\varepsilon F}(m)\}_{m\ge 1}$ is centered in F). Therefore

$$\exists y_{\varepsilon F} \in G_{\varepsilon F}: \quad \|y_{\varepsilon F}\|_X \le R \quad \text{and} \quad [\mathscr{B}_\varepsilon(y_{\varepsilon F}), v_i]_+ \ge 0 \quad \forall i \ge 1.$$

Due to the density of the system $\{v_i\}_{i\ge 1}$ in the space F we have the validity of (2.39) for $y_{\varepsilon F} \in F$. Proposition 2.6 is proved. □

We continue the proof of Proposition 2.5. For arbitrary $\varepsilon > 0$ and $F \in \mathscr{F}$ let us consider the nonempty set

$$G_{\varepsilon F} = \{y_{\varepsilon F} \in F \mid y_{\varepsilon F} \text{ satisfies(2.39) and } \|y_{\varepsilon F}\|_X \le R\}$$

and prove its boundness in $D(L)$ uniformly by F for any fixed $\varepsilon > 0$.

From inequality (2.39) and equalities (2.31), (2.32) and (2.28) we obtain that for any $\varepsilon > 0$, $F \in \mathscr{F}$, $y_{\varepsilon F} \in G_{\varepsilon F}$

$$\begin{aligned}&\varepsilon\|(Ly_{\varepsilon F})\|_{X^*} + \langle Ly_{\varepsilon F}, y_{\varepsilon F}\rangle_X + [A(y_{\varepsilon F}), y_{\varepsilon F}]_+ + [B(y_{\varepsilon F}), y_{\varepsilon F}]_+ \\ &-\langle f, y_{\varepsilon F}\rangle_X \ge 0 \ge \varepsilon\|(Ly_{\varepsilon F})\|_{X^*} + \langle Ly_{\varepsilon F}, y_{\varepsilon F}\rangle_X \\ &+[A(y_{\varepsilon F}), y_{\varepsilon F}]_- + [B(y_{\varepsilon F}), y_{\varepsilon F}]_- - \langle f, y_{\varepsilon F}\rangle_X.\end{aligned}$$

Hence due to the definition of upper and lower support functions the existence of

$$d'(y_{\varepsilon F}) \in A(y_{\varepsilon F}) \quad \text{and} \quad d''(y_{\varepsilon F}) \in B(y_{\varepsilon F}),$$

such that

$$\left.\begin{aligned}\varepsilon\|Ly_{\varepsilon F}\|^2_{X^*} + \langle Ly_{\varepsilon F}, y_{\varepsilon F}\rangle_X + \langle d'(y_{\varepsilon F}), y_{\varepsilon F}\rangle_{X_1} \\ + \langle d''(y_{\varepsilon F}), y_{\varepsilon F}\rangle_{X_2} = \langle f, y_{\varepsilon F}\rangle_X\end{aligned}\right\} \tag{2.43}$$

follows. Due to $L \geq 0$ and (2.43) we obtain that

$$\langle d'(y_{\varepsilon F}), y_{\varepsilon F}\rangle_{X_1} + \langle d''(y_{\varepsilon F}), y_{\varepsilon F}\rangle_{X_2} \leq \|f\|_{X^*} R.$$

Since for the map $A + B : X \to C_v(X^*)$ Condition (Π) is valid, the existence of $C_1 > 0$ such that

$$\|d'(y_{\varepsilon F}) + d''(y_{\varepsilon F})\|_{X^*} \leq C_1 \quad \forall \varepsilon > 0,\ F \in \mathscr{F},\ y_{\varepsilon F} \in G_{\varepsilon F} \tag{2.44}$$

follows. In virtue of validity of estimations (2.43) and (2.44) we obtain also that for each $\varepsilon > 0$

$$\varepsilon\|Ly_{\varepsilon F}\|^2_{X^*} \leq (C_1 + \|f\|_{X^*})R$$

for all $F \in \mathscr{F}$ and $y_{\varepsilon F} \in G_{\varepsilon F}$. Hence, for each fixed $\varepsilon > 0$

$$\sup_{F\in\mathscr{F}} \|G_{\varepsilon F}\|_+^{(D(L))} \leq R + \frac{1}{\varepsilon}(C_1 + \|f\|_{X^*})R =: C_2, \tag{2.45}$$

where

$$\|G_{\varepsilon F}\|_+^{(D(L))} = \sup_{y\in G_{\varepsilon F}} \|y\|_{D(L)}.$$

Remark that C_2 depends on $\varepsilon > 0$.

From the inequality (2.39) and from Proposition 2 for any $\varepsilon > 0$, $F \in \mathscr{F}$, $y_{\varepsilon F} \in G_{\varepsilon F}$ the existence of $\beta(y_{\varepsilon F}) \in \mathscr{B}_\varepsilon(y_{\varepsilon F})$ such that

$$\langle \beta(y_{\varepsilon F}), h_F\rangle_{D(L)} = 0 \quad \forall h_F \in F \tag{2.46}$$

follows. Now we prove boundness of $\{\beta(y_{\varepsilon F})\}_{F\in\mathscr{F}}$ in the space $(D(L))^*$ $\forall\varepsilon > 0$. From (2.32), (2.31), (2.28), (2.44), (2.45), from the definitions $G_{\varepsilon F}$ and $\|\cdot\|_{D(L)}$ for all $F \in \mathscr{F}$, $y_{\varepsilon F} \in G_{\varepsilon F}$ and $\omega \in D(L)$ it follows that:

$$\begin{aligned}
\langle \beta(y_{\varepsilon F}), \omega\rangle_{D(L)} &\leq [\mathscr{B}_\varepsilon(y_{\varepsilon F}), \omega]_+ \leq \varepsilon\|L\omega\|_{X^*}\|Ly_{\varepsilon F}\|_{X^*} \\
&\quad + \|Ly_{\varepsilon F}\|_{X^*}\|\omega\|_X + C_1\|\omega\|_X + \|f\|_{X^*}\|y_{\varepsilon F}\|_X \\
&\leq (1+\varepsilon)\|\omega\|_{D(L)}\|y_{\varepsilon F}\|_{D(L)} + C_1\|\omega\|_{D(L)} + \|f\|_{X^*}\|y_{\varepsilon F}\|_{D(L)} \\
&\leq (1+\varepsilon)\|\omega\|_{D(L)}\|G_{\varepsilon F}\|_+ + C_1\|\omega\|_{D(L)} + \|f\|_{X^*}\|G_{\varepsilon F}\|_+.
\end{aligned}$$

Hence for every $\omega \in D(L)$

$$\begin{aligned}
&\sup_{F\in\mathscr{F}} \sup_{y_{\varepsilon F}\in G_{\varepsilon F}} \langle \beta(y_{\varepsilon F}), \omega\rangle_{D(L)} \leq \sup_{F\in\mathscr{F}} [\mathscr{B}_\varepsilon(G_{\varepsilon F}), \omega]_+ \\
&\leq (1+\varepsilon)\|\omega\|_{D(L)}\|G_{\varepsilon F}\|_+ + C_1\|\omega\|_{D(L)} + \|f\|_{X^*}\|G_{\varepsilon F}\|_+ < +\infty.
\end{aligned}$$

Therefore, in virtue of Banach–Steinhaus Theorem there exists $C_3 > 0$ such that

$$\forall F \in \mathscr{F},\ y_{\varepsilon F} \in G_{\varepsilon F} \quad \|\beta(y_{\varepsilon F})\|_{(D(L))^*} \leq C_3. \tag{2.47}$$

Remark that C_3 depends on $\varepsilon > 0$.

For each $F_0 \in \mathscr{F}$ we define the following set:

$$K_{\varepsilon F_0} = \bigcup_{F \in \mathscr{F} : F \supset F_0} G_{\varepsilon F}.$$

Due to (2.45) we have

$$\sup_{F \in \mathscr{F}} \| K_{\varepsilon F} \|_+^{(D(L))} \leq C_3$$

and for any finite set of subspaces $\{F_j\}_{j=1}^n \subset \mathscr{F}$ and any $F \in \mathscr{F}$ such that $F \supset \bigcup_{j=1}^n F_j$ we have:

$$\emptyset \neq K_{\varepsilon F} \subset \bigcap_{j=1}^n K_{\varepsilon F_j}.$$

Hence, in virtue of the Banach–Alaoglu Theorem (the $D(L)$ due to Proposition 1.17 is reflexive) $\{\overline{K}_{\varepsilon F}^{\omega}\}_{F \in \mathscr{F}}$ is the centered system of weakly compact sets where $\overline{K}_{\varepsilon F}^{\omega}$ is a weak closure of $K_{\varepsilon F}$ in $D(L)$. So, due to [RS80, Proposition, p. 98] there exists

$$y_0 \in \bigcap_{F \in \mathscr{F}} \overline{K}_{\varepsilon F}^{\omega}.$$

Now let us show that y_0 is a solution of problem (2.38). We prove this Proposition by contradiction. Is $\bar{0} \notin \mathscr{B}_\varepsilon(y_0)$ then due to [R73, Theorem 3.4, p. 70] there exists $\omega_0 \in D(L)$ such that

$$[\mathscr{B}_\varepsilon(y_0), \omega_0]_- > 0. \tag{2.48}$$

We take $F_0 \in \mathscr{F}$ from the condition $y_0, \omega_0 \in F_0$. Then there exist subsequences

$$\{y_n\}_{n \geq 1} \subset K_{\varepsilon F_0} \quad \text{and} \quad \{F_n\}_{n \geq 1} \subset \mathscr{F} \quad (F_0 \subset \bigcap_{n \geq 1} F_n),$$

such that

$$G_{\varepsilon F_n} \ni y_n \rightharpoonup y_0 \quad \text{in } D(L) \quad \text{as } n \to \infty. \tag{2.49}$$

Due to (2.47) and in virtue of the Banach–Alaoglu Theorem (the space $D(L)$ is reflexive due to Proposition 1.17) we can assume that

$$\mathscr{B}_\varepsilon(y_n) \ni \beta(y_n) \rightharpoonup \beta \quad \text{in} (D(L))^* \quad \text{as } n \to \infty \tag{2.50}$$

fore some $\beta \in (D(L))^*$.

Setting in (2.46) $h_{F_n} = y_n - y_0 \in F_n$ we obtain

$$0 = \langle \beta(y_n), y_n - y_0 \rangle_{D(L)} \to 0 \quad \text{as } n \to \infty.$$

Therefore, due to (2.49) and (2.50) we can apply λ_0-pseudomonotony for $\mathscr{B}_\varepsilon$ on $D(L)$. It implies the existence of subsequences $\{y_{n_m}\}_{m \geq 1} \subset \{y_n\}_{n \geq 1}$ and

$\{\beta_{n_m}\}_{m\geq 1} \subset \{\beta_n\}_{n\geq 1}$ such that

$$[\mathscr{B}_\varepsilon(y_0), y_0-\omega]_- \leq \varliminf_{m\to\infty} \langle \beta(y_{n_m}), y_{n_m}-\omega\rangle_{D(L)} \leq \langle \beta, y_0-\omega\rangle_{D(L)} \quad \forall \omega \in D(L).$$

Substituting in the last relation $\omega = y_0 - \omega_0$ we obtain:

$$[\mathscr{B}_\varepsilon(y_0), \omega_0]_- \leq \langle \beta, \omega_0\rangle_{D(L)} = \lim_{m\to\infty} \langle \beta(y_{n_m}), \omega_0\rangle_{D(L)} = \lim_{m\to\infty} 0 = 0.$$

The latter contradicts with (2.48). Hence, $y_0 \in D(L)$ is the solution of problem (2.38).

To complete the proof let us remark that $n \geq 1$ $\|y_n\|_X \leq R$. Due to (2.49) and the continuity of the embedding of $D(L)$ in X it follows that $y_n \rightharpoonup y_0$ in X. Hence,

$$R \geq \varliminf_{n\to\infty} \|y_n\|_X \geq \|y_0\|_X.$$

Proposition 2.5 is proved. □

Let us continue the proving Theorem 2.2. Due to (2.31)–(2.33) and Proposition 3 we have that for every $\varepsilon > 0$ there exist $d'(y_\varepsilon) \in A(y_\varepsilon)$, $d''(y_\varepsilon) \in B(y_\varepsilon)$ and $m_\varepsilon(y_\varepsilon) \in \mathscr{M}_\varepsilon(y_\varepsilon)$ such that

$$\langle m_\varepsilon(y_\varepsilon), h\rangle_{D(L)} + \langle d'(y_\varepsilon), h\rangle_{X_1} + \langle d''(y_\varepsilon), h\rangle_{X_2} - \langle f, h\rangle_X = 0 \quad \forall h \in D(L), \tag{2.51}$$

moreover, due to (2.28),

$$\langle m_\varepsilon(y_\varepsilon), y_\varepsilon\rangle_{D(L)} = \varepsilon\|Ly_\varepsilon\|^2_{X^*} + \langle Ly_\varepsilon, y_\varepsilon\rangle_X.$$

Hence from (2.51) we obtain:

$$\varepsilon\|Ly_\varepsilon\|^2_{X^*} + \langle Ly_\varepsilon, y_\varepsilon\rangle_X + \langle d'(y_\varepsilon), y_\varepsilon\rangle_{X_1} + \langle d''(y_\varepsilon), y_\varepsilon\rangle_{X_2} = \langle f, y_\varepsilon\rangle_X. \tag{2.52}$$

Due to $L \geq 0$ and (2.52) we have:

$$\langle d'(y_\varepsilon), y_\varepsilon\rangle_{X_1} + \langle d''(y_\varepsilon), y_\varepsilon\rangle_{X_2} \leq \|f\|_{X^*}R \quad \forall \varepsilon > 0.$$

Since the map A (and B respectively) satisfies Condition (Π) on X_1 (on X_2 respectively) then there exist $C_4 > 0$ and $C_5 > 0$ such that

$$\|d'(y_\varepsilon)\|_{X_1^*} \leq C_4, \quad \|d''(y_\varepsilon)\|_{X_2^*} \leq C_5 \quad \forall \varepsilon > 0. \tag{2.53}$$

From (2.51) and (2.33) it follows that for any $h \in D(L)$

$$\varepsilon[Lh, J^{-1}(Ly_\varepsilon)]_+ + \langle Ly_\varepsilon, h\rangle_X + \langle d'(y_\varepsilon), h\rangle_{X_1} + \langle d''(y_\varepsilon), h\rangle_{X_2} - \langle f, h\rangle_X \geq 0. \tag{2.54}$$

For every $\varepsilon > 0$ due to Proposition 3 (the space X is reflexive)we take $h_\varepsilon \in J^{-1}(Ly_\varepsilon)$ from (2.54) such that

$$\varepsilon\langle Lh, h_\varepsilon\rangle_X + \langle Ly_\varepsilon, h\rangle_X + \langle d'(y_\varepsilon), h\rangle_{X_1} + \langle d''(y_\varepsilon), h\rangle_{X_2} - \langle f, h\rangle_X = 0 \quad \forall h \in D(L). \tag{2.55}$$

The form

$$D(L) \ni h \to \langle Lh, h_\varepsilon\rangle_X = \frac{1}{\varepsilon}(\langle f, h\rangle_{X_2} - \langle Ly_\varepsilon, h\rangle_X - \langle d'(y_\varepsilon), h\rangle_{X_1} - \langle d''(y_\varepsilon), h\rangle_{X_2})$$

is continuous in the topology induced from X on $D(L)$. Due to the definition of the operator L^* adjoint with L it follows that

$$h_\varepsilon \in D(L^*) \quad \text{and} \quad \langle Lh_\varepsilon, h_\varepsilon\rangle_X = \langle L^*h_\varepsilon, h_\varepsilon\rangle_X.$$

Hence and from (2.55), specifically, it follows:

$$\varepsilon\langle L^*h_\varepsilon, h\rangle_X + \langle Ly_\varepsilon, h\rangle_X + \langle d'(y_\varepsilon), h\rangle_{X_1} + \langle d''(y_\varepsilon), h\rangle_{X_2} - \langle f, h\rangle_X$$
$$= \langle \varepsilon L^*h_\varepsilon + Ly_\varepsilon + d'(y_\varepsilon) + d''(y_\varepsilon) - f, h\rangle_X = 0 \quad \forall h \in D(L),$$

namely, due to density of $D(L)$ in Y, y_ε satisfies the inclusion (2.29).

Now we set in (2.55) $h = h_\varepsilon$. Hence we obtain:

$$\varepsilon\langle Lh_\varepsilon, h_\varepsilon\rangle_X + \langle Ly_\varepsilon, h_\varepsilon\rangle_X + \langle d'(y_\varepsilon), h_\varepsilon\rangle_{X_1} + \langle d''(y_\varepsilon), h_\varepsilon\rangle_{X_2} = \langle f, h_\varepsilon\rangle_X.$$

Due to $L \geq 0$, (2.28) and (2.53) we have:

$$\|Ly_\varepsilon\|^2_{X^*} \leq (C_4 + C_5 + \|f\|_{X^*}) \cdot \|Ly_\varepsilon\|_{X^*} \quad \forall \varepsilon > 0.$$

Hence, there exists $C_6 > 0$ such that

$$\|Ly_\varepsilon\|_{X^*} \leq C_6 \quad \forall \varepsilon > 0.$$

Therefore due to Proposition 1, estimations (2.53) and previous inequality, in virtue of the Banach–Alaoglu Theorem we can affirm that up to a subsequence

$$y_\varepsilon \rightharpoonup y_0 \text{ in } D(L),\ A(y_\varepsilon) + B(y_\varepsilon) \ni d'(y_\varepsilon) + d''(y_\varepsilon) =: d(y_\varepsilon) \rightharpoonup d_0 \text{ in } X^* \tag{2.56}$$

for some $y_0 \in D(L)$ and $d_0 \in X^*$. We will denote the given subsequences as $\{y_\varepsilon\}$ and $\{d_\varepsilon\}$ respectively.

Let us prove that

$$\varlimsup_{\varepsilon\to 0+} \langle d(y_\varepsilon), y_\varepsilon - y_0\rangle_X \leq 0. \tag{2.57}$$

Indeed, due to (2.54)–(2.56) (since $L \geq 0$) we obtain that

$$\overline{\lim_{\varepsilon\to 0+}} \langle d(y_\varepsilon), y_\varepsilon - y_0\rangle_X \leq \overline{\lim_{\varepsilon\to 0+}} 2C_6 R\varepsilon + \overline{\lim_{\varepsilon\to 0+}} \langle Ly_0, y_\varepsilon - y_0\rangle_X + \overline{\lim_{\varepsilon\to 0+}} \langle f, y_\varepsilon - y_0\rangle_X \leq 0.$$

Now we show that $y_0 \in D(L)$ is a solution of problem (2.27). Due to λ_0-pseudomonotony of $A + B$ on $D(L)$, (2.56) and (2.57) by passing (if it necessary) to the subsequence we obtain:

$$\begin{aligned}\langle d_0, y_0 - \omega\rangle_X &\geq \lim_{\varepsilon\to 0+} \langle d(y_\varepsilon), y_\varepsilon - y_0\rangle_X + \lim_{\varepsilon\to 0+} \langle d(y_\varepsilon), y_0 - \omega\rangle_X \\ &\geq \underline{\lim_{\varepsilon\to 0+}} \langle d(y_\varepsilon), y_\varepsilon - \omega\rangle_X \\ &\geq [(A + B)(y_0), y_0 - \omega]_- \quad \forall \omega \in D(L). \end{aligned} \tag{2.58}$$

On the other hand setting in (2.54) $h = \omega - y_0$, we obtain:

$$\begin{aligned}\langle d_0, y_0 - \omega\rangle_X &= \lim_{\varepsilon\to 0+} \langle d(y_\varepsilon), y_0 - \omega\rangle_X \\ &\leq \overline{\lim_{\varepsilon\to 0+}} (C_6 + \|\omega\|_X) R\varepsilon + \overline{\lim_{\varepsilon\to 0+}} \langle Ly_\varepsilon, \omega - y_0\rangle_X + \overline{\lim_{\varepsilon\to 0+}} \langle f, y_\varepsilon - \omega\rangle_X \\ &= \langle f - Ly_0, y_0 - \omega\rangle_X \quad \forall \omega \in D(L).\end{aligned}$$

Hence, due to (2.58), for any $\omega \in D(L)$

$$\langle f - Ly_0, \omega\rangle_X \leq [(A + B)(y_0), \omega]_+.$$

This means that for any $\omega \in D(L)$

$$\langle Ly_0, \omega\rangle_X + [A(y_0), \omega]_+ + [B(y_0), \omega]_+ \geq \langle f, \omega\rangle_X,$$

namely, due to density of $D(L)$ in X, $y_0 \in D(L)$ is a solution of problem (2.27). The Theorem is proved. □

Corollary 2.1. *Let X be a reflexive Banach space,*

$$L : D(L) \subset X \to X^*$$

be a linear, densely defined, maximal monotony on $D(L)$ operator, $A{:}X_1 \to C_v(X_1^*)$ and $B{:}X_2 \to C_v(X_2^*)$ be finite-dimensionally locally bounded, λ_0-pseudomonotone on $D(L)$ multivalued maps, which satisfy Condition (Π). We assume that the embedding of $D(L)$ in some Banach space Y is compact and dense, and the embedding of X in Y is dense and continuous, and let $N : Y \rightrightarrows Y^*$ be locally bounded multivalued map such that the graph N is closed in $Y \times Y_w^*$ (namely with respect to strong topology in Y and weakly star one in Y^*) and which satisfies Condition (Π). Moreover, for $f \in X^*$ there exists $R > 0$ such that

$$[A(y), y]_+ + [B(y), y]_+ + [N(y), y]_+ - \langle f, y\rangle_X \geq 0 \quad \forall y \in X : \|y\|_X = R. \tag{2.59}$$

Then there exists at least one solution of the problem

$$Ly + A(y) + B(y) + N(y) \ni f, \quad y \in D(L), \tag{2.60}$$

turned out by the singular perturbations method.

Proof. We set $C(y) = B(y) + N(y)$ for any $y \in X \subset Y$. Due to continuity of the embedding $X \subset Y$ and Lemma 2 it follows that C satisfies Condition (Π) on X. Finite-dimensional local boundness of C is obvious. In virtue of Proposition 1.26 the map C is λ_0-pseudomonotone on $D(L)$. Therefore, we apply Theorem 2.2 to A, C, L. Consequently, problem (2.60) has at least one solution, turned out by singular perturbations method. □

Example 2.1. The symmetrical hyperbolic systems. Let $\Omega \subset \mathbb{R}^n$ be a bounded domain with the boundary $\partial\Omega$ of the class C^1 and Ω locally lies on one side of $\partial\Omega$. Consider the controlled system

$$\frac{\partial y}{\partial t} + \sum_{i=1}^{n} B_i(t,x)\frac{\partial y}{\partial x_i} + U(t,x)y + a(t,x,y) = u(t,x) \tag{2.61}$$

where $(t;x) \in (0,T)\times\Omega = Q$; $y(t,x) = \{y_1(t,x), \ldots, y_n(t,x)\}$, $B_i(t,x)$ are the real symmetrical $m\times m$-matrixes of the class $C^1(\overline{Q})$, $U(t,x)$ is a real $m\times m$-matrix of the class $C^1(\overline{Q})$. The operator

$$\Lambda(t) = \sum_{i}^{n} B_i(t,x)\frac{\partial}{\partial x_i} + U(t,x)$$

is formally positive, i.e.

$$U(t,x) + U^*(t,x) - \sum_{i=1}^{n}\frac{\partial B_i}{\partial x_i}(t,x) \geq 0, \quad (t,x) \in Q.$$

Consider $m \times l$-matrix $L(t,x)$ having the constant rank $l \leq m$ and the boundary conditions generated by it

$$L(t,x)y(t,x) = 0, \quad (t,x) \in (0,T)\times\partial\Omega. \tag{2.62}$$

Let $\nu(x)$ be an external normal to $\partial\Omega$ and suppose that the matrix $B^\nu(t,x) = \sum_{i=1}^{n} B_i(t,x)\nu_x(x)$ is invertible $\forall (t,x) \in (0,T)\times\partial\Omega$. In addition, let

$$\langle\xi, B^\nu(t,x)\xi\rangle \geq 0 \quad \forall\, \xi \in \operatorname{Ker} L(t,x)$$

and the maximality condition is fulfilled: for any subspace $\mathbb{E} \subset \mathbb{R}^n$, strictly containing Ker $L(t,x)$, there may be found $\xi \in \mathbb{E}$ that $\langle \xi, B^\nu(t,x)\xi \rangle \geq 0$.

Define $X = (L_p(Q))^m$, the operator

$$\Lambda y = \frac{\partial y}{\partial t} + \sum_{i=1}^{n} B_i(t,x)\frac{\partial y}{\partial x_i} + U(t,x)y$$

with the definition domain $D(\Lambda) = \{\xi \in X \,|\, \Lambda\xi \in X^*,\ L(t,x)\xi = 0 \text{ on } \sum = (0,T) \times \partial\Omega,\ \xi(0,x) = 0\ \forall\ x \in \Omega\}$. In these conditions the differentiable operator $\Lambda : D(\Lambda) \subset X \to X^*$ is maximally monotone for $p \geq 2$ [ZM04].

Afterwards suppose that $a : (0,T) \times \Omega \times \mathbb{R}^n \to \mathbb{R}^m$ is a Caratheodory mapping, and here

(1) $|a(t,x,\xi)| \leq C_1|\xi|^{p-1} + C_2,\ (t,x,\xi) \in (0,T) \times \Omega \times \mathbb{R}^m$.
(2) $(a(t,x,\xi),\xi) \leq \alpha|\xi|^p + \beta,\ (t,x,\xi) \in (0,T) \times \Omega \times \mathbb{R}^m$.
(3) $(a(t,x,\xi_1) - a(t,x,\xi_2), \xi_1 - \xi_2) \geq 0\ \forall\ \xi_1,\xi_2 \in \mathbb{R}^m;\ (t,x) \in (0,T) \times \Omega$. Here the constants $C_1 > 0,\ \alpha > 0,\ C_2, \beta \in \mathbb{R}$ does not depend on $(t;x;\xi)$.

For a fixed U the corresponding Nemitskii operator $A(y)$ for a and Λ satisfy all the requirements of Theorem 2.2.

Example 2.2. The nonlinear transfer equation. Let Θ be a locally compact space in $\mathbb{R}^n$ with the Radon measure μ and the following controlled system is considered

$$\frac{\partial y}{\partial t} + \sum_{i=1}^{n} \theta_i \frac{\partial y}{\partial x_i} + \int_{\theta} K(t,x,\theta,\tilde{\theta})y(t,x,\theta)\mu(d\tilde{\theta}) + a(t,x,\theta) = U(t,x,y) \quad (2.63)$$

where $\theta \in \Theta,\ (t,x) \in (0,T) \times \Omega = Q$, the kernel K is such that $y \to Ky = \int_{\theta} K(t,x,\theta,\tilde{\theta})y(t,x,\tilde{\theta})\mu(d\tilde{\theta})$ is the linear continuous operator in $L_p(Q \times \Theta) = X$. And here the function a satisfies the requirements of the previous example. Equation (2.63) is studied under the initial-bounded conditions $y(t,x,\theta) = 0$ if $(t,x) \in \sum,\ \theta \in \Theta$ and

$$\sum_{i=1}^{n} \theta_i \cos(\nu, x_i) < 0, \quad y(0,x,\theta) = 0. \quad (2.64)$$

On the operator $\Lambda y = \frac{\partial y}{\partial t} + \sum\limits_{i=1}^{n} \theta_i \frac{\partial y}{\partial x_i} + Ky$ with the definition domain

$$D(\Lambda) = \{\xi \in X \,|\, \Lambda\xi \in X^* \text{ and (2.64) is valid}\}$$

the closure Λ is constructed. As it is known [ZM04], the operator Λ constructed in such a way is maximally monotone. Therefore, adequately choosing $u \in U \subset X^*$ we obtain the conditions of Theorem 2.2.

Example 2.3. The nonlinear Schrödinger equation (complex case). Consider the controlled system

$$\frac{\partial y}{\partial t} - i\Delta y + |y|^{p-2}y = u \in L_3(Q)$$

with the initial-bounded conditions $y|_\Sigma = 0$, $y(0,x) = 0$, $x \in \Omega$. Passing to the real case we obtain

$$\frac{\partial y_1}{\partial t} + \Delta y_2 + |y|^{p-2}y_1 = u_1, \quad \frac{\partial y_2}{\partial t} - \Delta y_1 + |y|^{p-2}y_2 = u_2$$

where $y = y_1 + iy_2$. Suppose $X = (L_p(Q))^2$,

$$\Lambda y = \left\{\frac{\partial y_1}{\partial t} + y_2; \frac{\partial y_2}{\partial t} + y_1\right\};$$
$$A(u,y) = \{|y|^{p-2}y_1 - u_1; |y|^{p-2}y_2 - u_2\},$$
$$D(\Lambda)0 = \{y \in X \,|\, \Lambda y \in X^*,\ y_{10} = y_{20} = 0,\ y_{1\Sigma} = y_{2\Sigma} = 0\}.$$

The operator $\Lambda: D(\Lambda) \subset X \to X^*$ is maximally monotone [ZM04] and it is easy to describe the subset of the admissible controls for the operator A to satisfy Theorem 2.2.

2.3 Dubinskii Method for Noncoercive Operators

Differential-operator equations, inclusions and evolutional variational inequalities have been extensively studied for last several decades. Interest in these objects is caused, first of all, by their wide applications. As a rule, they are associated with problems in mathematical physics, partial differential equations the differential operators of which admit a discontinuity with respect to the phase variable, differential equations with discontinuous right-hand side, problems in the theory of control and optimization, etc. In physics and mechanics, applied problems related to phase transitions, unidirectional conductivity of boundaries of substances, propagation of electromagnetic, acoustic, vibroacoustic, hydroacoustic, and seismoacoustic waves, quantum mechanical effects, etc., simulated the investigation of evolution equations and inclusions [DL76, ZMN04, SY02]. The latest investigations in this direction have been devoted to quasilinear equations with homogeneous boundary conditions and linearized equations with non-linear conditions on the boundary of the domain that can be reduced to non-linear operator differential equations and inclusions. However, linearized objects do not always adequately describe the investigated processes. For this reason, it is necessary to consider inclusions and variational inequalities with substantially narrower set of properties. For monotone noncoercive maps, such objects were studied in [GGZ74]. The differential-operator inclusions with noncoercive maps of pseudomonotone type multimaps have not been systematically studied yet.

In this section we consider the first order evolutional inclusions with noncoercive maps of the Volterra type. Our aim is to develop a noncoercive theory for these objects and to obtain new results related to solvability and to validate constructive methods of investigation. The results obtained in this section are new for both evolutional equations and evolutional inclusions.

As before let $(V_1, \|\cdot\|_{V_1})$ and $(V_2, \|\cdot\|_{V_2})$ be reflexive Banach spaces, continuously embedded in a Hilbert space $(H, (\cdot,\cdot))$ such that for a numerable set $\Phi \subset V = V_1 \cap V_2$

$$\Phi \text{ is dense in } V,\ V_1,\ V_2 \text{ and in } H.$$

After identifying $H \equiv H^*$ we obtain

$$V_1 \subset H \subset V_1^*, \quad V_2 \subset H \subset V_2^*,$$

with continuous and dense embedding, $(V_i^*, \|\cdot\|_{V_i^*})$, $i = 1, 2$ be the topologically adjoint with V_i space with respect to the form

$$\langle \cdot,\cdot \rangle_{V_i} : V_i^* \times V_i \to \mathbb{R},$$

that coincides (on $H \times \Phi$) with the inner product in H.

Let us consider functional spaces $X_i = L_{r_i}(S; H) \cap L_{p_i}(S; V_i)$, where S is finite time interval, $1 < p_i \le r_i < +\infty$. The spaces X_i are reflexive Banach spaces with the norms

$$\|y\|_{X_i} = \|y\|_{L_{p_i}(S;V_i)} + \|y\|_{L_{r_i}(S;H)},$$

$X = X_1 \cap X_2$, $\|y\|_X = \|y\|_{X_1} + \|y\|_{X_2}$. Let X_i^* $(i = 1, 2)$ be topologically adjoint with X_i. Then

$$X^* = X_1^* + X_2^* = L_{q_1}(S; V_1^*) + L_{q_2}(S; V_2^*) + L_{r_1'}(S; H) + L_{r_2'}(S; H),$$

where $r_i^{-1} + r_i'^{-1} = p_i^{-1} + q_i^{-1} = 1$ $(i = 1, 2)$. Let us define the pairing on $X^* \times X$

$$\begin{aligned}\langle f, y \rangle &= \int_S (f_{11}(\tau), y(\tau))_H \, d\tau + \int_S (f_{12}(\tau), y(\tau))_H \, d\tau + \int_S \langle f_{21}(\tau), y(\tau)\rangle_{V_1} \, d\tau \\ &\quad + \int_S \langle f_{22}(\tau), y(\tau)\rangle_{V_2} \, d\tau \\ &= \int_S (f(\tau), y(\tau)) \, d\tau,\end{aligned}$$

where $f = f_{11} + f_{12} + f_{21} + f_{22}$, $f_{1i} \in L_{r_i'}(S; H)$, $f_{2i} \in L_{q_i}(S; V_i^*)$ $(i = 1, 2)$. We recall that $\langle \cdot,\cdot \rangle$ coincides with the inner product in $\mathcal{H} = L_2(S; H)$ on $\mathcal{H}$.

Let $A : X \rightrightarrows X^*$ be the multivalued noncoercive map. We consider solutions in the class $W = \{y \in X \mid y' \in X^*\}$ of the following problem:

$$\begin{cases} \langle y',\xi\rangle + [A(y),\xi]_+ \geq \langle f,\xi\rangle \quad \forall \xi \in W, \\ y(0)=\bar{0}, \end{cases}$$

where $f \in X^*$ is arbitrary, y' is the derivative of an element $y \in X$ considered in the sense of the scalar distributions space $\mathscr{D}^*(S;V^*) = \mathscr{L}(\mathscr{D}(S);V_w^*)$, with $V = V_1 \cap V_2$.

On the reflexive Banach space W let us introduce the graph norm $\|y\|_W = \|y\|_X + \|y'\|_{X^*}$, where

$$\|f\|_{X^*} = \inf_{\substack{f=f_{11}+f_{12}+f_{21}+f_{22}:\\ f_{1i}\in L_{r_i'}(S;H),\ f_{2i}\in L_{q_i}(S;V_i^*)\ (i=1,2)}} \max\Big\{\|f_{11}\|_{L_{r_1'}(S;H)};\ \|f_{12}\|_{L_{r_2'}(S;H)};\ \|f_{21}\|_{L_{q_1}(S;V_1^*)};\ \|f_{22}\|_{L_{q_2}(S;V_2^*)}\Big\}.$$

Remark that the embedding $W \subset C(S;H)$ is continuous.

Now let us prove the Theorem about resolvability of differential-operator inclusions with nonlinear noncoercive maps of w_λ-pseudomonotone type. Theorem 2.3 and its Corollary deal with the solvability of evolutional inequalities with non-coercive multi-valued Volterra operators with $(X;W)$-semi-bounded variation. The proof of these statements is based on the results of Theorem 2.1. For differential-operator equations, similar results were obtained for monotone single-valued maps in [GGZ74]. For single-valued maps with semi-bounded variation, the corresponding results are presented in [ZMN04]. For the analysis of the solvability of differential-operator inclusions and evolutional variational inequalities with coercive multivalued maps, see [D65], [K94]–[VM00]. We remark that there are no similar results for differential-operator inclusions and evolutional inequalities.

Theorem 2.3. *Let* $\max\{r_1,r_2\} \geq 2$, $A : X \rightrightarrows X^*$ *be the radial lower semicontinuous multivalued operator and for some* $\lambda > 0$

$$\sup_{\zeta(y)\in A(y)} \int_S e^{-2\lambda t}\big(\zeta(y)(t)+\lambda y(t), y(t)\big)dt \geq \gamma(\|y\|_X)\|y\|_X \quad \forall y \in X, \tag{2.65}$$

where $\gamma : \mathbb{R}_+ \to \mathbb{R}$ *is bounded below on bounded in* $\mathbb{R}_+$ *sets function such that* $\gamma(r) \to +\infty$ *as* $r \to \infty$, *and* $\forall R > 0$, $\forall y,\xi \in X$: $\|y\|_X \leq R$, $\|\xi\|_X \leq R$,

$$\begin{aligned} &\inf_{\zeta(y)\in A(y)} \int_S e^{-2\lambda t}\big(\zeta(y)(t)+\lambda y(t), y(t)-\xi(t)\big)dt \\ &\geq \sup_{\zeta(\xi)\in A(\xi)} \int_S e^{-2\lambda t}\big(\zeta(\xi)(t)+\lambda\xi(t), y(t)-\xi(t)\big)dt \\ &\quad - C_A(R;\|y-\xi\|_{L_{p_0}(S;V_0)}), \end{aligned} \tag{2.66}$$

where $C_A \in \Phi_0$, $p_0 = \min\{p_1, p_2\}$, V is compactly embedded in Banach space V_0 and $V_0 \subset V^*$ is continuous. Then the statement of Theorem 2.1 is valid.

Proof. Let us introduce the following notation: $y_\lambda(t) = e^{-\lambda t} y(t)$, $f_\lambda(t) = e^{-\lambda t} f(t)$, for any $y \in X$ the set $\mathscr{A}_\lambda(y_\lambda) \in C_v(X^*)$ is well-defined by the relation

$$[\mathscr{A}_\lambda(y_\lambda), \omega]_+ = [\mathscr{A}(y) + \lambda y, \omega_\lambda]_+ \quad \forall \omega \in X$$

(we remark that the functional $\omega \longmapsto [\mathscr{A}(y) + \lambda y, \omega_\lambda]_+$ is semiadditive, positively homogenous and lower semicontinuous as supremum of linear continuous functionals). Then $y \in W$ is a solution of problem (2.2) if and only if y_λ satisfies

$$\langle y_\lambda', \xi\rangle + [A_\lambda y_\lambda, \xi]_+ \geq \langle f_\lambda, \xi\rangle \quad \forall \xi \in W, \quad y_\lambda(0) = \bar{0}.$$

Therefore we must make sure that A_λ satisfies all conditions of Theorem 2.1. Radial lower semicontinuity is obvious. Further, since $\|y_\lambda\|_X \leq \|y\|_X$ and $\|y_\lambda\|_X^{-1}$ $[A_\lambda y_\lambda, y_\lambda]_+$

$$\geq \|y\|_X^{-1} \sup_{\zeta(y)\in A(y)} \int_S e^{-2\lambda t}\big(\zeta(y)(t) + \lambda y(t), y(t)\big)dt \geq \gamma(\|y\|_X)\|y\|_X,$$

then the operator A_λ is $+$-coercive. By definition

$$\begin{aligned}
[A_\lambda y_\lambda, y_\lambda - \xi_\lambda]_- &= \inf_{\zeta(y)\in A(y)} \int_S \big(e^{-\lambda t}\zeta(y)(t) + \lambda y(t)e^{-\lambda t}, e^{-\lambda t}(y(t) - \xi(t))\big)dt \\
&= \inf_{\zeta(y)\in A(y)} \int_S e^{-2\lambda t}\big(\zeta(y)(t) + \lambda y(t), y(t) - \xi(t)\big)dt \\
&\geq \sup_{\zeta(\xi)\in A(\xi)} \int_S e^{-2\lambda t}\big(\zeta(\xi)(t) + \lambda \xi(t), y(t) - \xi(t)\big)dt \\
&\quad - C_A(R; \|y - \xi\|_{L_{p_0}(S;V_0)}) \\
&= \sup_{\zeta(\xi)\in A(\xi)} \int_S \big(e^{-\lambda t}\zeta(\xi)(t) + \lambda \xi(t)e^{-\lambda t}, e^{-\lambda t}(y(t) - \xi(t))\big)dt \\
&\quad - C_A(R; \|y - \xi\|_{L_{p_0}(S;V_0)}) = [A_\lambda \xi_\lambda, y_\lambda - \xi_\lambda]_+ \\
&\quad - C_A(R; \|y - \xi\|_{L_{p_0}(S;V_0)})
\end{aligned} \tag{2.67}$$

Let us consider the weighting space $L_{p_0,\lambda}(S; V_0)$ consisting of measurable functions $y_\lambda : S \to V_0$ for which the integral $\int_S e^{\lambda t p_0} \|y_\lambda(t)\|_{V_0}^{p_0} dt$ is finite. Then

$$\|y - \xi\|_{L_{p_0}(S;V_0)} = \left(\int_S e^{\lambda t p_0} \|y_\lambda(t) - \xi_\lambda(t)\|_{V_0}^{p_0} dt\right)^{1/p_0} = \|y_\lambda - \xi_\lambda\|_{L_{p_0,\lambda}(S;V_0)}.$$

Therefore due to (2.67) we obtain

$$[A_\lambda y_\lambda, y_\lambda - \xi_\lambda]_- \geq [A_\lambda \xi_\lambda, y_\lambda - \xi_\lambda]_+ - C_A(R; \|y_\lambda - \xi_\lambda\|_{L_{p_0,\lambda}(S;V_0)}).$$

The Proposition about compactness of the embedding $W \subset L_{p_0,\lambda}(S; V_0)$ completes the proof of the Theorem. This fact directly follows from continuity of the embedding

$$W \subset \left\{ y \in L_{p_0}(S; V) \mid y' \in L_{\min\{r_1', r_2'\}}(S; V^*) \right\}$$

and from Theorem 1.14 with $B_0 = V$, $B = V_0$, $B_1 = V^*$, $p_0 = p_0$ and $p_1 = \min\{r_1', r_2'\}$. □

Definition 2.2. The set B belongs to the class $\mathscr{H}(X^*)$ if and only if, when for arbitrary $n \geq 1$, $\{d_i\}_{i=1}^n \subset B$ and $E_j \subset S$, $j = \overline{1,n}$: $\forall j = \overline{1,n}$ E_j is measurable, $\cup_{j=1}^n E_j = S$, $E_i \cap E_j = \emptyset$ $\forall i \neq j$, $i, j = \overline{1,n}$ the element $d(\cdot) = \sum_{j=1}^n d_j(\cdot)\chi_{E_j}(\cdot) \in \overset{*}{\overline{\text{co}}}\, B$ where

$$\chi_{E_j}(\tau) = \begin{cases} 1, & \tau \in E_j, \\ 0, & \text{elsewhere.} \end{cases}$$

We remark that $\emptyset, X^* \in \mathscr{H}(X^*)$; $\forall f \in X^*$ $\{f\} \in \mathscr{H}(X^*)$; if $K : S \rightrightarrows V^*$ is a multivalued map, then

$$\{f \in X^* \mid f(t) \in K(t) \text{ for a.e. } t \in S\} \in \mathscr{H}(X^*).$$

On another hand, as for any $v \in V^*$ $(v \neq 0)$ $g(\cdot) = v \cdot \chi_{[0;T/2]}(\cdot) \notin B$, then the convex closed set $B = \{f \in X^* \mid f(\cdot) \equiv \alpha v \in V^*, \alpha \in [0,1]\} \notin \mathscr{H}(X^*)$.

Further let us assume the validity of the following condition for A:

Definition 2.3. The multivalued map $A : X \rightrightarrows X^*$ satisfies *Condition* (H), if for arbitrary $y \in X$ $A(y) \in \mathscr{H}(X^*)$.

Definition 2.4. The multivalued map $A : X \rightrightarrows X^*$ be the *Volterra type operator*, if for any $u, v \in X$, $t \in S$ from the equality $u(s) = v(s)$ for a.e. $s \in [0,t]$ it follows that $[A(u), \xi_t]_+ = [A(v), \xi_t]_+$ $\forall \xi_t \in X$: $\xi_t(s) = 0$ for a.e. $s \in S \backslash [0,t]$.

Corollary 2.2. *If* $\max\{r_1, r_2\} \geq 2$ *and for some* $\lambda \geq 0$ $A + \lambda I : X \rightrightarrows X^*$ *is radial lower semicontinuous* $+$*-coercive operator of Volterra type with* $(X; W)$*-semibounded variation with* $\|\cdot\|_W' = \|\cdot\|_{L_{p_0}(S;V_0)}$ *(*p_0*,* V_0 *satisfy conditions of Theorem 2.3) and also satisfy Condition* (H)*. Then the statement of Theorem 2.3 is valid.*

Remark 2.6. In Corollary 2.2 Condition (H) for A and $+$-coercivity of $A + \lambda I$ on X can de replaced by $-$-coercivity of $A + \lambda I$ on X.

Proof. Let us prove that while the conditions of given corollary hold true the relations (2.65) and (2.66) are valid. Radial lower semicontinuity is easily verified. We prove the semiboundness of variation. Let for all $R > 0$, $y, \xi \in X$: $\|y\|_X \le R$, $\|\xi\|_X \le R$ the following relation be true:

$$[A(y) - A(\xi) + \lambda y - \lambda \xi, y - \xi]_- + C_A(R; \|y - \xi\|'_W) \ge 0.$$

We set $\hat{C_A}(R;\cdot) = \max\limits_{\tau\in[0,t]} C_A(R;\tau)$ for all $R, t \ge 0$ ($\hat{C_A} \in \Phi_0$),

$$z_t(\tau) = \begin{cases} z(\tau), & 0 \le \tau \le t, \\ \bar{0}, & t < \tau \le T, \end{cases} \qquad t \in S,\ z \in X.$$

Let $\zeta, \eta \in A$ are fixed selectors. Since A is the Volterra type operator then $\forall t \in S$

$$\int_0^t (\zeta(y)(\tau) + \lambda y(\tau) - \eta(\xi)(\tau) - \lambda\xi(\tau), y(\tau) - \xi(\tau))d\tau + \hat{C_A}\left(R; \|y - \xi\|'_{W_t}\right)$$

$$= \int_0^T (\zeta(y_t)(\tau) + \lambda y_t(\tau) - \eta(\xi_t)(\tau) - \lambda\xi_t(\tau), y_t(\tau) - \xi_t(\tau))d\tau$$
$$+ \hat{C_A}\left(R; \|y_t - \xi_t\|'_W\right)$$
$$\ge [(A + \lambda I)(y_t) - (A + \lambda I)(\xi_t), y_t - \xi_t]_- + \hat{C_A}(R; \|y_t - \xi_t\|'_W) \ge 0,$$

since $\|y_t\|_X \le \|y\|_X$ and $\|\xi_t\|_X \le \|\xi\|_X$. Here $\|\cdot\|'_{W_t} = \|\cdot\|_{L_{p_0}([0,t];V_0)}$.

We set

$$g(\tau) = \left(\zeta(y)(\tau) + \lambda y(\tau) - \eta(\xi)(\tau) - \lambda\xi(\tau), y(\tau) - \xi(\tau)\right), \quad \tau \in S,$$

$$h(t) = \hat{C_A}\left(R; \|y - \xi\|'_{W_t}\right), \quad t \in S.$$

Remark that the function $S \ni t \to h(t)$ is monotone nondecreasing and

$$\int_0^t g(\tau)d\tau \ge -h(t) \quad \forall t \in S.$$

Hence,

$$\int_0^T e^{-2\lambda\tau}(\zeta(y)(\tau) + \lambda y(\tau) - \eta(\xi)(\tau) - \lambda\xi(\tau), y(\tau) - \xi(\tau))d\tau$$

$$= \int_0^T e^{-2\lambda\tau} g(\tau)d\tau = e^{-2\lambda T}\int_0^T g(\tau)d\tau + 2\lambda\int_0^T e^{-2\lambda\tau}\int_0^\tau g(s)ds\,d\tau$$

$$\geq -e^{-2\lambda T}h(T) - 2\lambda \int_0^T e^{-2\lambda\tau}h(\tau)d\tau \geq -h(T)\left[e^{-2\lambda T} + 2\lambda \int_0^T e^{-2\lambda\tau}d\tau\right]$$
$$= -h(T) = -\hat{C}_A\left(R; \|y-\xi\|'_W\right),$$

and this implies (2.66).

Let us verify (2.65). For any $r > 0$ let us set

$$\tilde{\gamma}(r) = \inf_{y\in X,\, \|y\|_X=r} \sup_{d\in A(y)} \frac{\langle d+\lambda y, y\rangle_X}{\|y\|_X}; \quad \tilde{\gamma}(0) := 0.$$

(a) Let us prove that $\tilde{\gamma}(r) > -\infty$. If it is not true, then $\exists\{y_n\}_{n\geq 1} \subset \{y \in X \mid \|y\|_X = r\}$, $\exists\{d_n\}_{n\geq 1} \subset X^*$, $d_n \in A(y_n)$ $\forall n \geq 1$:

$$\frac{\langle d_n + \lambda y_n, y_n\rangle_X}{\|y_n\|_X} \to -\infty \quad \text{as} \quad n \to +\infty.$$

Then $\langle d_n, y_n\rangle_X \leq \langle d_n, y_n\rangle_X + \lambda\|y_n\|_Y^2 = \langle d_n + \lambda y_n, y_n\rangle_X = r\cdot\frac{\langle d_n+\lambda y_n, y_n\rangle_X}{\|y_n\|_X} \to -\infty$ as $n \to +\infty$. Due to the boundedness of A we have that $\exists c_1 > 0 : \|d_n\|_{X^*} \leq c_1$ $\forall n \geq 1$. We obtain the contradiction with the assumption. Thus, $\tilde{\gamma}(r) > -\infty$.

(b) We remark, that $\forall y \in X$

$$[A(y) + \lambda y, y]_+ \geq \tilde{\gamma}(\|y\|_X)\|y\|_X. \tag{2.68}$$

Due to the boundednes of A (similarly to the case a) it follows that $\tilde{\gamma}$ is lower bounded on bounded sets.

(c) Let us show that $\tilde{\gamma}(r) \to +\infty$ as $r \to +\infty$. If it is not true, then $\exists\{r_n\}_{n\geq 1} \subset (0, +\infty) : r_n \nearrow +\infty$ as $n \to +\infty$ and $\tilde{\gamma}(r_n) \leq c$ $\forall n \geq 1$, where c is not depends on $n \geq 1$. So, $\forall n \geq 1$ $\exists y_n \in X : \|y_n\|_X = r_n$ and

$$\frac{[A(y_n) + \lambda y_n, y_n]_+}{\|y_n\|_X} \leq c + 1 \quad \forall n \geq 1.$$

Due to $\|y_n\|_X \to +\infty$ as $n \to +\infty$, then we have the contradiction with +-coercivity of $A + \lambda I$.

Finally, $[A(y) + \lambda y, y]_+ \geq \tilde{\gamma}(\|y\|_X)\|y\|_X$ $\forall y \in X$, where $\tilde{\gamma} : \mathbb{R}_+ \to \mathbb{R}$ is a lower bounded on bounded in $\mathbb{R}_+$ sets function such that $\tilde{\gamma}(r) \to +\infty$ as $r \to +\infty$. So, $\inf_{r\geq 0} \tilde{\gamma}(r) =: a > -\infty$. For any $b > a$ we consider the nonempty bounded set $A_b = \{c \geq 0 \mid \tilde{\gamma}(c) \leq b\}$. Let $c_b = \sup A_b$, $b > a$. We remark that $\forall b_1 > b_2 > a$ $c_{b_2} \leq c_{b_1} < +\infty$ and $c_b \to +\infty$ as $b \to +\infty$. We set

$$\widehat{\gamma}(t) = \begin{cases} a, & t \in [0, c_{a+1}], \\ a+k, & t \in (c_{a+k}, c_{a+k+1}],\ k \geq 1. \end{cases}$$

Then, $\widehat{\gamma} : \mathbb{R}_+ \to \mathbb{R}$ is a lower bounded on bounded sets nondecreasing function such that $\widehat{\gamma}(r) \to +\infty$ as $r \to \infty$ and $\widehat{\gamma}(t) \le \tilde{\gamma}(t)\ \forall t \ge 0$.

Let us fix an arbitrary $y \in X$. As A is the Volterra type, then

$$\begin{aligned}\forall t \in S \quad & \sup_{d \in A(y)} \int_0^t \big(d(\tau) + \lambda y(\tau), y(\tau)\big) d\tau \\ &= \sup_{d \in A(y)} \int_0^T \big(d(\tau) + \lambda y_t(\tau), y_t(\tau)\big) d\tau \ge \widehat{\gamma}(\|y_t\|_X)\|y_t\|_X \\ &= \widehat{\gamma}(\|y\|_{X_t})\|y\|_{X_t},\end{aligned}$$

where $\|y\|_{X_t} = \|y_t\|_X$, $y_t(\tau) = \begin{cases} y(\tau), & \tau \in [0, t], \\ \bar{0}, & \text{elsewhere.}\end{cases}$

Let for any $d \in A(y)$

$$\begin{aligned} g_d(\tau) &= \big(d(\tau) + \lambda y(\tau), y(\tau)\big), \quad \text{for a.e.} \quad \tau \in S, \\ h(t) &= \widehat{\gamma}(\|y\|_{X_t})\|y\|_{X_t}, \quad t \in S.\end{aligned}$$

We remark that for all $t \in S$ $h(t) \ge \min\{\widehat{\gamma}(0), 0\}\|y\|_X$ and

$$\forall t \in S \qquad \sup_{d \in A(y)} \int_0^t g_d(\tau) d\tau \ge h(t).$$

Let us show that

$$\begin{aligned}\sup_{d \in A(y)} \int_0^T e^{-2\lambda\tau}(d(\tau) + \lambda y(\tau), y(\tau)) d\tau &\ge e^{-2\lambda T} \sup_{d \in A(y)} \int_0^T (d(\tau) + \lambda y(\tau), y(\tau)) d\tau \\ &+ \sup_{d \in A(y)} \int_0^T (e^{-2\lambda\tau} - e^{-2\lambda T})(d(\tau) + \lambda y(\tau), y(\tau)) d\tau. \qquad (2.69)\end{aligned}$$

Let us set $\varphi(\tau) = e^{-2\lambda(T-\tau)}$, $\tau \in [0, T]$, ($\varphi \in (0, 1]$). $\forall n \ge 1$ $\varphi_n(\tau) = \sum_{i=0}^{n-1} \varphi(\frac{iT}{n}) \chi_{[\frac{iT}{n}, \frac{(i+1)T}{n})}(\tau)$, $\tau \in [0, T]$. Then, $\forall d_1 \in A(y)$ $\forall d_2 \in A(y)$ $\forall i = \overline{0, n-1}$ $\varphi(\frac{iT}{n})d_1 + (1 - \varphi(\frac{iT}{n}))d_2 \in A(y)$. We remark that $\forall \tau \in [0, T]$ $|\varphi_n(\tau) - \varphi(\tau)| \le \frac{2\lambda T}{n}$. Due the Condition (H) we obtain that $d = \sum_{i=0}^{n-1} (\varphi(\frac{iT}{n})d_1 + (1 - \varphi(\frac{iT}{n}))d_2) \chi_{[t_i, t_{i+1})}(\tau) \in A(y)$. So,

$$\sup_{d\in A(y)}\int_0^T e^{-2\lambda\tau}(d(\tau)+\lambda y(\tau),y(\tau))d\tau \ge \int_0^T (d(\tau)+\lambda y(\tau),y(\tau)e^{-2\lambda\tau})d\tau$$

$$=\int_0^T \varphi_n(\tau)(d_1(\tau)+\lambda y(\tau),y(\tau)e^{-2\lambda\tau})d\tau$$

$$+\int_0^T (1-\varphi_n(\tau))(d_2(\tau)+\lambda y(\tau),y(\tau)e^{-2\lambda\tau})d\tau$$

$$\ge e^{-2\lambda T}\int_0^T (d_1(\tau)+\lambda y(\tau),y(\tau))d\tau$$

$$+\int_0^T (e^{-2\lambda\tau}-e^{-2\lambda T})(d_2(\tau)+\lambda y(\tau),y(\tau))d\tau$$

$$-\frac{2\lambda T}{n}(\|A(y)\|_+\|y\|_X+\lambda\|y\|_Y^2+\|A(y)\|_+\|y\|_X+\lambda\|y\|_Y^2).$$

When $n\to+\infty$, taking sup by $d_1\in A(y)$, $d_2\in A(y)$ ia last inequality we will obtain (2.69). From (2.69) it follows that

$$\sup_{d\in A(y)}\int_0^T e^{-2\lambda\tau}(d(\tau)+\lambda y(\tau),y(\tau))d\tau$$

$$\ge e^{-2\lambda T}h(T)+2\lambda\sup_{d\in A(y)}\int_0^T e^{-2\lambda s}\int_0^s g_d(\tau)d\tau ds$$

$$\ge e^{-2\lambda T}h(T)+2\lambda T\sup_{d\in A(y)}\inf_{s\in S}e^{-2\lambda s}\int_0^s (d(\tau)+\lambda y(\tau),y(\tau))d\tau.$$

Let us show that

$$\sup_{d\in A(y)}\inf_{s\in S}e^{-2\lambda s}\int_0^s (d(\tau)+\lambda y(\tau),y(\tau))d\tau\ge -c_1\|y\|_X,$$

where $c_1=\max\{-\widehat{\gamma}(0),0\}\ge 0$, that does not depend on $y\in X$.

Let $y \in X$ be fixed. Let us set

$$\varphi(s,d) = e^{-2\lambda s}\int_0^s \big(d(\tau)+\lambda y(\tau), y(\tau)\big)d\tau, \quad a = \sup_{d\in A(y)} \inf_{s\in S}\varphi(s,d),$$

$$A_d = \{s \in S \mid \varphi(s,d) \le a\}, \quad s \in S,\ d \in A(y).$$

In virtue of the continuity of $\varphi(\cdot, d)$ on S it follows that A_d is nonempty closed set for any $d \in A(y)$. Really let for any fixed $d \in A(y)$ there is $s_d \in S$ such that

$$\varphi(s_d, d) = \min_{\hat{s}\in S}\varphi(\hat{s}, d) \le a.$$

The closureness of A_d follows from the continuity of $\varphi(\cdot, d)$ on S.

Let us prove that $\{A_d\}_{d\in A(y)}$ is a centered system. For fixed $\{d_i\}_{i=1}^n \subset A(y)$, $n \ge 1$ let us set

$$\psi_i(\cdot) = \big(d_i(\cdot)+\lambda y(\cdot), y(\cdot)\big), \quad \psi(\cdot) = \max_{i=1}^{n}\psi_i(\cdot),$$

$$E_0 = \emptyset, \quad E_j = \left\{\tau \in S \setminus \left(\cup_{i=0}^{j-1} E_i\right) \mid \psi_j(\tau) = \psi(\tau)\right\}, \quad j = \overline{1,n},$$

$$d(\cdot) = \sum_{j=1}^{n} d_j(\cdot)\chi_{E_j}(\cdot).$$

We remark that $\forall j = \overline{1,n}$ E_j is measurable set, $\cup_{j=1}^n E_j = S$, $E_i \cap E_j = \emptyset$ $\forall i \ne j, i, j = \overline{1,n}$, $d \in X^*$. Furthermore,

$$\varphi(s,d_i) = e^{-2\lambda s}\int_0^s \psi_i(\tau)d\tau \le e^{-2\lambda s}\int_0^s \psi(\tau)d\tau = \varphi(s,d), \quad s \in S,\ i = \overline{1,n}.$$

So, due to Condition (H), $d \in A(y)$ and for some $s_d \in S$

$$\varphi(s_d, d_i) \le \varphi(s_d, d) = \min_{\hat{s}\in S}\varphi(\hat{s}, d) \le a, \quad i = \overline{1,n}.$$

Thus, $s_d \in \cap_{i=1}^n A_{d_i} \ne \emptyset$.

As S is a compact and the family of closed subsets $\{A_d\}_{d\in A(y)}$ is centered, then $\exists s_0 \in S$: $s_0 \in \cap_{d\in A(y)} A_d$. This means that

$$\sup_{d\in A(y)}\inf_{s\in S} e^{-2\lambda s}\int_0^s \big(d(\tau)+\lambda y(\tau), y(\tau)\big)d\tau$$

$$\geq \sup_{d\in A(y)} e^{-2\lambda s_0}\int_0^{s_0} \big(d(\tau)+\lambda y(\tau), y(\tau)\big)d\tau$$

$$= e^{-2\lambda s_0}\sup_{d\in A(y)}\int_0^{s_0} g_d(\tau)d\tau \geq e^{-2\lambda s_0}h(s_0)$$

$$\geq e^{-2\lambda s_0}\min\{\widehat{\gamma}(0),0\}\|y\|_X \geq -\max\{-\widehat{\gamma}(0),0\}\|y\|_X = -c_1\|y\|_X.$$

So, $\forall y\in X$

$$\sup_{d\in A(y)}\int_0^T e^{-2\lambda\tau}(d(\tau)+\lambda y(\tau), y(\tau))d\tau \geq \big(e^{-2\lambda T}\widehat{\gamma}(\|y\|_X)-2\lambda c_1 T\big)\|y\|_X.$$

If we set $\gamma(r)=e^{-2\lambda T}\widehat{\gamma}(r)-2\lambda c_1 T$, then we obtain (2.65). □

Remark 2.7. The condition (2.65) in Theorem 2.3 can be replaced by similar one (2.20), (2.21), namely for each $y\in X$

$$[y]_{X_1}+[y]_{X_2}+\lambda_0\|y\|_{L_{p_0}(S;H)} \geq \beta\|y\|_X,$$

$$\inf_{\zeta(y)\in A(y)}\int_S e^{-2\lambda t}\big(\zeta(y)(t)+\lambda y(t), y(t)\big)dt \geq \gamma_1[y]_{X_1}^{p_1}+\gamma_2[y]_{X_2}^{p_2}+\alpha.$$

Proof. Indeed (see the proof of Theorem 2.3), $\|y\|_X\geq\|y_\lambda\|_X$ and $[y]_{X_i}\geq[y_\lambda]_{X_i}$, $i=1,2$. Therefore

$$[A_\lambda y_\lambda, y_\lambda]_- = \inf_{\zeta(y)\in A(y)}\int_S e^{-2\lambda t}\big(\zeta(y)(t)+\lambda y(t), y(t)\big)dt$$
$$\geq \gamma_1[y]_{X_1}^{p_1}+\gamma_2[y]_{X_2}^{p_2}+\alpha \geq \gamma_1[y_\lambda]_{X_1}^{p_1}+\gamma_2[y_\lambda]_{X_2}^{p_2}+\alpha.$$

Then using the identity of weighting norms it is easy to show the validity of the inequality

$$[y_\lambda]_{X_1}+[y_\lambda]_{X_2}+\widehat{\lambda}_0\|y_\lambda\|_{L_{p_0}(S;H)} \geq \widehat{\beta}\|y_\lambda\|_X.$$

□

Corollary 2.3. *Let $V_2=H$, $r_1\geq 2$, $p_2=r_2=2$, $\lambda_0>0$, $A+\lambda_0 I: X_1\rightrightarrows X_1^*$ is +-coercive, radial lower semicontinuous multivalued map with (X_1,W)-semibounded variation, $\varphi: X_2\to\mathbb{R}$ is convex lower semicontinuous functional. Then for every $f\in X^*$ there exists at least one solution $y\in W$ of the problem:*

$$\langle y', \xi - y\rangle + [Ay, \xi - y]_+ + \varphi(\xi) - \varphi(y) \geq \langle f, \xi - y\rangle \quad \forall \xi \in W, \quad y(0) = \bar{0}, \tag{2.70}$$

under the condition that A and $\partial\varphi$ are operators of Volterra type which satisfy Condition (H).

Remark 2.8. In Corollary 2.3 Condition (H) for A and $\partial\varphi$ and $+$-coercivity for $A + \lambda_0 I$ can be replaced by $-$-coercivity for $A + \lambda_0 I$ on X_1.

Remark 2.9. The inequality in (2.70) is equivalent to the following one:

$$\langle y', \xi - y\rangle + [Ay, \xi - y]_+ + [\partial\varphi(y), \xi - y]_+ \geq \langle f, \xi - y\rangle \quad \forall \xi \in W, \tag{2.71}$$

where

$$\partial\varphi(y) = \{p \in X_2^* \mid \varphi(\xi) - \varphi(y) \geq \langle p, \xi - y\rangle \quad \forall \xi \in X_2\}$$

is a subdifferential of the convex functional φ.

The inequality (2.71) follows from the formula: $\forall u, v \in X_2$

$$D_+\varphi(u; v - u) := \lim_{t \to 0+} \frac{\varphi(u + t(v - u)) - \varphi(u)}{t} = [\partial\varphi(u), v - u]_+.$$

Remark 2.10. In the last corollary we do not demand coercivity of φ (respectively, of $\partial\varphi$) on X_2.

Proof. Now we show that the operator $C = \lambda I + \lambda_0 I + A + \partial\varphi : X \rightrightarrows X^*$ satisfies all conditions of Corollary 2.2 for some $\lambda > 0$.

To prove this it is sufficient to show the same for the multivalued operator

$$B(y) = \lambda y + \partial\varphi(y) \quad \forall y \in X_2,$$

where $\lambda > 0$ is arbitrary fixed. Radial lower semicontinuity follows from upper semicontinuity of $\partial\varphi$ on X_2. Due to monotony of $\partial\varphi$ on X_2 for each $y \in X_2$,

$$\begin{aligned}[B(y), y]_+ &= [(\lambda I + \partial\varphi)(y), y]_+ = \lambda\langle y, y\rangle + [\partial\varphi(y), y]_+ \\ &= \lambda(y, y)_{\mathcal{H}} + [\partial\varphi(y), y - \bar{0}]_+ \geq \lambda\|y\|^2_{\mathcal{H}} + [\partial\varphi(\bar{0}), y - \bar{0}]_+ \\ &\geq \lambda\|y\|^2_{X_2} - \|\partial\varphi(\bar{0})\|_+\|y\|_{X_2}.\end{aligned}$$

Hence, for all $y \in X_2$

$$\frac{[B(y), y]_+}{\|y\|_{X_2}} \geq \lambda\|y\|_{X_2} - \|\partial\varphi(\bar{0})\|_+ \to +\infty \text{ as } \|y\|_{X_2} \to +\infty.$$

So, B is $+$-coercive. From monotony of $\partial\varphi$, since $\lambda > 0$ it follows that B is also monotone and therefore has semibounded variation on W. Hence problem (2.71) has at least one solution $y \in W$ such that $y(0) = \bar{0}$. So y is a solution of (2.70). □

2.4 Faedo-Galerkin Method for Differential-Operator Inclusions

Differential-operator equations and inclusions appear at an analysis and control of mathematical models of non-linear technological and industrial processes and fields [ZN96, ZMN04]. Since the monotonous and continuous relationship between determinative parameters is not natural [P85], we must take into account such non-linear effects as unsmoothness, unmonotony etc [P85] when studying more exact geophysical models. All of it leads to the necessity of qualitative and numerical investigation of models which can be described, in particular, by partial differential equations that contain the unmonotony non-linearity in the main part [S94]. One of the main classes of such operators is the class of operators of w_{λ_0}-pseudomonotone type [CVM04, K91, S94]. The stationary operator inclusions with such maps were systematically studied in works [KKK08a, KKK08b]. The goal of the given section is to validate the Faedo-Galerkin method for periodic solutions and Cauchi problem solutions of the class of the evolutional inclusions with maps of w_{λ_0}-pseudomonotone type in infinite-dimensional spaces. This is the new direction of investigations in comparison with well-known results (see for example [KKK08a, KM05a, Pap94, P85, S94, ZN96] and references there).

2.4.1 Faedo-Galerkin Method I

2.4.1.1 Problem Definition

Again let $(V_i; H; V_i^*)$ – be evolutionary triples such that the space $V = V_1 \cap V_2$ is continuously and densely embedded in H, $\{h_i\}_{i\geq 1} \subset V$ – is complete in V countable vectors system, $H_n = \text{span}\{h_i\}_{i=1}^n$, $n \geq 1$;

$$\begin{aligned} X &= L_{r_1}(S;H) \cap L_{r_2}(S;H) \cap L_{p_1}(S;V_1) \cap L_{p_2}(S;V_2), \\ X^* &= L_{q_1}(S;V_1^*) + L_{q_2}(S;V_2^*) + L_{r_2'}(S;H) + L_{r_1'}(S;H), \\ X_i &= L_{r_i}(S;H) \cap L_{p_i}(S;V_i), \quad X_i^* = L_{q_i}(S;V_i^*) + L_{r_i'}(S;H), \\ W &= \{y \in X \mid y' \in X^*\}, \quad W_i = \{y \in X_i \mid y' \in X^*\} \quad i = 1,2. \end{aligned}$$

with the norms corresponding, we assume that $p_0 := \max\{r_1, r_2\} < +\infty$ (see Remark 1.3).

For the multivalued map $C : X \rightrightarrows X^*$ we consider the problem:

$$\begin{cases} y' + C(y) \ni f \\ y(0) = y_0. \end{cases} \tag{2.72}$$

Here $f \in X^*$, $y_0 \in H$ is arbitrary fixed, y' is the derivative of an element $y \in X$ considered in the sense of scalar distributions space $\mathscr{D}^*(S; V^*)$, $S = [0; T]$.

2.4.1.2 Faedo-Galerkin Method

For each $n \geq 1$ let us consider Banach spaces

$$X_n = L_{p_0}(S; H_n), \quad X_n^* = L_{q_0}(S; H_n), \quad W_n = \{y \in X_n \mid y' \in X_n^*\},$$

where $1/p_0 + 1/q_0 = 1$. Let us also remind, that for any $n \geq 1$ I_n – the canonical embedding of X_n in X, $I_n^* : X^* \to X_n^*$ – is adjoint with I_n.

Let us introduce the following maps:

$$C_n := I_n^* C I_n : X_n \rightrightarrows X_n^*, \quad f_n := I_n^* f \in X_n^*.$$

Let us consider the sequence $\{y_{0n}\}_{n \geq 0} \subset H$:

$$\forall n \geq 1 \quad H_n \ni y_{0n} \to y_0 \quad \text{in } H \quad \text{as } n \to +\infty. \tag{2.73}$$

Together with problem (2.72) $\forall n \geq 1$ we consider the following class of problems:

$$\begin{cases} y_n' + C_n(y_n) \ni f_n \\ y_n(0) = y_{0n}. \end{cases} \tag{2.74}$$

Definition 2.5. We will say, that the solution $y \in W$ of (2.72) turns out by *Faedo-Galerkin method*, if y is a weak limit of a subsequence $\{y_{n_k}\}_{k \geq 1}$ from $\{y_n\}_{n \geq 1}$ in W, which satisfies the following conditions:

(a) $\forall n \geq 1 \quad W_n \ni y_n$ is a solution of (2.74).
(b) $y_{0n} \to y_0$ in H as $n \to \infty$.

2.4.1.3 Main Results

Theorem 2.4. *Let the multivalued map $C : X \to C_v(X^*)$ satisfies the following conditions:*

(1) C is λ_0-pseudomonotone on W.
(2) C is finite-dimensionally locally bounded.
(3) C satisfies Property (Π) on X.
(4) C satisfies the following coercive property:

$$\exists c > 0 : \quad \frac{[C(y), y]_+ - c\|C(y)\|_+}{\|y\|_X} \to +\infty \quad \textit{as} \quad \|y\|_X \to \infty.$$

Besides, let the system of vectors $\{h_j\}_{j \geq 1} \subset V_1 \cap V_2$ exists, it is complete in V_1, V_2 and such that for $i = 1, 2$ the triple $(\{h_j\}_{j \geq 1}; V_i; H)$ satisfies Condition (γ) with *constant C_i. Then for arbitrary $f \in X^*$, $y_0 \in H$ the set*

$$K_H(f) := \Big\{ y \in W \mid y \text{ is the solution of problem (2.72), which turns out by Faedo-Galerkin method} \Big\}$$

is nonempty. Moreover the representation

$$K_H(f) = \bigcup_{\{y_{0n}\}_{n\geq 1}\subset H \text{ satisfies (2.73)}} \bigcap_{n\geq 1} \Big[\bigcup_{m\geq n} K_m(f_m)(y_{0m}) \Big]_{X_w}, \tag{2.75}$$

is true, where for each $n \geq 1$

$$K_n(f_n)(y_{0n}) = \Big\{ y_n \in W_n \mid y_n \text{ is a solution of problem (2.74)} \Big\},$$

$[\cdot]_{X_w}$ is closuring operator in the space $(X; \sigma(X; X^*))$.

Proof. Let us consider the arbitrary sequence $\{y_{0n}\}_{n\geq 1} \subset H$, which satisfies (2.73). Then $\exists \delta > 0$:

$$\sup_{n\geq 1} \|y_{0n}\|_H \leq \delta. \tag{2.76}$$

Proposition 2.8. *There exists a sequence $\{x_n\}_{n\geq 1} \subset W$ such that*

(a) $\forall n \geq 1 \quad x_n \in W_n \subset W$.
(b) $\forall n \geq 1 \quad \|x_n\|_X \leq c$, $\|x_n'\|_{X^*} \leq \frac{\max^2\{C_1;C_2\}\cdot\delta}{2c} =: \bar{\delta}$,
where $c > 0$ is the constant from the condition (4) of Theorem 2.4.
(c) $\forall n \geq 1 \quad x_n(0) = y_{0n}$.

Proof. Let us fix the arbitrary $n \geq 1$ and prove the solvability of the problem:

$$\left.\begin{aligned} u_n' + \varepsilon^2 J(u_n) &\ni \bar{0}, \\ u_n(0) &= y_{0n}, \end{aligned}\right\} \tag{2.77}$$

in class W, where u_n' is the derivative of the element $u_n \in X$ considered in the sense of $\mathscr{D}(S; V^*)$,

$$\varepsilon = \frac{\delta \cdot \max\{C_1, C_2\}}{c\sqrt{2}} > 0,$$

C_i, $i = 1, 2$ are the constants from the condition of Theorem 2.4,

$$J(u) = \partial\left(\frac{1}{2}\|\cdot\|_X^2\right)(u) = \{p \in X^* \mid \langle p, u\rangle_X = \|u\|_X^2 = \|p\|_{X^*}^2\} \in C_v(X^*)$$

for all $u \in X$ (see Proposition 8).

Lemma 2.2. *Problem (2.77) has at least one solution* $u_n \in W$ *such that* $\|u_n\|_X \le \frac{c}{\max\{C_1,C_2\}}$, $\|u_n'\|_{X^*} \le \frac{\max\{C_1,C_2\}\cdot\delta^2}{2c}$.

Proof. Let us consider the auxiliary problem:

$$\left.\begin{aligned} Lv_n + \varepsilon^2 Z_n(v_n) \ni \bar{0},\\ v_n \in D(L), \end{aligned}\right\} \tag{2.78}$$

where $L : D(L) \subset X \to X^*$ is taken in the following way:

$$Lv = v' \quad \forall v \in D(L) = \{v \in W \mid v(0) = \bar{0}\},$$

v' is the derivative of an element $v \in X$ considered in the sense of $\mathscr{D}(S; V^*)$,

$$Z_n(v) = J(v - z_n) \in C_v(X^*) \quad \forall v \in X, \quad z_n(\cdot) \equiv y_{0n} \in W.$$

Let us prove the solvability of problem (2.78) using Theorem 2.2 with

$$X_1 = X_2 = X, \quad A = Z_n, \quad B \equiv \bar{0}, \quad f = \bar{0}, \quad L = L, \quad D(L) = D(L).$$

At first let us make sure that maps L and Z_n satisfy the following conditions:

(i_1) $L{:}D(L) \subset X \to X^*$ is linear maximal monotone on $D(L)$.
(i_2) $Z_n{:}X \to C_v(X^*)$ is monotone, radial lower semicontinuous, bounded, coercive.
(i_3) the set $D(L)$ is dense in X.

Now we consider (i_1). Linearity of L follows from linearity of the set $D(L)$ and from linearity of the derivative. Now we prove the monotony. For each $y \in D(L)$, using equality (1.16) we have:

$$\langle Ly, y\rangle_X = \langle y', y\rangle_X = \frac{1}{2}(\|y(T)\|_H^2 - \|y(0)\|_H^2) = \frac{1}{2}\|y(T)\|_H^2 \ge 0.$$

Hence and using linearity of L on $D(L)$ monotony of L on $D(L)$ follows. Let us prove maximal monotony of L on $D(L)$. Let $v \in X$, $w \in X^*$ be such that

$$\forall u \in D(L) \quad \langle w - Lu, v - u\rangle_X \ge 0.$$

Firstly we are going to prove that $v \in W$ and $v' = w$. Indeed, let $u = h\varphi x \in D(L)$, where $\varphi \in \mathscr{D}(S)$, $x \in H$, $h > 0$. Then

$$\begin{aligned} 0 \le \langle w - \varphi' hx, v - \varphi hx\rangle_X &= \langle w, v\rangle_X - \left(\int_S (\varphi'(s)v(s) + \varphi(s)w(s))ds, hx\right)\\ &\quad + \langle \varphi' hx, \varphi hx\rangle_X\\ &= \langle w, v\rangle_X + h\langle v'(\varphi) - w(\varphi), x\rangle_X, \end{aligned}$$

where $v'(\varphi)$, $w(\varphi)$ are the values of distribution v', w on the element $\varphi \in \mathscr{D}(S)$. Therefore,

$$\forall \varphi \in \mathscr{D}(S)\ \forall x \in X \quad \langle v'(\varphi) - w(\varphi), x\rangle_X \geq 0.$$

Hence,

$$v'(\varphi) - w(\varphi) = \bar{0} \quad \forall \varphi \in \mathscr{D}(S).$$

This means that $v' = w \in X^*$. To prove that $v(0) = \bar{0}$ let us take

$$u(t) = v(T) \cdot \frac{t}{T} \quad \forall t \in S, \quad u \in D(L).$$

Using (1.16) we obtain, that

$$\begin{aligned} 0 &\leq \langle v' - Lu, v - u\rangle_X = \langle v' - u', v - u\rangle_X \\ &= \frac{1}{2}\left(\|v(T) - u(T)\|_H^2 - \|v(0) - u(0)\|_H^2\right) \\ &= -\frac{1}{2}\|v(0)\|_H^2 \leq 0. \end{aligned}$$

Hence $\|v(0)\|_H = 0$. So, $v \in D(L)$ and $v' = w$.

Let us consider (i_2). Monotony and boundness of Z_n on X follow from the same properties as those ones of J, which in their turn follow from Theorem 2 and Proposition 8. Let us prove radial lower semicontinuity. Let y, $\xi \in X$ be arbitrary fixed. From Proposition 1

$$\begin{aligned} \underline{\lim}_{t\to 0+} [Z_n(y + t\xi), \xi]_+ &\geq \underline{\lim}_{t\to 0+} [Z_n(y + t\xi), \xi]_- \\ &= -\overline{\lim}_{t\to 0+} [J(y - z_n + t\xi - tz_n), -\xi]_+, \end{aligned}$$

but since $J = \partial\left(\|\cdot\|_X^2/2\right)$, from Theorem 1 and the identity (15) we have:

$$\begin{aligned} \overline{\lim}_{t\to 0+} [J(y - z_n + t\xi - tz_n), -\xi]_+ &\leq [J(y - z_n), -\xi]_+ \\ &= -[J(y - z_n), \xi]_- = -[Z_n(y), \xi]_-. \end{aligned}$$

Therefore, radial lower semicontinuity of Z_n on X is proved. At last we prove coercivity of Z_n. From Proposition 1 and identity (15) we have:

$$\begin{aligned} &\forall u \in X : \|u\|_X \geq 2\|z_n\|_X \quad [Z_n(u), u]_+ \geq [J(u - z_n), u - z_n]_+ \\ &\quad + [J(u - z_n), z_n]_- \geq \|u - z_n\|_X^2 - \|u - z_n\|_X\|z_n\|_X \\ &\geq \|u - z_n\|_X(\|u_X\| - 2\|z_n\|_X) \geq (\|u\|_X - \|z_n\|_X)(\|u\|_X - 2\|z_n\|_X) \\ &= \|u\|_X^2 - 3\|z_n\|_X\|u\|_X + \|z_n\|_X^2 \geq \|u\|_X(\|u\|_X - 3\|z_n\|_X). \end{aligned}$$

Hence, for all $u \in X: \|u\|_X \geq 2\|z_n\|_X$

$$\frac{[Z_n(u), u]_+}{\|u\|_X} \geq \|u\|_X - 3\|z_n\|_X \to +\infty \quad \text{as} \quad \|u\|_X \to +\infty.$$

Coercivity of Z_n on X is proved.

Let us prove (i_3). From Proposition 1.8 it is sufficient to prove that for each $n \geq 1$ the set $D(L) \cap X_n$ is dense in X_n.

Lemma 2.3. *For each $n \geq 1$*

$$D(L) \cap X_n = \{y \in W_n \mid y(0) = \bar{0}\} =: \Omega_n.$$

Proof. "$\subset$" Let $y \in D(L) \cap X_n \subset X_n$. Then from Proposition 1.9 and Corollary 1.2 we have:

$$y' = (P_n y)' = P_n y' \in P_n X^* = I_n^* X^* = X_n^*.$$

Hence, $y \in W_n$ and $y(0) = \bar{0}$.

"$\supset$" Let $y \in W_n$: $y(0) = \bar{0}$. This means that $y \in X_n$, $y' \in X_n^* \subset X^*$ and $y(0) = \bar{0}$, namely $y \in D(L) \cap X_n$. Lemma 2.3 is proved. □

Let us continue proving the condition (i_3). From [GGZ74, Lemma VI.1.5, p. 249] it is sufficient to approximate an arbitrary function from $C^1(S; H_n)$ by functions from Ω_n by the norm of the space $\|\cdot\|_{X_n}$. Let $y \in C^1(S; H_n)$ be an arbitrary function,

$$c := \max_{t \in S} \|y(t)\|_{H_n} + \max_{t \in S} \|y'(t)\|_{H_n}.$$

For each $m \geq 1$ let us set

$$y_m(t) = \begin{cases} \left[my'(1/m) - m^2 y(1/m)\right] t^2 \\ \qquad + [2my(1/m) - y'(1/m)]\, t, & t \in [0, 1/m], \\ y(t), & t \in (1/m, T]. \end{cases}$$

Note that $y_m \in \Omega_n$. Indeed,

$$y_m(0) = \bar{0}, \quad y_m(1/m-) = y(1/m), \quad y_m'(1/m-) = y_m'(1/m+) = y_m'(1/m).$$

From the fact that for each $m \geq 1$

$$\|y - y_m\|_{X_n} = \left(\int_0^{1/m} \|y(t) - y_m(t)\|_H^{p_0} dt \right)^{1/p_0}$$
$$\leq \frac{c}{m^{1/p_0}} + \frac{mc + m^2 c}{m^{2+1/p_0}} + \frac{2mc + c}{m^{1+1/p_0}} \leq \frac{6c}{m^{1/p_0}} \to 0 \quad \text{as} \quad m \to \infty,$$

the validity of (i_3) follows.

From the properties (i_1)–(i_3) it follows that all conditions of Theorem 2.2 hold true, and therefore the auxiliary problem (2.78) has at least one solution $v_n \in W$: $v_n(0) = \bar{0}$. Hence, keeping in mind the substitution $u = v + z_n$ (noticing that $Lz_n = \bar{0}$) we have that $y_n := v_n + z_n \in W$ is a solution of problem (2.77). Therefore the solvability of problem (2.77) is proved.

Now let $u_n \in W$ be one of the solutions of problem (2.77). Then from Proposition 1, Proposition 8 and formula (1.16) we obtain:

$$\frac{\delta^2}{2} \geq \frac{\|y_{0n}\|_H^2}{2} \geq \left(\frac{\|u_n(0)\|_H^2}{2} - \frac{\|u_n(T)\|_H^2}{2}\right) = \langle -u_n', u_n\rangle \geq \varepsilon^2[J(u_n), u_n]_-$$
$$= \varepsilon^2\|u_n\|_X^2 = \varepsilon^2\|J(u_n)\|_{+(-)}^2 = \frac{1}{\varepsilon^2}\|u_n'\|_{X^*}^2.$$

Hence,

$$\|u_n\|_X \leq \frac{c}{\max\{C_1; C_2\}}, \quad \|u_n'\|_{X^*} \leq \frac{\max\{C_1, C_2\}\cdot\delta^2}{2c}.$$

Lemma 2.2 is proved. □

Let us continue proving Proposition 2.8. We take $u_n \in W$ from the Lemma just proved and set $x_n(\cdot) = P_n u_n(\cdot)$. From Proposition 1.6 it follows that $x_n(0) = y_{0n}$. From Proposition 1.6 and Lemma 2.2 it follows that $x_n \in W_n$ and

$$\|x_n\|_X = \|P_n u_n(\cdot)\|_X \leq \max\{C_1, C_2\}\|u_n\|_X \leq c.$$

From Proposition 1.5 and Lemma 2.2 it follows that $x_n \in W_n$ and

$$\|x_n'\|_{X^*} = \|P_n u_n'(\cdot)\|_{X^*} \leq \max\{C_1, C_2\}\|u_n'\|_{X^*} \leq \frac{\max\{C_1, C_2\}^2\cdot\delta^2}{2c} = \bar{\delta}.$$

From arbitrariness of $n \geq 1$ it follows that Proposition 2.8 is proved. □

Let us continue to prove Theorem 2.4. From Proposition 2.8 for each $n \geq 1$ we take $x_n \in W_n$ in such way that $x_n(0) = y_{0n}$, $\|x_n\|_X \leq c$ and $\|x_n'\|_{X^*} \leq \bar{\delta}$. We use the coercivity condition (4). Let $\gamma : \mathbb{R}_+ \mapsto \mathbb{R}$ be defined in the following way:

$$\gamma(r) = \inf_{\|y\|_X = r} \|y\|_X^{-1}\cdot([C(y), y]_+ - c\|C(y)\|_+) \quad \forall r \geq 0.$$

From condition (4) of the current Theorem it follows that

$$\gamma(r) \longrightarrow +\infty \quad \text{as} \quad r \longrightarrow +\infty,$$

also from Proposition 1 for all $n \geq 1$ and $y \in X$ it follows:

$$
\begin{aligned}
&[C(y+x_n)-f+x_n',y]_+ \\
&\quad \geq [C(y+x_n),y+x_n]_+ + \langle x_n'-f,y\rangle_X + [C(y+x_n),-x_n]_- \\
&\quad \geq [C(y+x_n),y+x_n]_+ - \left(\|x_n'\|_{X^*}+\|f\|_{X^*}\right)\|y\|_X - \|C(y+x_n)\|_+\|x_n\|_X \\
&\quad \geq [C(y+x_n),y+x_n]_+ - \left(\bar{\delta}+\|f\|_{X^*}\right)\|y\|_X - c\|C(y+x_n)\|_+.
\end{aligned}
$$

Therefore, for all $y \in X$: $\|y\|_X > c$

$$
\begin{aligned}
\frac{[C(y+x_n)-f+x_n',y]_+}{\|y\|_X} &\geq \frac{[C(y+x_n),y+x_n]_+ - c\|C(y+x_n)\|_+}{\|y\|_X} - \bar{\delta} - \|f\|_{X^*} \\
&\geq \gamma(\|y+x_n\|_X)\frac{\|y\|_X - c}{\|y\|_X} - \|f\|_{X^*} - \bar{\delta} \to +\infty
\end{aligned}
$$

uniformly on $n \geq 1$ as $\|y\|_X \to +\infty$, since $\|y+x_n\|_X \geq \|y\|_X - c$. Hence, there exists $r_0 > \delta$ such that

$$
[C(y+x_n)-f+x_n',y]_+ \geq 0 \qquad \forall n \geq 1,\ y \in X:\ \|y\|_X \geq r_0. \tag{2.79}
$$

We set $R = 3r_0$. Then $\forall z \in \overline{B}_{\bar{\delta}}(\bar{0}) \subset \overline{B}_{r_0}(\bar{0})$, in particular for $z = x_n$,

$$
\overline{B}_{r_0}(\bar{0}) \subset \overline{B}_{2r_0}(z) = \left\{y \in X \;\middle|\; \|y-z\|_X \leq 2r_0\right\} \subset \overline{B}_R(\bar{0}). \tag{2.80}
$$

Resolvability of approximating problems

Lemma 2.4. *For each $n \geq 1$ there exists a solution $y_n \in W_n$ of problem (2.74) such that $\|y_n\|_X \leq R$.*

Proof. Let for every $n \geq 1$

$$
D_n(\cdot) := C_n(\,\cdot\, + x_n) : X_n \rightrightarrows X_n^*.
$$

Given maps satisfy the following properties (see. [KM05a, p. 115–117]:

(i_1) $C_n, D_n : X_n \to C_v(X_n^*)$.
(i_2) C_n, D_n is λ_0-pseudomonotone on W_n, locally finite-dimensionally bounded; Moreover,
(i_3) $[D_n(y) - f_n + x_n', y]_+ \geq 0 \qquad \forall y \in X_n :\ \|y\|_{X_n} = 2r_0$.
(i_4) D_n satisfies Property (Π) on X_n.

Simultaneously with problem (2.74) we consider the following one:

$$
\begin{cases} z_n' + D_n(z_n) \ni f_n - x_n' \\ z_n(0) = \bar{0}. \end{cases} \tag{2.81}
$$

to find a solution z_n in W_n.

Let us introduce the operator $L_n : D(L_n) \subset X_n \to X_n^*$ with definitional domain

$$D(L_n) = \{y \in W_n \mid y(0) = \bar{0}\} = W_n^0,$$

that maps by the rule:

$$\forall y \in W_n^0 \quad L_n y = y',$$

where the derivative y' is considered in the sense of distributions space $\mathscr{D}^*(S; H_n)$. Note that L_n well-defined, since $W_n^0 \subset W_n$.

Lemma 2.5. *[KM05a, Lemma 5, p. 117] For the operator L_n the following properties take place:*

(i_5) *L_n is linear.*
(i_6) $\forall y \in W_n^0 \; \langle L_n y, y\rangle \geq 0.$
(i_7) *L_n is maximal monotone.*

Let us get back to the proof of Lemma 2.4. To prove the solvability of problem (2.81) we apply Theorem 2.2 with $X_1 = X_2 = X = X_n$, $A = D_n$, $B \equiv \bar{0}$, $L = L_n$, $D(L) = W_n^0$, $f = f_n - x_n'$, $R = 2r_0$. Due to properties (i_1)–(i_7) The existence of a solution $z_n \in W_n$ of problem (2.81) follows: $\|z_n\|_X \leq 2r_0$.

To complete the proof of Lemma 2.4 it is necessary to note that

$$y_n := z_n + x_n \in W_n$$

is a solution of problem (2.74), if

$$z_n \in W_n : \quad \|z_n\|_X \leq 2r_0$$

is a solution of problem (2.81). From $\|z_n\|_X \leq 2r_0$, due to (2.80), it follows that $\|y_n\|_X \leq R$. Lemma 2.4 is proved. □

Passing to limit Due to Lemma 2.4 we have a sequence of Galerkin approximate solutions $\{y_n\}_{n\geq 1}$, that satisfies following conditions:

$$\text{(a) } \forall n \geq 1 : \quad \|y_n\|_X \leq R. \tag{2.82}$$

$$\text{(b) } \forall n \geq 1 : \quad y_n \in W_n \subset W, \quad y_n' + C_n(y_n) \ni f_n; \tag{2.83}$$

$$\text{(c) } \forall n \geq 1 : \quad y_n(0) = y_{0n} \to y_0 \text{ in } H \text{ as } n \to \infty. \tag{2.84}$$

From inclusion (2.83) we have that

$$\forall n \geq 1 \, \exists d_n \in C(y_n) : \; I_n^* d_n =: d_n^1 = f_n - y_n' \in C_n(y_n) = I_n^* C(y_n). \tag{2.85}$$

Lemma 2.6. *From sequences $\{y_n\}_{n\geq 1}$, $\{d_n\}_{n\geq 1}$, that satisfy (2.82)–(2.85), subsequences $\{y_{n_k}\}_{k\geq 1} \subset \{y_n\}_{n\geq 1}$ and $\{d_{n_k}\}_{k\geq 1} \subset \{d_n\}_{n\geq 1}$ can be isolated in such way that for some $y \in W$, $d \in X^*$, $z \in H$ following types of convergence will take place:*

$$(1)\ y_{n_k} \rightharpoonup y \quad \text{in } X \quad \text{as } k \to \infty; \tag{2.86}$$

$$(2)\ y'_{n_k} \rightharpoonup y' \quad \text{in } X^* \quad \text{as } k \to \infty; \tag{2.87}$$

$$(3)\ d_{n_k} \rightharpoonup d \quad \text{in } X^* \quad \text{as } k \to \infty; \tag{2.88}$$

$$(4)\ y_{n_k}(T) \rightharpoonup z \quad \text{in } H \quad \text{as } k \to \infty. \tag{2.89}$$

Moreover in (2.86)–(2.89):

$$(i)\ y(0) = y_0, \quad (ii)\ z = y(T). \tag{2.90}$$

Proof. 1°. Let us prove boundness of $\{d_n\}_{n\geq 1}$ in X^*. Due to (2.85), (1.16) and (2.76), $\forall n \geq 1$

$$\begin{aligned}+\infty > \|f\|_{X^*} R \geq \|f\|_{X^*} \|y_n\|_X \geq \langle f, y_n\rangle = \langle f_n, y_n\rangle \\ = \langle y'_n, y_n\rangle + \langle d_n^1, y_n\rangle \geq \frac{1}{2}\left(\|y_n(T)\|_H^2 - \|y_n(0)\|_H^2\right) + \langle d_n, y_n\rangle\end{aligned}$$

Therefore, $\forall n \geq 1$

$$\langle d_n, y_n\rangle \leq \|f\|_{X^*} R + \delta^2/2 =: c_2 < +\infty.$$

From here, and also from the estimate (2.82) and Property (Π) for C it follows that

$$\exists c_4 > 0: \quad \forall n \geq 1 \quad \|d_n\|_{X^*} \leq c_4. \tag{2.91}$$

2°. Let us prove boundness of $\{y'_n\}_{n\geq 1}$ in X^*. Due to (2.85) it follows that $\forall n \geq 1$ $y'_n = I_n^*(f - d_n)$, and therefore keeping in mind (2.82), (2.91), we have:

$$\|y_n\|_{X^*} \leq \|y_n\|_W \leq R + \max\{C_1, C_2\}\left(\|f\|_{X^*} + c_4\right) =: c_5 < +\infty, \tag{2.92}$$

where $C_i \geq 1$ are the constant from Condition (γ). Hence from the continuity of the embedding W in $C(S; H)$ (Corollary 1.1) the existence of $c_6 > 0$ follows:

$$\forall n \geq 1 \quad \text{foe all } t \in S \qquad \|y_n(t)\|_H \leq c_6 < +\infty,$$

In particular,

$$\forall n \geq 1 \quad \|y_n(T)\|_H \leq c_6. \tag{2.93}$$

3°. From estimates (2.92), (2.91) and (2.93), due to the Banach–Alaoglu Theorem (Theorem 1.2), keeping in mind reflexivity of X, the existence of subsequences

$$\{y_{n_k}\}_{k\geq 1} \subset \{y_n\}_{n\geq 1}, \quad \{d_{n_k}\}_{k\geq 1} \subset \{d_n\}_{n\geq 1}$$

and of elements $y \in W$, $d \in X^*$ and $z \in H$, for which consequence types (2.86)–(2.89) follows.

4°. Let us prove that

$$y' = f - d. \tag{2.94}$$

Let $\varphi \in D(S)$, $n \in \mathbb{N}$ and $h \in H_n$. Then $\forall k : n_k \geq n$ we have:

$$\begin{aligned}\Big(\int_S \varphi(\tau)(y'_{n_k}(\tau) + d_{n_k}(\tau))d\tau, h\Big) &= \int_S \Big(\varphi(\tau)(y'_{n_k}(\tau) + d_{n_k}(\tau)), h\Big)d\tau \\ &= \int_S \Big(y'_{n_k}(\tau) + d_{n_k}(\tau), \varphi(\tau)h\Big)d\tau \\ &= \langle y'_{n_k} + d_{n_k}, \psi\rangle,\end{aligned}$$

where $\psi(\tau) = h \cdot \varphi(\tau) \in X_n \subset X$.

Note that we use the property of Bochner's integral here [GGZ74, Theorem IV.1.8, c.153]. Since for $n_k \geq n$ $H_{n_k} \supset H_n$, then

$$\begin{aligned}\langle y'_{n_k} + d_{n_k}, \psi\rangle &= \langle y'_{n_k} + d_{n_k}, I_{n_k}\psi\rangle = \langle I^*_{n_k}(y'_{n_k} + d_{n_k}), \psi\rangle \\ &= \langle y'_{n_k} + d^1_{n_k}, \psi\rangle = \langle f_{n_k}, \psi\rangle.\end{aligned}$$

Due to Proposition 1.5 we have:

$$\begin{aligned}&\forall k \geq 1\ (n_k \geq n)\langle f_{n_k}, \psi\rangle = \langle f, I_{n_k}\psi\rangle \\ &= \int_S (f(\tau), \varphi(\tau)h)d\tau = \int_S (\varphi(\tau)f(\tau), h)d\tau = \Big(\int_S \varphi(\tau)f(\tau)d\tau, h\Big).\end{aligned}$$

Therefore, for all $k : n_k \geq n$

$$\begin{aligned}\Big(\int_S \varphi(\tau)y'_{n_k}(\tau)d\tau, h\Big) &= \Big(\int_S \varphi(\tau)(f(\tau) - d_{n_k}(\tau))d\tau, h\Big) \\ &= \int_S ((f(\tau) - d_{n_k}(\tau)), \varphi(\tau)h)d\tau = \langle f - d_{n_k}, \psi\rangle \\ \to \langle f - d, \psi\rangle &= \Big(\int_S \varphi(\tau)(f(\tau) - d(\tau))d\tau, h\Big) \quad \text{as } k \to \infty. \end{aligned} \tag{2.95}$$

The latter follows from the weak convergence of d_{n_k} to d in X^*.

From convergence (2.87) we have:

$$\Big(\int_S \varphi(\tau)y'_{n_k}(\tau)d\tau, h\Big) \to \Big(\int_S \varphi(\tau)y'(\tau)d\tau, h\Big) = (y'(\varphi), h) \text{ as } k \to +\infty, \tag{2.96}$$

where

$$\forall \varphi \in \mathscr{D}(S) \quad y'(\varphi) = -y(\varphi') = -\int_S y(\tau)\varphi'(\tau)d\tau$$

is the derivative of an element y considered in the sense of $\mathscr{D}^*(S, V^*)$.

Hence, due to (2.95) and (2.96) it follows that

$$\forall \varphi \in \mathscr{D}(S)\ \forall h \in \bigcup_{n\geq 1} H_n \quad (y'(\varphi), h) = \left(\int_S \varphi(\tau)(f(\tau) - d(\tau))d\tau, h\right).$$

Since $\bigcup_{n\geq 1} H_n$ in dense in V then

$$\forall \varphi \in \mathscr{D}(S) \quad y'(\varphi) = \int_S \varphi(\tau)(f(\tau) - d(\tau))d\tau.$$

So, $y' = f - d$.

5°. Let us prove that $y(0) = y_0$. Let $h \in H_n$, $\varphi \in \mathscr{D}(S)$, $n \in \mathbb{N}$, $\psi(\tau) := (T - \tau)h \in X_n$. From (2.94) it follows:

$$\begin{aligned}
\langle y', \psi\rangle &= \int_S (y'(\tau), \psi(\tau))d\tau = \int_S (f(\tau) - d(\tau), \psi(\tau))d\tau \\
&= \lim_{k\to\infty} \int_S (f(\tau) - d_{n_k}(\tau), \psi(\tau))d\tau = \lim_{k\to\infty} \langle f - d_{n_k}, I_{n_k}\psi\rangle \\
&= \lim_{k\to\infty} \langle I_{n_k}^*(f - d_{n_k}), \psi\rangle = \lim_{k\to\infty} \langle (f_{n_k} - d_{n_k}^1), \psi\rangle = \lim_{k\to\infty} \langle y_{n_k}', \psi\rangle.
\end{aligned}$$

Now we use the formula (1.15). Noting that $\psi'(\tau) = -h$, $\tau \in S$, we obtain:

$$\begin{aligned}
\lim_{k\to\infty} \langle y_{n_k}', \psi\rangle &= \lim_{k\to\infty} \left\{ -\langle \psi', y_{n_k}\rangle + \big(y_{n_k}(T), \psi(T)\big) - \big(y_{n_k 0}, Th\big) \right\} \\
&= \lim_{k\to\infty} \left\{ \int_S (y_{n_k}(\tau), h)d\tau - \big(y_{n_k 0}, Th\big) \right\} \\
&= \lim_{k\to\infty} \int_S (y_{n_k}(\tau), h)d\tau - \lim_{k\to\infty} \big(y_{n_k 0}, Th\big) \\
&= \int_S (y(\tau), h)d\tau - (y_0, Th) = -\langle \psi', y\rangle - (y_0, Th).
\end{aligned}$$

The latter holds true due to $y_{n_k} \rightharpoonup y$ in X and $y_{n_k 0} \to y_0$ in H. Again we apply the formula (1.15) to the last expression:

$$\begin{aligned}-\langle \psi', y\rangle - (y_0, Th) &= \langle y', \psi\rangle - (y(T), \psi(T)) + (y(0), \psi(0)) - (y_0, Th)\\ &= \langle y', \psi\rangle + T(y(0) - y_0, h).\end{aligned}$$

Hence, $\forall h \in \bigcup_{n\ge1} H_n$

$$\langle y', \psi\rangle = \langle y', \psi\rangle + T(y(0) - y_0, h) \Leftrightarrow (y(0) - y_0, h) = 0.$$

Due to density of $\bigcup_{n\ge1} H_n$ in H it follows that $y(0) = y_0$, and the statement (i) from (2.90) is proved.

6°. To complete the proof we must show that $y(T) = z$. The proof is similar to that of 5°. Indeed, we take $\psi \equiv h \in \bigcup_{n\ge1} H_n$. Hence $\psi \in X_{n_0}$ for some n_0. Again we use the formula (1.15):

$$\begin{aligned}(y(T) - y(0), h) &= \int_S (y'(\tau), h)d\tau = \lim_{k\to\infty} \int_S (y'_{n_k}(\tau), h)d\tau\\ &= \lim_{k\to\infty} (y_{n_k}(T) - y_{n_k}(0), h) = (z - y(0), h).\end{aligned}$$

The last equality holds true due to (2.89) and (i) from (2.90). Therefore,

$$\forall h \in \bigcup_{n\ge1} H_n \quad (y(T) - z, h) = 0,$$

that is equivalent to $y(T) = z$.

Lemma 2.6 is proved. □

Now, to prove that y is a solution of problem (2.72) it is necessary to show that y satisfies the inclusion from (2.72). Due to identity (2.94), it is sufficient to prove that $d \in C(y)$.

At first let us make sure that

$$\overline{\lim_{k\to\infty}} \langle d_{n_k}, y_{n_k} - y\rangle \le 0. \tag{2.97}$$

Indeed, due to (2.94), $\forall k \ge 1$ we have:

$$\begin{aligned}\langle d_{n_k}, y_{n_k} - y\rangle &= \langle d_{n_k}, y_{n_k}\rangle - \langle d_{n_k}, y\rangle = \langle d^1_{n_k}, y_{n_k} - y\rangle - \langle d_{n_k}, y\rangle\\ &= \langle f_{n_k} - y'_{n_k}, y_{n_k}\rangle - \langle d_{n_k}, y\rangle\\ &= \langle f_{n_k}, y_{n_k}\rangle - \langle y'_{n_k}, y_{n_k}\rangle - \langle d_{n_k}, y\rangle\\ &= \langle f, y_{n_k}\rangle - \langle d_{n_k}, y\rangle + \frac{1}{2}\left(\|y_{n_k}(0)\|^2_H - \|y_{n_k}(T)\|^2_H\right).\end{aligned} \tag{2.98}$$

The latter is obtained from (1.16). Further in left and right sides of the equality (2.98) we pass to upper limit as $k \to \infty$. We have:

$$\begin{aligned}
\overline{\lim_{k\to\infty}} \langle d_{n_k}, y_{n_k} - y\rangle &\le \overline{\lim_{k\to\infty}} \langle f, y_{n_k}\rangle + \overline{\lim_{k\to\infty}} \langle d_{n_k}, -y\rangle \\
&\quad + \overline{\lim_{k\to\infty}} \frac{1}{2}\left(\|y_{n_k}(0)\|_H^2 - \|y_{n_k}(T)\|_H^2\right) \\
&\le \langle f, y\rangle_X - \langle d, y\rangle + \frac{1}{2}\left(\|y(0)\|_H^2 - \|y(T)\|_H^2\right) \\
&= \langle f - d, y\rangle - \langle y', y\rangle = \langle y' - y', y\rangle = 0.
\end{aligned}$$

The latter holds true due to Lemma 2.6, formulas (1.16), (2.94), the inequality

$$\overline{\lim_{k\to\infty}} (-\|y_{n_k}(T)\|_H) \le -\|y(T)\|_H,$$

due to [GGZ74, Lemma I.5.3], (2.89) and (ii) from (2.90). The inequality (2.97) is proved.

From the conditions (2.86), (2.87), (2.88), (2.97) and λ_0-pseudomonotony of C on W it follows that there exist $\{d_m\} \subset \{d_{n_k}\}_{k\ge 1}$, $\{y_m\} \subset \{y_{n_k}\}_{k\ge 1}$, such that

$$\forall \omega \in X \quad \underline{\lim_{m\to\infty}} \langle d_m, y_m - \omega\rangle \ \ge\ [C(y), y - \omega]_- . \tag{2.99}$$

If we prove that

$$\langle d, y\rangle \ \ge\ \overline{\lim_{m\to\infty}} \langle d_m, y_m\rangle, \tag{2.100}$$

hence from (2.99) and from convergence of (2.91) we will have:

$$\forall \omega \in X \quad [C(y), y - \omega]_- \le \langle d, y - \omega\rangle,$$

and using Proposition 2 we obtain that this is equivalent to the inclusion $y \in C(y)$. Therefore, y is a solution of problem (2.72).

Let us prove (2.100):

$$\begin{aligned}
\overline{\lim_{m\to\infty}} \langle d_m, y_m\rangle &= \overline{\lim_{m\to\infty}} \langle d_m, I_m y_m\rangle = \overline{\lim_{m\to\infty}} \langle d_m^1, y_m\rangle = \overline{\lim_{m\to\infty}} \langle f_m - y_m', y_m\rangle \\
&\le \overline{\lim_{m\to\infty}} \langle f_m, y_m\rangle + \overline{\lim_{m\to\infty}} (-\langle y_m', y_m\rangle) = \overline{\lim_{m\to\infty}} \langle f, y_m\rangle \\
&\quad + \frac{1}{2} \overline{\lim_{m\to\infty}} \left(\|y_m(0)\|_H^2 - \|y_m(T)\|_H^2\right) \le \langle f, y\rangle \\
&\quad - \frac{1}{2}\left(\|y(T)\|_H^2 - \|y(0)\|_H^2\right) = \langle f, y\rangle - \langle y', y\rangle = \langle d, y\rangle.
\end{aligned}$$

So, $y \in W$ is a solution of problem (2.72).

The representation (2.75) follows immediately from passing to limit of the current Theorem, from the property of upper topological limits [K66, Property 2.29.IV.8] and from Definition 2.5.

The Theorem is proved. □

2.4.2 *Faedo-Galerkin Method II*

2.4.2.1 Problem Definition

Again let $(V_i; H; V_i^*)$ be evolutionary triples such that the space $V = V_1 \cap V_2$ is continuously and densely embedded in H, $\{h_i\}_{i\geq 1} \subset V$ – is complete in V countable vectors system, $H_n = \text{span}\{h_i\}_{i=1}^n, n \geq 1$;

$$\begin{aligned} X &= L_{r_1}(S;H) \cap L_{r_2}(S;H) \cap L_{p_1}(S;V_1) \cap L_{p_2}(S;V_2), \\ X^* &= L_{q_1}(S;V_1^*) + L_{q_2}(S;V_2^*) + L_{r_2'}(S;H) + L_{r_1'}(S;H), \\ X_i &= L_{r_i}(S;H) \cap L_{p_i}(S;V_i), \quad X_i^* = L_{q_i}(S;V_i^*) + L_{r_i'}(S;H), \\ W &= \{y \in X \mid y' \in X^*\}, \quad W_i = \{y \in X_i \mid y' \in X^*\} \quad i = 1,2. \end{aligned}$$

with the norms corresponding, we assume that $p_0 := \max\{r_1, r_2\} < +\infty$ (see Remark 1.3).

For the multivalued map $A : X \rightrightarrows X^*$ and linear dense defined map $L : D(L) \subset X \to X^*$ we consider the problem:

$$\begin{cases} Ly + A(y) \ni f \\ y \in D(L). \end{cases} \tag{2.101}$$

Here $f \in X^*$ is arbitrary fixed. On $D(L)$ we consider the graph norm

$$\|y\|_{D(L)} = \|y\|_X + \|Ly\|_{X^*} \quad \forall y \in D(L).$$

2.4.2.2 Faedo-Galerkin Method

Foe each $n \geq 1$ let us consider Banach spaces

$$X_n = L_{p_0}(S;H_n), \quad X_n^* = L_{q_0}(S;H_n), \quad W_n = \{y \in X_n \mid y' \in X_n^*\},$$

where $1/p_0 + 1/q_0 = 1$. Let us also remind, that for any $n \geq 1$ I_n – the canonical embedding of X_n in X, $I_n^* : X^* \to X_n^*$ – is adjoint with I_n.

Let us introduce the following maps: $A_n := I_n^* A I_n : X_n \rightrightarrows X_n^*$,

$$L_n := I_n^* L I_n : D(L_n) = X_n \cap D(L) \subset X_n \rightrightarrows X_n^*, \quad f_n := I_n^* f \in X_n^*.$$

Together with problem (2.101) $\forall n \geq 1$ we consider the following class of problems:

$$\begin{cases} L_n y_n + A_n(y_n) \ni f_n \\ y_n \in D(L_n). \end{cases} \tag{2.102}$$

Definition 2.6. We will say, that the solution $y \in W$ of (2.101) turns out by the *Faedo-Galerkin method*, if y is the weak limit of a subsequence $\{y_{n_k}\}_{k\geq 1}$ from $\{y_n\}_{n\geq 1}$ in $D(L)$, where for each $n \geq 1$ y_n is a solution of problem (2.102).

2.4.2.3 Main Results

Theorem 2.5. *Let $L : D(L) \subset X \to X^*$ be a linear operator, $A : X_1 \to C_v(X_1^*)$ and $B : X_2 \to C_v(X_2^*)$ be multivalued maps such that*

(1) L is maximal monotone on $D(L)$ and satisfies

- *Condition L_1: for each $n \geq 1$ and $x_n \in D(L_n)$ $Lx_n \in X_n^*$.*
- *Condition L_2: for each $n \geq 1$ the set $D(L_n)$ is dense in X_n.*
- *Condition L_3: for each $n \geq 1$ L_n is maximal monotone on $D(L)$.*

(2) There exist Banach spaces W_1 and W_2 such that $W_1 \subset X_1$, $W_2 \subset X_2$ and $D(L) \subset W_1 \cap W_2$ with continuous embedding.
(3) A is λ_0-pseudomonotone on W_1 and satisfies Condition (Π).
(4) B is λ_0-pseudomonotone on W_2 and satisfies Condition (Π).
(5) The sum $C = A + B : X \overset{\to}{\to} X^*$ is finite-dimensionally locally bounded and weakly $+$-coercive.

Furthermore, let $\{h_j\}_{j\geq 1} \subset V$ be a complete vectors system in V_1, V_2, H such that $\forall i = 1, 2$ the triple $\left(\{h_j\}_{j\geq 1}; V_i; H\right)$ satisfies Condition (γ).

Then for each $f \in X^*$ the set

$$K_H(f) := \{y \in D(L) \mid y \text{ is the solution of (2.101), obtained by Faedo-Galerkin method}\}$$

is nonempty and the representation

$$K_H(f) = \bigcap_{n\geq 1} \Big[\bigcup_{m\geq n} K_m(f_m)\Big]_{X_w} \tag{2.103}$$

is true, where for each $n \geq 1$

$$K_n(f_n) = \{y_n \in D(L_n) \mid y_n \text{ is the solution of (2.102)}\}$$

and $[\cdot]_{X_w}$ is the closure operator in the space X with respect to the weak topology.

Moreover, if the operator $A + B : X \overset{\to}{\to} X^*$ is $-$-coercive, then $K_H(f)$ is weakly compact in X and in $D(L)$ concerning the graph norm.

Remark 2.11. The sufficient condition to get the weak +-coercivity of $A + B$ is that: A is +-coercive and it satisfies Condition (κ) on X_1, B is +-coercive and it satisfies Condition (κ) on X_2 (see Lemma 1).

Remark 2.12. From Condition L_2 on the operator L and from Proposition 1.8 it follows that L is dense defined.

Proof. By Lemma 1.15 and Lemma 2 we consider the λ_0-pseudomonotone on $W_1 \cap W_2$ (and hence on $D(L)$), finite-dimensionally locally bounded, weakly +-coercive map

$$X \ni y \to C(y) := A(y) + B(y) \in C_v(X^*),$$

which satisfies Condition (Π).

Let $f \in X^*$ be fixed. Now let us use the weak +-coercivity condition for C. There exists $R > 0$ such that

$$[C(y) - f, y]_+ \geq 0 \qquad \forall y \in X : \ \|y\|_X = R. \tag{2.104}$$

Resolvability of Approximating Problems

Lemma 2.7. *For each $n \geq 1$ there exists a solution of problem (2.102) $y_n \in D(L_n)$ such that $\|y_n\|_X \leq R$.*

Proof. Let us prove that for each $n \geq 1$

$$C_n := A_n + B_n = I_n^*(A + B) : X_n \to C_v(X_n^*)$$

satisfies the next hypothesis:

(i_1) C_n satisfies Condition (Π).
(i_2) C_n is λ_0-pseudomonotone on $D(L_n)$, locally finite-dimensionally bounded.
(i_3) $[C_n(y_n) - f_n, y_n]_+ \geq 0 \qquad \forall y_n \in X_n : \ \|y_n\|_{X_n} = R.$

Let us consider (i_1). Let $B \subset X_n$ be some nonempty bounded subset, $k > 0$ be a constant and $d_n = I_n^* d \in C_n$ (where $d \in C$ is a selector) is such that

$$\langle d_n(y), y\rangle \leq k \quad \text{for each } y \in B.$$

Since for each $y \in X_n$ $\langle d_n(y), y\rangle = \langle I_n^* d(y), y\rangle = \langle d(y), y\rangle$ then

$$\langle d(y), y\rangle \leq k \quad \text{for each } y \in B.$$

Since C satisfies Condition (Π) there exists $K > 0$ such that

$$\|d(y)\|_{X^*} \leq K \quad \text{for all } y \in B.$$

Consequently,

$$\sup_{y\in B} \|d_n(y)\|_{X^*} \le K\|I_n^*\|_{\mathscr{L}(X^*;X_n^*)} < +\infty.$$

Now we consider (i_2). Since the boundness of

$$I_n \in \mathscr{L}(X_n; X), \quad I_n^* \in \mathscr{L}(X^*; X_n^*)$$

and the locally finite-dimensional boundness of $C : X \to C_v(X^*)$ then the locally finite-dimensional boundness of C_n on X_n follows.

Now we prove the λ_0-pseudomonotony of C_n on $D(L_n)$. Let $\{y_m\}_{m\ge 0} \subset D(L_n)$ be an arbitrary sequence such that

$$y_m \rightharpoonup y_0 \text{ in } D(L_n),\ d_n(y_m) = I_n^* d(y_m) \in C_n(y_m) \rightharpoonup d \in X_n^* \quad \text{as } m \to +\infty,$$

where $d(y_m) \in C(y_m)$ is a selector, and inequality (1.55) holds. Since $D(L_n) \subset D(L)$ with continuous embedding then

$$y_m \rightharpoonup y_0 \quad \text{in} \quad D(L) \quad \text{as} \quad m \to +\infty. \tag{2.105}$$

Since $\forall m \ge 1$

$$\langle I_n^* d(y_m), y_m - y_0\rangle = \langle d(y_m), y_m - y_0\rangle$$

then

$$\overline{\lim_{m\to\infty}} \langle d(y_m), y_m - y_0\rangle = \overline{\lim_{m\to\infty}} \langle d_n(y_m), y_m - y_0\rangle \le 0. \tag{2.106}$$

Hence

$$\overline{\lim_{m\to\infty}} \langle d(y_m), y_m\rangle \le \overline{\lim_{m\to\infty}} \langle d_n(y_m), y_m - y_0\rangle + \overline{\lim_{m\to\infty}} \langle d_n(y_m), y_0\rangle \le \langle d, y_0\rangle < +\infty.$$

Since C satisfies Condition (Π) we have that the sequence $\{d(y_m)\}_{m\ge 1}$ is bounded in X^*. Hence, up to a subsequence

$$d(y_m) \rightharpoonup g \quad \text{in} \quad X^* \text{ as } m \to \infty$$

for some $g \in X^*$. Consequently from (2.105) and (2.106), we get the existence of the subsequence $\{y_{m_k}\}_{k\ge 1} \subset \{y_m\}_{m\ge 1}$ such that $\forall w \in X$

$$\underline{\lim_{k\to\infty}} \langle d(y_{m_k}), y_{m_k} - w\rangle \ge [C(y_0), y_0 - w]_-.$$

This means that for each $w \in X_n$

$$\underline{\lim}_{k\to\infty} \langle d_n(y_{m_k}), y_{m_k} - w\rangle \geq [C_n(y_0), y_0 - w]_-.$$

So, C_n is λ_0-pseudomonotone on $D(L_n)$.

The condition (i_3) holds thanks to (2.104).

Now let us continue the proof of the given Lemma. From Theorem 2.2 with $V = W = X = X_n$, $A = C_n$, $B \equiv \bar{0}$, $L = L_n$, $D(L) = D(L_n)$, $f = f_n$, $r = R$ and with the properties (i_1)–(i_3) for C_n, $L_2 - L_3$ for L_n, it follows that problem (2.102) has at least one solution $y_n \in D(L_n)$ such that $\|y_n\|_X \leq R$.

Let us remark that under condition (Π) on C_n it is easy to find the next estimate (2.109) from which it is possible to use the λ_0-pseudomonotony for C on $D(L_n)$.

The Lemma is proved. □

Passing to Limit

Due to Lemma 2.7 we have a sequence of Galerkin's approximate solutions $\{y_n\}_{n\geq 1}$ that satisfies the next conditions

$$\forall n \geq 1 \quad \|y_n\|_X \leq R; \tag{2.107}$$

$$\forall n \geq 1 \quad y_n \in D(L_n) \subset D(L), \quad L_n y_n + d_n(y_n) = f_n, \tag{2.108}$$

where $d_n(y_n) = I_n^* d(y_n)$, $d(y_n) \in C(y_n)$ is a selector.

In order to prove the given Theorem we need to obtain the next important

Lemma 2.8. *Let for some subsequence $\{n_k\}_{k\geq 1}$ from the natural scale the sequence $\{y_{n_k}\}_{k\geq 1}$ satisfy the next conditions:*

- $\forall k \geq 1$ $y_{n_k} \in D(L_{n_k}) = D(L) \cap X_{n_k}$.
- $\forall k \geq 1$ $L_{n_k} y_{n_k} + d_{n_k}(y_{n_k}) = f_{n_k}$, $d_{n_k}(y_{n_k}) = I_{n_k}^* d(y_{n_k})$, $d(y_{n_k}) \in C(y_{n_k})$.
- $y_{n_k} \rightharpoonup y$ *in* X *as* $k \to +\infty$ *for some* $y \in X$.

Then, $y \in K_H(f)$.

Proof. From the definitions of L_{n_k}, d_{n_k} and f_{n_k} for each $k \geq 1$

$$\begin{aligned}\langle d_{n_k}(y_{n_k}), y_{n_k}\rangle &= \langle f_{n_k} - L_{n_k} y_{n_k}, y_{n_k}\rangle = \langle f - L y_{n_k}, y_{n_k}\rangle \\ &\leq \|f\|_{X^*} \sup_{k\geq 1} \|y_{n_k}\|_X =: K_1 < +\infty,\end{aligned}$$

where K_1 is a constant which does not depend on $k \geq 1$. Hence, due to Property (Π) for operator C it follows that there exists $K_2 > 0$ such that for each $k \geq 1$

$$\|d(y_{n_k})\|_{X^*} \leq K_2 < +\infty. \tag{2.109}$$

Since Condition L_1 for L and Proposition 1.7 it follows that for all $k \geq 1$

$$\left.\begin{aligned} \sup_{k\geq 1} \|Ly_{n_k}\|_{X^*} &= \sup_{k\geq 1} \|L_{n_k} y_{n_k}\|_{X^*} \\ &\leq \max\{C_1, C_2\}(D + \|f\|_{X^*}) =: C_3 < +\infty. \end{aligned}\right\} \tag{2.110}$$

where K_3 is a constant which does not depend on $k \geq 1$. Hence, for each $k \geq 1$

$$\|y_{n_k}\|_{D(L)} = \|y_{n_k}\|_X + \|Ly_{n_k}\|_{X^*} \leq \sup_{k\geq 1} \|y_{n_k}\|_X + K_3 =: K_4 < +\infty,$$

where K_4 is a constant which does not depend on $k \geq 1$. Consequently, due to (2.109), Corollary 1.8 and the Banach–Alaoglu Theorem there exists a subsequence $\{y_m\}$ from $\{y_{n_k}\}$ such that for some $y \in D(L)$ and $d \in X^*$ the next convergence takes place:

$$y_m \rightharpoonup y \text{ in } D(L), \quad d(y_m) \rightharpoonup d \text{ in } X^*. \tag{2.111}$$

(a) Let us prove that

$$\lim_{m\to\infty} \langle Ly_m + d(y_m), y_m - y\rangle = 0. \tag{2.112}$$

Since the set $\bigcup_{n\geq 1} X_n$ is dense in X then for each m there exists $u_m \in X_m$ (for example $u_m \in \operatorname{argmin}_{v_m\in X_m} \|y - v_m\|_X$) such that $u_m \to y$ in X. So, due to (2.110), (2.109) we obtain that for each m

$$\begin{aligned} &|\langle Ly_m + d(y_m), y_m - y\rangle| \leq |\langle Ly_m + d(y_m), y_m - u_m\rangle| + |\langle Ly_m + d(y_m), u_m - y\rangle| \\ &\leq |\langle f, y_m - u_m\rangle| + (K_3 + K_2)\cdot \|y - u_m\|_X \to |\langle f, y - y\rangle| = 0. \end{aligned}$$

(b) Now we obtain that

$$\overline{\lim_{m\to\infty}} \langle d(y_m), y_m - y\rangle \leq 0. \tag{2.113}$$

From (2.112), (2.111) and from the monotony of L we have

$$\begin{aligned} &\overline{\lim_{m\to\infty}} \langle d(y_m), y_m - y\rangle \\ &\quad = \lim_{m\to\infty} \langle Ly_m + d(y_m), y_m - y\rangle - \lim_{m\to\infty} \left(\langle Ly_m - Ly, y_m - y\rangle + \langle Ly, y_m - y\rangle\right) \\ &\quad \leq 0 + \overline{\lim_{m\to\infty}} \left(-\langle Ly_m - Ly, y_m - y\rangle\right) + \overline{\lim_{m\to\infty}} \langle Ly, y - y_m\rangle \leq 0. \end{aligned}$$

From (2.111) and (2.113) we can use the λ_0-pseudomonotony of C on $D(L)$. Hence, there exists a subsequence $\{y_k\}_k$ from $\{y_m\}_m$ such that

$$\forall \omega \in X \quad \varliminf_{k\to\infty} \langle d(y_k), y_k - \omega\rangle \geq [C(y), y-\omega]_{-}. \tag{2.114}$$

In particular, from (2.113) and (2.114) it follows that

$$\lim_{k\to\infty} \langle d(y_k), y_k - y\rangle = 0.$$

(c) Let us prove that

$$\forall u \in D(L)\bigcap\big(\bigcup_{n\geq 1} X_n\big) \quad \langle f - d - Ly + Lu, u\rangle \geq 0. \tag{2.115}$$

In order to prove (2.115) it is necessary to obtain that

$$\forall u \in D(L)\bigcap\big(\bigcup_{n\geq 1} X_n\big) \quad \varliminf_{k\to\infty} \langle Ly_k - Ly + Lu, u\rangle \geq 0. \tag{2.116}$$

From the monotony of L and from (2.111), for each $u \in D(L)\bigcap\big(\bigcup_{n\geq 1} X_n\big)$ we have

$$\varliminf_{k\to\infty} \langle Ly_k - Ly + Lu, u\rangle \geq \lim_{k\to\infty} \langle Ly_k - Ly, u\rangle = 0.$$

Further let $u \in D(L)\bigcap\big(\bigcup_{n\geq 1} X_n\big)$ be arbitrary fixed. Then there exists $n_0 \geq 1$ such that $u \in D(L) \cap X_{n_0}$ and for each $k \geq n_0$

$$\left.\begin{aligned}\forall m \geq n_0 \quad \langle Ly_m, u\rangle_X &= \langle L_m y_m, u\rangle_{X_m} = \langle f_m - I_m^* d_m, u\rangle_{X_m} \\ &= \langle f - d_m, u\rangle_X \to \langle f - d, u\rangle_X.\end{aligned}\right\} \tag{2.117}$$

So, (2.115) directly follows from (2.116) and (2.117).
(d) Now we prove that $Ly = f - d$. Let us use (2.115). We obtain that for each $t > 0$ and $u \in D(L)\bigcap\big(\bigcup_{n\geq 1} X_n\big)$

$$\langle f - d - Ly, t\cdot u\rangle \geq -\langle t\cdot Lu, t\cdot u\rangle.$$

that is equivalent to

$$\langle f - d - Ly, u\rangle \geq -t\cdot\langle Lu, u\rangle.$$

Hence,

$$\forall u \in D(L)\bigcap\big(\bigcup_{n} X_n\big) \quad \langle f - d - Ly, u\rangle \geq 0$$

and, by Proposition 1.8, the last relation is equivalent to $Ly = f - d$.
(e) In order to prove that $y \in D(L)$ is the solution of (2.101) it is enough to show that $d \in C(y)$. Since (2.113), (2.114) and (2.111) it follows that for each $\omega \in X$

$$[C(y), y-\omega]_- \le \underline{\lim}_{k\to\infty} \langle d(y_k), y_k - \omega\rangle$$

$$\le \overline{\lim}_{k\to\infty} \langle d(y_k), y_k - y\rangle + \lim_{k\to\infty} \langle d(y_k), y-\omega\rangle \le \langle d, y-\omega\rangle,$$

that is equivalent to the required statement. So, $y \in K_H(f)$.

The Lemma is proved. □

Since (2.107), (2.108), Lemma 2.8, the Banach–Alaoglu Theorem and the topological property of the upper limit [K66, Property 2.29.IV.8] it follows that

$$\emptyset \ne \bigcap_{n\ge 1}\Big[\bigcup_{m\ge n} K_m(f_m)\Big]_{X_w} \subset K_H(f).$$

The converse inclusion is obviously; it follows from the same topological property of upper limit and from $D(L) \subset X$ with continuous embedding.

Now let us prove that $K_H(f)$ is weakly compact in X and in $D(L)$ under the – -coercivity condition on the operator $C = A + B : X \to C_v(X^*)$. Since (2.103) and $D(L) \subset X$ with continuous embedding it is enough to show that the given set is bounded in $D(L)$. Let $\{y_n\}_{n\ge 1} \subset K_H(f)$ be an arbitrary sequence. Then for some $d_n \in C(y_n)$

$$Ly_n + d(y_n) = f.$$

If $\{y_n\}_{n\ge 1}$ is such that

$$\|y_n\|_X \to +\infty \quad \text{as} \quad n \to \infty,$$

we obtain the contradiction

$$\begin{aligned} +\infty \leftarrow \frac{1}{\|y_n\|_X}[C(y_n), y_n]_- &\le \frac{1}{\|y_n\|_X}\langle d(y_n), y_n\rangle \\ &\le \frac{1}{\|y_n\|_X}\langle Ly_n + d(y_n), y_n\rangle \\ &= \frac{1}{\|y_n\|_X}\langle f, y_n\rangle \le \|f\|_{X^*} < +\infty. \end{aligned} \tag{2.118}$$

Hence, for some $k > 0$

$$\|y_n\|_X \le k \quad \forall n \ge 1. \tag{2.119}$$

Due to Condition (Π) for C, from (2.118) to (2.119) it follows the existence of $K > 0$:

$$\|d_n\|_{X^*} \le K.$$

Hence,

$$\|Ly_n\|_{X^*} \le K + \|f\|_{X^*} \quad \text{and} \quad \|y_n\|_{D(L)} \le k + K + \|f\|_{X^*}.$$

The Theorem is proved. □

2.4.3 On Searching the Periodic Solutions for Differential-Operator Inclusions by Faedo-Galerkin Method

Let $A : X_1 \to C_v(X_1^*)$ and $B : X_2 \to C_v(X_2^*)$ be multivalued maps. We consider the next problem:

$$\begin{cases} y' + A(y) + B(y) \ni f, \\ y(0) = y(T), \end{cases} \tag{2.120}$$

in order to find the solutions by FG method in the class

$$W = \{y \in X \mid y' \in X^*\},$$

where the derivative y' of an element $y \in X$ is considered in the sense of scalar distributions space $\mathscr{D}^*(S; V^*) = \mathscr{L}(\mathscr{D}(S); V_w^*)$, with $V = V_1 \cap V_2$, V_w^* equals to V^* with the topology $\sigma(V^*, V)$ [RS80]. We consider the norm on W

$$\|y\|_W = \|y\|_X + \|y'\|_{X^*} \text{ for each } y \in W.$$

We also consider the spaces $W_i = \{y \in X_i \mid y' \in X^*\}, i = 1, 2.$

Remark 2.13. It is clear that the space W is continuously embedded in $C(S; V^*)$. Hence, the condition from (2.120) has sense.

In parallel with problem (2.120) we consider the next class of problems in order to search the solutions in $W_n = \{y \in X_n \mid y' \in X_n^*\}$:

$$\begin{cases} y_n' + A_n(y_n) + B_n(y_n) \ni f_n, \\ y(0) = y(T), \end{cases} \tag{2.121}$$

where the maps A_n, B_n, f_n were introduced later, the derivative y_n' of an element $y_n \in X_n$ is considered in the sense of $\mathscr{D}^*(S; H_n)$.

Let

$$W_{per} := \{y \in W \mid y(0) = y(T)\},$$

let us introduce the map

$$L : D(L) = W_{per} \subset X \to X^*$$

in such way $Ly = y'$ for each $y \in W_{per}$.

From the main resolvability Theorem it follows the next Corollary:

Corollary 2.4. *Let $A : X_1 \to C_v(X_1^*)$ and $B : X_2 \to C_v(X_2^*)$ be multivalued maps such that*

(1) A is λ_0-pseudomonotone on W_1 and it satisfies Condition (Π).
(2) B is λ_0-pseudomonotone on W_2 and it satisfies Condition (Π).

(3) *The sum* $C = A + B : X \overset{\to}{\to} X^*$ *is finite-dimensionally locally bounded and weakly* $+$*-coercive.*

Furthermore, let $\{h_j\}_{j\geq 1} \subset V$ *be a complete vector system in* V_1, V_2, H *such that as* $i = 1, 2$ *the triple* $\big(\{h_j\}_{j\geq 1}; V_i; H\big)$ *satisfies Condition* (γ).

Then for each $f \in X^*$ *the set*

$$K_H^{per}(f) := \{y \in W \mid y \textit{ is the solution of (2.120), obtained by Faedo-Galerkin method}\}$$

is nonempty and the representation

$$K_H^{per}(f) = \bigcap_{n\geq 1}\Big[\bigcup_{m\geq n} K_m^{per}(f_m)\Big]_{X_w}$$

is true, where for each $n \geq 1$

$$K_n^{per}(f_n) = \{y_n \in W_n \mid y_n \textit{ is the solution of (2.121)}\}.$$

Moreover, if the operator $A + B : X \overset{\to}{\to} X^*$ *is* $-$*-coercive, then* $K_H^{per}(f)$ *is weakly compact in* X *and in* W.

Proof. At first let us prove the maximal monotony of L on W_{per}. For $v \in X$, $w \in X^*$ such that for each $u \in W_{per}$ $\langle w - Lu, v - u\rangle \geq 0$ is true, let us prove that $v \in W_{per}$ and $v' = w$. If we take $u = h\varphi x \in W_{per}$ with $\varphi \in \mathscr{D}(S)$, $x \in V$ and $h > 0$ we get

$$\begin{aligned}0 \leq \langle w - \varphi' h x, v - \varphi h x\rangle &= \langle w, v\rangle - \left\langle \int_S (\varphi'(s)v(s) + \varphi(s)w(s))ds, hx\right\rangle \\ &\quad + \langle \varphi' h x, \varphi h x\rangle \\ &= \langle w, v\rangle + h\langle v'(\varphi) - w(\varphi), x\rangle,\end{aligned}$$

where $v'(\varphi)$ and $w(\varphi)$ are the values of the distributions v' and w on $\varphi \in \mathscr{D}(S)$. So, for each $\varphi \in \mathscr{D}(S)$ and $x \in V$ $\langle v'(\varphi) - w(\varphi), x\rangle \geq 0$ is true. Thus we obtain $v'(\varphi) = w(\varphi)$ for all $\varphi \in \mathscr{D}(S)$. It means that $v' = w \in X^*$. Now we prove $v(0) = v(T)$. If we use [GGZ74, Theorem VI.1.17] with $u(t) \equiv v(T) \in W_{per}$, we obtain that

$$\begin{aligned}0 \leq \langle v' - Lu, v - u\rangle &= \langle v' - u', v - u\rangle \\ &= \frac{1}{2}\Big(\|v(T) - v(T)\|_H^2 - \|v(0) - v(T)\|_H^2\Big) \\ &= -\frac{1}{2}\|v(0) - v(T)\|_H^2 \leq 0\end{aligned}$$

and then $v(0) = v(T)$.

In order to prove the given Proposition, it is enough to show that L satisfies Conditions L_1–L_3. Condition L_1 follows from Proposition 1.9.

Condition L_2 follows from [GGZ74, Lemma VI.1.5] and from the fact that the set $C^1(S; H_n)$ is dense in $L_{p_0}(S, H_n) = X_n$. Condition L_3 follows from [GGZ74, Lemma VI.1.7] with $V = H = H_n$ and $X = X_n$. □

2.4.4 On the Solvability for One Cauchi Problem by FG Method

Let $A : X_1 \to C_v(X_1^*)$ and $B : X_2 \to C_v(X_2^*)$ be multivalued maps. We consider the next problem:

$$\begin{cases} y' + A(y) + B(y) \ni f, \\ y(0) = \bar{0}, \end{cases} \tag{2.122}$$

in order to find the solutions by FG method in the class

$$W = \{y \in X \mid y' \in X^*\},$$

where the derivative y' of an element $y \in X$ is considered in the sense $\mathscr{D}^*(S; V^*)$. We consider the norm on W

$$\|y\|_W = \|y\|_X + \|y'\|_{X^*} \text{ for each } y \in W.$$

We also consider the spaces $W_i = \{y \in X_i \mid y' \in X^*\}, i = 1, 2$.

Remark 2.14. It is clear that the space W is continuously embedded in $C(S; V^*)$. Hence, the condition (2.122) has sense.

In parallel with problem (2.122) we consider the next class of problems in order to search the solutions in $W_n = \{y \in X_n \mid y' \in X_n^*\}$:

$$\begin{cases} y_n' + A_n(y_n) + B_n(y_n) \ni f_n, \\ y(0) = \bar{0}, \end{cases} \tag{2.123}$$

where the maps A_n, B_n, f_n were introduced latter, the derivative y_n' of an element $y_n \in X_n$ is considered in the sense of $\mathscr{D}^*(S; H_n)$.

Let

$$W_{\bar{0}} := \{y \in W \mid y(0) = \bar{0}\},$$

let us introduce the map

$$L : D(L) = W_{\bar{0}} \subset X \to X^*$$

in such way $Ly = y'$ for each $y \in W_{\bar{0}}$.

From the main resolvability Theorem it follows the next

Corollary 2.5. *Let $A : X_1 \to C_v(X_1^*)$ and $B : X_2 \to C_v(X_2^*)$ be multivalued maps such that*

(1) A is λ_0-pseudomonotone on W_1 and it satisfies Condition (Π).
(2) B is λ_0-pseudomonotone on W_2 and it satisfies Condition (Π).
(3) The sum $C = A + B : X \overset{\rightarrow}{\rightarrow} X^*$ is finite-dimensionally locally bounded and weakly $+$-coercive.

Furthermore, let $\{h_j\}_{j\geq 1} \subset V$ be a complete vector system in V_1, V_2, H such that as $i = 1, 2$ the triple $(\{h_j\}_{j\geq 1}; V_i; H)$ satisfies Condition (γ).
Then for each $f \in X^*$ the set

$$K_H^{\bar{0}}(f) := \{y \in W \mid y \text{ is the solution of (2.122), obtained by Faedo-Galerkin method}\}$$

is nonempty and the representation

$$K_H^{\bar{0}}(f) = \bigcap_{n\geq 1}\Big[\bigcup_{m\geq n} K_m^{\bar{0}}(f_m)\Big]_{X_w}$$

is true, where for each $n \geq 1$

$$K_n^{\bar{0}}(f_n) = \{y_n \in W_n \mid y_n \text{ is the solution of (2.123)}\}.$$

Moreover, if the operator $A + B : X \overset{\rightarrow}{\rightarrow} X^*$ is $-$-coercive, then $K_H^{\bar{0}}(f)$ is weakly compact in X and in W.

Proof. At first let us prove the maximal monotony of L on $W_{\bar{0}}$. For $v \in X$, $w \in X^*$ such that for each $u \in W_{\bar{0}}$ $\langle w - Lu, v - u\rangle \geq 0$ is true, let us prove that $v \in W_{\bar{0}}$ and $v' = w$. By the analogy with the proof of Corollary 2.4, we obtain $v' = w \in X^*$. Now we prove $v(0) = \bar{0}$. If we use [GGZ74, Theorem IV.1.17] with $u(t) = v(T)\frac{t}{T} \in W_{\bar{0}}$, we obtain that

$$\begin{aligned} 0 &\leq \langle v' - Lu, v - u\rangle = \langle v' - u', v - u\rangle \\ &= \frac{1}{2}\left(\|v(T) - v(T)\|_H^2 - \|v(0)\|_H^2\right) = -\frac{1}{2}\|v(0)\|_H^2 \leq 0 \end{aligned}$$

and then $v(0) = \bar{0}$.

In order to prove the given Proposition, it is enough to show that L satisfies Conditions L_1–L_3. It follows from the proof of Lemma 2.2. □

Example 2.4. Let us consider the bounded domain $\Omega \subset \mathbb{R}^n$ with rather smooth boundary $\partial\Omega$, $S = [0, T]$, $Q = \Omega \times (0; T)$, $\Gamma_T = \partial\Omega \times (0; T)$. Let, as $i = 1, 2$, $m_i \in \mathbb{N}$, N_1^i(respectively N_2^i) the number of the derivatives respect to the variable x of order $\leq m_i - 1$ (respectively m_i) and $\{A_\alpha^i(x, t, \eta, \xi)\}_{|\alpha|\leq m_i}$ be a family of real functions defined in $Q \times \mathbb{R}^{N_1^i} \times \mathbb{R}^{N_2^i}$. Let

$$D^k u = \{D^\beta u, |\beta| = k\} \text{ be the differentiations by } x,$$
$$\delta_i u = \{u, Du, \dots, D^{m_i - 1} u\},$$
$$A^i_\alpha(x,t,\delta_i u, D^{m_i} v) : x,t \to A^i_\alpha(x,t,\delta_i u(x,t), D^{m_i} v(x,t)).$$

Let $\psi : \mathbb{R} \to \mathbb{R}$ be some locally Lipschitzean real function and its Clarke generalized gradient $\Phi = \partial_{Cl}\psi : \mathbb{R} \rightrightarrows \mathbb{R}$ satisfies the growth condition

$$\exists p_3 \geq 2, C > 0 : \qquad \|\Phi(t)\|_+ \leq C(1 + |t|^{p_3 - 1}) \quad \forall t \in \mathbb{R}. \tag{2.124}$$

Let us consider the next problem with Dirichlet boundary conditions:

$$\frac{\partial y(x,t)}{\partial t} + \sum_{|\alpha| \leq m_1} (-1)^{|\alpha|} D^\alpha (A^1_\alpha(x,t,\delta_1 y, D^{m_1} y))$$
$$+ \sum_{|\alpha| \leq m_2} (-1)^{|\alpha|} D^\alpha (A^2_\alpha(x,t,\delta_2 y, D^{m_2} y)) + \Phi(y(x,t)) \ni f(x,t) \text{ in } Q, \tag{2.125}$$

$$D^\alpha y(x,t) = 0 \quad \text{on } \Gamma_T \text{ as } |\alpha| \leq m_i - 1 \text{ and } i = 1,2 \tag{2.126}$$

$$\text{and} \quad y(x,0) = 0 \quad \text{in } \Omega. \tag{2.127}$$

Let us assume for $i = \overline{1,3}$ $q_i > 1$: $p_i^{-1} + q_i^{-1} = 1$, $H = L_2(\Omega)$, $V_3 = L_{p_3}(\Omega)$ and $V_i = W_0^{m_i,p_i}(\Omega)$ with $p_i > 1$, $m_i = 0,1,2,\dots$ such that $V_i \subset H$ with continuous embedding, $i = 1,2$. Consider the function $\varphi : L_{p_3}(Q) \to \mathbb{R}$ defined by

$$\varphi(y) = \int_Q \psi(y(x,t))dxdt \qquad \forall y \in L_{p_3}(Q).$$

Using the growth condition (2.124) and Lebourgs mean value Theorem, we note that the function φ is well-defined and Lipschitz continuous on bounded sets in $L_{p_3}(Q)$, thus locally Lipschitz so that Clarkes generalized gradient $\partial_{Cl}\varphi : L_{p_3}(Q) \rightrightarrows L_{q_3}(Q)$ is well-defined. Moreover, the Aubin–Clarke Theorem (see [C90, p. 83]) ensures that, for each $y \in L_{p_3}(Q)$ we have

$$p \in \partial_{Cl}\varphi(y) \Rightarrow p \in L_{q_3}(Q) \text{ with } p(x,t) \in \partial_{Cl}\psi(y(x,t)) \text{ for a.e. } (x,t) \in Q.$$

Under suitable conditions on the coefficients A^i_α, the given problems can be written again:

$$y' + A_1(y) + A_2(y) + \partial_{Cl}\varphi(y) \ni f, \qquad y(0) = \bar{0}, \tag{2.128}$$

where

$$f \in X^* = L_{q_1}(S; W^{-m_1,q_1}(\Omega)) + L_{q_2}(S; W^{-m_2,q_2}(\Omega)) + L_{q_3}(Q).$$

Each element $y \in W$ that satisfies (2.128) is called a generalized solution of problem (2.125)–(2.127).

Choice of basis. We assume that there is a complete vectors system $\{h_j\}_{j\geq 1} \subset W_0^{m_1,p_1}(\Omega) \cap W_0^{m_2,p_2}(\Omega)$ such that the triples

$$\left(\{h_j\}_{j\geq 1}; W_0^{m_i,p_i}(\Omega); L_2(\Omega)\right), \quad i = 1,2$$

satisfy Condition (γ).

For example, when $n = 1$ as $\{h_j\}_{j\geq 1}$ we may take the "special" basis for the pair $(H_0^{\max\{m_1;m_2\}+\varepsilon}(\Omega); L_2(\Omega))$ with a suitable $\varepsilon \geq 0$ [KMP06a, KMT06]. As it is well-known, the triple $(\{h_j\}_{j\geq 1}; L_p(\Omega); L_2(\Omega))$ satisfies Condition (γ) for $p > 1$. Then, using (for example) the results [KMP06a, KMT06], we obtain the necessary condition.

Definition of operators A_i. Let $A_\alpha^i(x,t,\eta,\xi)$, defined in $Q \times \mathbb{R}^{N_1^i} \times \mathbb{R}^{N_2^i}$, satisfy the conditions

for almost each $x,t \in$ Q the map $\eta,\xi \to A_\alpha^i(x,t,\eta,\xi)$ is continuous on $\mathbb{R}^{N_1^i} \times \mathbb{R}^{N_2^i}$;

$$\text{for all } \eta,\xi \text{ the map } x,t \to A_\alpha^i(x,t,\eta,\xi) \text{ is measurable on } Q, \tag{2.129}$$

$$\text{for all } u,v \in L^{p_i}(0,T;V_i) =: \mathcal{V}_i \quad A_\alpha^i(x,t,\delta_i u, D^{m_i}u) \in L^{q_i}(Q). \tag{2.130}$$

Then for each $u \in \mathcal{V}_i$ the map

$$w \to a_i(u,w) = \sum_{|\alpha|\leq m_i} \int_Q A_\alpha^i(x,t,\delta_i u, D^{m_i}u) D^\alpha w dx dt,$$

is continuous on $\mathcal{V}_i$ and then

$$\text{there exists } A_i(u) \in \mathcal{V}_i^* \text{ such that } a_i(u,w) = \langle A_i(u), w\rangle. \tag{2.131}$$

Conditions on A_i. Similarly to [L69, Sects. 2.2.5, 2.2.6, 3.2.1] we have

$$A_i(u) = A_i(u,u), \quad A_i(u,v) = A_{i1}(u,v) + A_{i2}(u),$$

where

$$\langle A_{i1}(u,v), w\rangle = \sum_{|\alpha|=m_i} \int_Q A_\alpha^i(x,t,\delta_i u, D^{m_i}v) D^\alpha w dx dt,$$

$$\langle A_{i2}(u), w\rangle = \sum_{|\alpha|\leq m_i-1} \int_Q A_\alpha^i(x,t,\delta_i u, D^{m_i}u) D^\alpha w dx dt.$$

We add the next conditions:

$$\langle A_{i1}(u,u), u-v\rangle - \langle A_{i1}(u,v), u-v\rangle \geq 0 \ \forall u, v \in \mathcal{V}_i; \tag{2.132}$$

$$\text{if } u_j \rightharpoonup u \text{ in } \mathcal{V}_i,\ u'_j \rightharpoonup u' \text{ in } \mathcal{V}_i^* \text{ and if } \langle A_{i1}(u_j,u_j) - A_{i1}(u_j,u), u_j - u\rangle \to 0,$$
$$\text{then } A^i_\alpha(x,t,\delta u_j, D^{m_i}u_j) \rightharpoonup A^i_\alpha(x,t,\delta u, D^{m_i}u) \text{ in} L^{q_i}(Q); \tag{2.133}$$

$$\text{“coercivity”}. \tag{2.134}$$

Remark 2.15. Similarly to [L69, Theorem 2.2.8] the sufficient conditions to get (2.132), (2.133) are:

$$\sum_{|\alpha|=m_i} A^i_\alpha(x,t,\eta,\xi)\xi_\alpha \frac{1}{|\xi| + |\xi|^{p_i-1}} \to +\infty \text{ as } |\xi| \to \infty$$

for almost each $x, t \in Q$ and $|\eta|$ bounded;

$$\sum_{|\alpha|=m_i} (A^i_\alpha(x,t,\eta,\xi) - A^i_\alpha(x,t,\eta,\xi^*))(\xi_\alpha - \xi^*_\alpha) > 0 \text{ as } \xi \neq \xi^*$$

for almost each $x, t \in Q$ and $\forall \eta$.

The next condition lets the coercivity:

$$\sum_{|\alpha|=m_i} A^i_\alpha(x,t,\eta,\xi)\xi_\alpha \geq c|\xi|^{p_i} \text{ for rather large } |\xi|.$$

A sufficient condition to get (2.130) (see [L69, p. 332]) is:

$$|A^i_\alpha(x,t,\eta,\xi)| \leq c[|\eta|^{p_i-1} + |\xi|^{p_i-1} + k(x,t)], \quad k \in L_{q_i}(Q). \tag{2.135}$$

By analogy with the proof of [L69, Theorem 3.2.1] and [L69, Proposition 2.2.6] we get the next

Proposition 2.9. *Let operator $A_i : \mathcal{V}_i \to \mathcal{V}_i^*$ $(i = 1, 2)$, defined in (2.131), satisfy (2.129), (2.130), (2.132), (2.133) and (2.134). Then A_i is pseudomonotone on W_i (in classical sense). Moreover it is bounded if (2.135) holds.*

Under the listed above conditions for each $f \in X^*$ there exists a generalized solution of problem (2.125)–(2.127) $y \in W$.

2.5 The Method of Finite Differences for Differential-Operator Inclusions in Banach Spaces

In the analysis of mathematical models of nonlinear processes and fields in physics, chemical kinetics, and geophysics, mathematical objects such as operator differential equations occur in problems of optimal control theory. In physics and mechanics, studying evolution equations and inclusions of second order was initiated by applied problems dealt with phase transitions and unilateral conductivity of boundaries of substances, propagation of electromagnetic, acoustic, vibro-, hydro-, and seismoacoustic waves, and quantum-mechanical effects. Studying equations that describe wave processes with "nonlinear friction" is rather difficult and needs a special technique [SY02]. Recent related studies cover quasilinear equations with homogeneous boundary conditions and linearized equations with nonlinear boundary conditions reduced to nonlinear differential-operator equations and inclusions. However, linearized objects not always adequately describe the process under study. Therefore, a need arises to consider evolution inclusions and variational inequalities with a much narrower set of properties. Recent developments on this subject are concerned with operator differential equations and inclusions with nonmonotonic nonlinearity globally bounded in the phase variable. We will develop the finite differences method for solutions of evolution inclusion with λ_0-pseudomonotonic mappings. The purpose of the study is to derive new theorems on the resolvability and substantiation of constructive methods for the approximation of such solutions. The results can be applied, for example, to analyze processes in the structural dynamics of a system of knowledge accumulated by large groups of people during purposeful learning [KZY09].

2.5.1 Setting of the Problem

Let Φ be a separable locally convex linear topological space, Φ^* be the topologically adjoint with Φ. By (f, ξ) we denote the canonical pairing of $f \in \Phi^*$ and $\xi \in \Phi$.

Let the spaces $\mathscr{V}$, $\mathscr{H}$ and $\mathscr{V}^*$ be given. Moreover

$$\Phi \subset \mathscr{V} \subset \Phi^*, \qquad \Phi \subset \mathscr{H} \subset \Phi^*, \qquad \Phi \subset \mathscr{V}^* \subset \Phi^*, \tag{2.136}$$

with continuous and dense embeddings. We assume that $\mathscr{H}$ is a Hilbert space with the scalar product $(h_1, h_2)_{\mathscr{H}}$ and norm $\|h\|_{\mathscr{H}}$, $\mathscr{V}$ is a reflexive separable Banach space with norm $\|v\|_{\mathscr{V}}$, $\mathscr{V}^*$ is the adjoint with $\mathscr{V}$ with the norm $\|f\|_{\mathscr{V}^*}$ associated with the bilinear form $(\cdot,\cdot)_{\mathscr{H}}$.

If $\xi, \psi \in \Phi$, then $(\xi, \psi) = (\xi, \psi)_{\mathscr{H}}$, i.e., it coincides with the scalar product in $\mathscr{H}$.

Let $\mathscr{V} = \mathscr{V}_1 \cap \mathscr{V}_2$ and $\|\cdot\|_{\mathscr{V}} = \|\cdot\|_{\mathscr{V}_1^*} + \|\cdot\|_{\mathscr{V}_2^*}$, where $(\mathscr{V}_i, \|\cdot\|_{\mathscr{V}_i})$, $i = \overline{1,2}$, are reflexive separable Banach spaces and the embeddings $\Phi \subset \mathscr{V}_i \subset \Phi^*$ and $\Phi \subset \mathscr{V}_i^* \subset \Phi^*$ are dense and continuous. The spaces $(\mathscr{V}_i^*, \|\cdot\|_{\mathscr{V}_i^*})$, $i = \overline{1,2}$, are the topologically adjoint with $(\mathscr{V}_i, \|\cdot\|_{\mathscr{V}_i})$. Then $\mathscr{V}^* = \mathscr{V}_1^* + \mathscr{V}_2^*$.

Let $\mathscr{A} : \mathscr{V}_1 \rightrightarrows \mathscr{V}_1^*$, $\mathscr{B} : \mathscr{V}_2 \rightrightarrows \mathscr{V}_2^*$ be a multivalued maps with nonempty convex closed bounded values, $\Lambda : \mathscr{V} \to \mathscr{V}^*$ be an unbounded operator with domain $D(\Lambda; \mathscr{V}, \mathscr{V}^*)$. We consider the following problem

$$u \in D(\Lambda; \mathscr{V}, \mathscr{V}^*), \tag{2.137}$$

$$\Lambda u + \mathscr{A}(u) + \mathscr{B}(u) \ni f, \tag{2.138}$$

where $f \in V^*$ is a fixed element.

Our aim consists in proving the existence of solutions by the method of finite differences (see [L69, Chap. 2.7]).

2.5.2 Preliminary Results

Let us assume that the space Φ is dense in $(\mathscr{V} \cap \mathscr{V}^*, \|v\|_{\mathscr{V}} + \|v\|_{\mathscr{V}^*})$. From this assumption it follows that

$$\mathscr{V} \cap \mathscr{V}^* \subset \mathscr{H}. \tag{2.139}$$

In fact, if $\xi \in \Phi$, then $\|v\|_{\mathscr{H}}^2 \le \|v\|_{\mathscr{V}^*}\|v\|_{\mathscr{V}}$, so that (2.139) follows.

Remark 2.16. If $\mathscr{V} \subset \mathscr{H}$, it is possible to avoid the space Φ. Identifying $\mathscr{H}$ and $\mathscr{H}^*$ we obtain the following embeddings:

$$\mathscr{V} \subset \mathscr{H} \subset \mathscr{V}^*. \tag{2.140}$$

Definition 2.7. The family of maps $\{G(s)\}_{s\ge 0}$ is said to be a semigroup of class C_0 in a Banach space X if $G(s) \in \mathscr{L}(X; X)$, for any $s \ge 0$, $G(0) = Id$, $G(s+t) = G(s) \circ G(t)$, $\forall s, t \ge 0$, and $G(t)x \to x$ as $t \to 0^+$, for all $x \in X$.

Let the family of maps $\{G(s)\}_{s\ge 0}$ be a semigroup of class C_0 on $\mathscr{V}, \mathscr{H}, \mathscr{V}^*$, that is, there are three semigroups, defined in the spaces $\mathscr{V}$, $\mathscr{H}$, and $\mathscr{V}^*$, respectively, which coincide on Φ. Each of them will be denoted by $\{G(s)\}_{s\ge 0}$. Moreover, we assume the following:

$$\begin{aligned} &\{G(s)\}_{s\ge 0} \text{ is a nonexpansive semigroup on } \mathscr{H}, \\ &\text{i.e. } \|G(s)\|_{\mathscr{L}(\mathscr{H};\mathscr{H})} \le 1, \ \forall s \ge 0. \end{aligned} \tag{2.141}$$

Further let $-\Lambda$ be the infinitesimal generator of the semigroup $\{G(s)\}_{s\ge 0}$ with $D(\Lambda; \mathscr{V})$ (resp. $D(\Lambda; \mathscr{H})$ or $D(\Lambda; \mathscr{V}^*)$) in $\mathscr{V}$ (resp. $\mathscr{H}$ or $\mathscr{V}^*$). It is well known [R73] that such generator exists. Moreover, it is a densely defined closed linear operator in the space $\mathscr{V}$ (resp. in $\mathscr{H}$ or $\mathscr{V}^*$).

Let $\{G^*(s)\}_{s\geq 0}$ be the semigroup adjoint with $G(s)$. Let $-\Lambda^*$ be the infinitesimal generator of the semigroup $\{G^*(s)\}_{s\geq 0}$ with domain $D(\Lambda^*;\mathscr{V})$ in $\mathscr{V}$, $D(\Lambda^*;\mathscr{H})$ in $\mathscr{H}$ and $D(\Lambda^*;\mathscr{V}^*)$ in $\mathscr{V}^*$. The operator Λ^* in $\mathscr{H}$ (resp. in $\mathscr{V}2$ or $\mathscr{V}^*$) is adjoint with the operator Λ in $\mathscr{H}$ (resp. in $\mathscr{V}$ or $\mathscr{V}^*$).

Lemma 2.9. *The sets $D(\Lambda;\mathscr{V}^*)\cap\mathscr{V}$ and $D(\Lambda^*;\mathscr{V}^*)\cap\mathscr{V}$ are dense in $\mathscr{V}$.*

Proof. In fact, for any $u\in\mathscr{V}$ and $\varepsilon>0$ there exists $\xi\in\Phi$ such that $\|u-\xi\|_{\mathscr{V}}<\varepsilon$. Then $\xi_n:=(I-\frac{1}{n}\Lambda)^{-1}\xi\in D(\Lambda;\mathscr{V}^*)\cap\mathscr{V}$, $\xi_n\to\xi$ in $\mathscr{V}$ as $n\to\infty$. □

Now we shall define Λ as an unbounded operator, which operates from $\mathscr{V}$ in $\mathscr{V}^*$ with domain $D(\Lambda;\mathscr{V},\mathscr{V}^*)$. Let us put

$$D(\Lambda;\mathscr{V},\mathscr{V}^*)=\left\{v\in\mathscr{V}\;\middle|\;\begin{array}{l}\text{the form } w\to(v,\Lambda^*w)\\ \text{is continuous in } D(\Lambda^*;\mathscr{V}^*)\cap\mathscr{V}\\ \text{with respect to the topology}\\ \text{induced by the space } \mathscr{V}.\end{array}\right\} \tag{2.142}$$

Then there is a unique element $\xi_v\in\mathscr{V}^*$ such that $\langle v,\Lambda^*w\rangle_{\mathscr{V}}=\langle\xi_v,w\rangle_{\mathscr{V}}$. If $v\in D(\Lambda;\mathscr{V}^*)\cap\mathscr{V}$, then $\xi_v=\Lambda v$. Thus, generally speaking we can put $\xi_v=\Lambda v$, whence

$$\langle v,\Lambda^*w\rangle_{\mathscr{V}}=\langle\Lambda v,w\rangle_{\mathscr{V}},\ \forall w\in D(\Lambda^*;\mathscr{V}^*)\cap\mathscr{V}. \tag{2.143}$$

Defining on $D(\Lambda;\mathscr{V},\mathscr{V}^*)$ the norm $\|v\|_{\mathscr{V}}+\|\Lambda v\|_{\mathscr{V}^*}$, we obtain Banach space. We define the space $D(\Lambda^*;\mathscr{V},\mathscr{V}^*)$ in a similar way.

Remark 2.17. If $\mathscr{V}\subset\mathscr{H}$, then

$$D(\Lambda;\mathscr{V},\mathscr{V}^*)=\mathscr{V}\cap D(\Lambda;\mathscr{V}^*)\quad\text{and}\quad D(\Lambda^*;\mathscr{V},\mathscr{V}^*)=\mathscr{V}\cap D(\Lambda^*;\mathscr{V}^*).$$

In the case when $\mathscr{V}$ is not contained in $\mathscr{H}$, we shall assume that

$$\begin{aligned}\mathscr{V}\cap D(\Lambda;\mathscr{V}^*)&\text{ is dense in } D(\Lambda;\mathscr{V},\mathscr{V}^*),\\ \mathscr{V}\cap D(\Lambda^*;\mathscr{V}^*)&\text{ is dense in } D(\Lambda^*;\mathscr{V},\mathscr{V}^*).\end{aligned} \tag{2.144}$$

Remark 2.18. It is known [L69, Chap. 2] that

$$\langle\Lambda v,v\rangle_{\mathscr{V}}\geq 0,\ \forall v\in D(\Lambda;\mathscr{V},\mathscr{V}^*),\ \ \langle\Lambda^*v,v\rangle_{\mathscr{V}}\geq 0,\ \forall v\in D(\Lambda^*;\mathscr{V},\mathscr{V}^*).$$

2.5.3 Method of Finite Differences

The natural approximation of inclusion (2.138) is the inclusion

$$\frac{I-G(h)}{h}u_h+\mathscr{A}(u_h)+\mathscr{B}(u_h)\ni f\quad(h>0). \tag{2.145}$$

However, if $\mathscr{V}$ is not contained in $\mathscr{H}$, then (2.145), generally speaking, has not solutions, and it is necessary to modify the given inclusion in an appropriate way. We shall choose a sequence $\theta_h \in (0, 1)$ such that

$$\frac{1-\theta_h}{h} \to 0 \text{ as } h \to 0. \tag{2.146}$$

We put $\theta_h = 1$ if $\mathscr{V} \subset \mathscr{H}$. Further, we define

$$\Lambda_h = \frac{I - \theta_h G(h)}{h}$$

and replace (2.145) by the inclusion

$$\Lambda_h u_h + \mathscr{A}(u_h) + \mathscr{B}(u_h) \ni f. \tag{2.147}$$

Definition 2.8. We will say, that the solution u of (2.137)–(2.138) turns out by *finite differences method*, if u is the weak limit of a subsequence $\{u_{h_{n_k}}\}_{k\geq 1}$ from $\{u_{h_n}\}_{n\geq 1}$ ($h_n \searrow 0+$ as $n \to \infty$) in $\mathscr{V}$, where for each $n \geq 1$ $u_{h_n} \in \mathscr{V}$ is a solution of problem (2.147).

2.5.4 Main Result

Let us validate the finite differences method for the class of differential-operator inclusions with $+$-coercive λ_0-pseudo-monotone maps. This method is the important method for the numerical investigation of such problems.

Theorem 2.6. *Assume the following conditions:*

1. $\mathscr{A} : \mathscr{V}_1 \to C_v(\mathscr{V}_1^*)$ *is a bounded, λ-pseudomonotone on $\mathscr{V}_1$ operator, which satisfies the $+$-coerciveness condition on $\mathscr{V}_1$;*
2. $\mathscr{B} : \mathscr{V}_2 \to C_v(\mathscr{V}_2^*)$ *is λ_0-pseudomonotone on $\mathscr{V}_2$ operator, which satisfies the $+$-coerciveness condition on $\mathscr{V}_2$ and Condition (Π)*
3. *The operator Λ satisfies all the conditions given in (2.141)–(2.144).*

Then for any $f \in \mathscr{V}^*$ there exists $u \in \mathscr{V}$ satisfying (2.137)–(2.138).

Remark 2.19. If $\mathscr{V} \subset \mathscr{H}$, inclusion (2.137) implies that $u \in \mathscr{V} \cap D(\Lambda; \mathscr{V}^*)$.

Proof. Let us use the coercitivity condition. From Lemma 1 it follows that $\mathscr{A} + \mathscr{B} : \mathscr{V} \rightrightarrows \mathscr{V}^*$ is $+$-coercive on $\mathscr{V}$. Let us define $\gamma : \mathbb{R}_+ \mapsto \mathbb{R}$ as

$$\gamma(r) = \inf_{\|y\|_X = r} \|y\|_{\mathscr{V}}^{-1} \cdot [\mathscr{A}(y) + \mathscr{B}(y), y]_+ \quad \forall r \geq 0,$$

at that,

$$\gamma(r) \longrightarrow +\infty \quad \text{as} \quad r \longrightarrow +\infty,$$

and for each $y \in \mathscr{V}$

$$[\mathscr{A}(y) + \mathscr{B}(y) - f, y]_+ \geq (\gamma(\|y\|_{\mathscr{V}}) - \|f\|_{\mathscr{V}^*})\|y\|_{\mathscr{V}}.$$

Hence, there exists $R > 0$ such that for all $u \in \mathscr{V}$ satisfying $\|u\|_{\mathscr{V}} = R$ we get

$$[\mathscr{A}(y) + \mathscr{B}(y) - f, y]_+ \geq 0. \tag{2.148}$$

Lemma 2.10. *Inclusion (2.147) has a solution $u_h \in \mathscr{V} \bigcap \mathscr{H}$ such that $\|u_h\|_{\mathscr{V}} \leq R$.*

Proof. Let us consider the map

$$\mathscr{D}_h = \Lambda_h + \mathscr{A} : \mathscr{H} \cap \mathscr{V}_1 \to C_v(\mathscr{H} + \mathscr{V}_1^*),$$

and also the following inclusion:

$$\mathscr{D}_h(u_h) + \mathscr{B}(u_h) \ni f.$$

The existence of a solution $u_h \in \mathscr{V} \bigcap \mathscr{H}$ of this inclusion such that $\|u_h\|_{\mathscr{V}} \leq R$ follows from Theorem 2.2 with $V = \mathscr{H} \cap \mathscr{V}_1$, $W = \mathscr{V}_2$, $A = D_h$, $B = \mathscr{B}$, $L \equiv \bar{0}$, $D(L) = V$, $f = f$, $R = R$, and the following Lemma.

Lemma 2.11. *The operator D_h satisfies the following conditions:*

$$[\mathscr{D}_h(u) + \mathscr{B}(u), u]_+ \geq 0 \quad \forall u \in \mathscr{V} \text{ such that } \|u\|_{\mathscr{V}} = R; \tag{2.149}$$

$$\mathscr{D}_h \text{ is } \lambda\text{-pseudomonotone on } \mathscr{H} \cap \mathscr{V}_1; \tag{2.150}$$

$$\mathscr{D}_h \text{ is bounded on } \mathscr{H} \cap \mathscr{V}_1. \tag{2.151}$$

Proof. As $G(s)$ is nonexpansive on $\mathscr{H}$, it follows that for any $v \in \mathscr{H}$,

$$\begin{aligned}(\Lambda_h v, v)_{\mathscr{H}} &= \frac{1}{h}\Big(v - \theta_h G(h)v, v\Big) \\ &\geq \frac{1}{h}\Big(\|v\|^2_{\mathscr{H}} - \theta_h \|G(h)v\|_{\mathscr{H}} \|v\|_{\mathscr{H}}\Big) \geq \frac{1-\theta_h}{h}\|v\|^2_{\mathscr{H}}.\end{aligned} \tag{2.152}$$

From here it follows the +-coercitivity for Λ_h on $\mathscr{H}$.

Using (2.148), (2.152) and Proposition 1, we get (2.149).

For (2.151) note that the boundness of $\mathscr{D}_h$ on $\mathscr{H} \cap \mathscr{V}_1$ follows from the boundness of Λ_h on $\mathscr{H}$ and the boundness of $\mathscr{A}$ on $\mathscr{V}_1$. The boundness of Λ_h on $\mathscr{H}$ follows immediately from the definition of Λ_h and estimate (2.141). Hence, it follows also that $\mathscr{D}_h$ satisfies the property $(\kappa)_+$.

Finally let us prove the λ-pseudomonotonicity of D_h on $\mathscr{H}\cap\mathscr{V}_1$. For this purpose Lemma 1.15 with $A = \Lambda_h$ on $\mathscr{V} = \mathscr{H}$ and $B = \mathscr{A}$ on $W = \mathscr{V}_1$ is used. From here, since $\mathscr{A}$ is λ-pseudomonotone and has bounded values on $\mathscr{V}_1$, it is enough to prove the λ-pseudomonotonicity of Λ_h on $\mathscr{H}$. Let us prove it. Indeed, let

$$y_n \rightharpoonup y \text{ in } \mathscr{H}, \quad \overline{\lim_{n\to\infty}}\,(\Lambda_h y_n, y_n - y)_{\mathscr{H}} \le 0.$$

Then, from estimate (2.152), we have

$$\underline{\lim_{n\to\infty}}\,(\Lambda_h y_n, y_n-y)_{\mathscr{H}} \ge \underline{\lim_{n\to\infty}}\,(\Lambda_h y_n-\Lambda_h y, y_n-y)_{\mathscr{H}}+\underline{\lim_{n\to\infty}}\,(\Lambda_h y, y_n-y)_{\mathscr{H}} \ge 0.$$

Hence, $\lim_{n\to\infty}(\Lambda_h y_n, y_n - y)_{\mathscr{H}} = 0$. Further, for any $u \in \mathscr{H}$, $s > 0$ let

$$w := y + s(u - y).$$

Then from

$$(\Lambda_h w - \Lambda_h y_n, y_n - y)_{\mathscr{H}} = (\Lambda_h w - \Lambda_h y_n, y_n - w)_{\mathscr{H}} + (\Lambda_h w - \Lambda_h y_n, w - y)_{\mathscr{H}} \le s(\Lambda_h w - \Lambda_h y_n, u - y)_{\mathscr{H}}$$

we have

$$s(\Lambda_h y_n, y - u)_{\mathscr{H}} \ge -(\Lambda_h y_n, y_n - y)_{\mathscr{H}} + (\Lambda_h w, y_n - y)_{\mathscr{H}} - s(\Lambda_h w, u - y)_{\mathscr{H}}, \quad \forall n \ge 1,$$

and

$$s\,\underline{\lim_{n\to\infty}}\,(\Lambda_h y_n, y - u)_{\mathscr{H}} \ge -s(\Lambda_h w, u - y)_{\mathscr{H}} \Leftrightarrow \underline{\lim_{n\to\infty}}\,(\Lambda_h y_n, y - u)_{\mathscr{H}} \ge -(\Lambda_h w, u - y)_{\mathscr{H}}.$$

Let $s \to 0^+$. Then

$$\underline{\lim_{n\to\infty}}\,(\Lambda_h y_n, y - u)_{\mathscr{H}} \ge -(\Lambda_h y, u - y)_{\mathscr{H}} = (\Lambda_h y, y - u)_{\mathscr{H}}$$

and

$$\underline{\lim_{n\to\infty}}\,(\Lambda_h y_n, y_n - u)_{\mathscr{H}} \ge \underline{\lim_{n\to\infty}}\,(\Lambda_h y_n, y_n - y)_{\mathscr{H}} + \underline{\lim_{n\to\infty}}\,(\Lambda_h y_n, y - u)_{\mathscr{H}} \ge (\Lambda_h y, y - u)_{\mathscr{H}}, \quad \forall u \in \mathscr{H}.$$

Thus we have the required statement. Lemma 2.11 is proved. □

Now Lemma 2.10 is also proved. □

Now we continue the proof of Theorem 2.6. We shall pass to the limit as $h \to 0^+$. From Lemma 2.10 for arbitrary $h > 0$ it follows the existence of $u_h \in \mathscr{H} \cap \mathscr{V}$, $d_h' \in A(u_h)$ and $d_h'' \in B(u_h)$ such that

$$\Lambda_h u_h + d_h' + d_h'' = f \tag{2.153}$$

and

$$\|u_h\|_{\mathscr{V}} \le R, \quad \text{for any } h > 0. \tag{2.154}$$

From estimate (2.154) and the boundness of the operator A on $\mathscr{V}_1$ it follows that

$$\mathscr{A}(u_h) \text{ is bounded in } \mathscr{V}_1^* \text{ as } h \to 0. \tag{2.155}$$

Let us prove that

$$d_h'' \text{ are bounded in } \mathscr{V}_2^* \text{ as } h \to 0. \tag{2.156}$$

First, from (2.152), (2.153), estimate (2.154), the boundness of the operator $\mathscr{A}$ and Proposition 1 we obtain that for any $\{h_n\} \subset (0, +\infty)$ such that $h_n \to 0$, as $n \to \infty$, we have

$$\begin{aligned}\sup_n \langle d_{h_n}'', u_{h_n}\rangle_{\mathscr{V}_2} &\le \sup_n \langle f, u_{h_n}\rangle_{\mathscr{V}_2} + \sup_n \langle -d_{h_n}', u_{h_n}\rangle_{\mathscr{V}_2} + \sup_n \langle -\Lambda_{h_n} u_{h_n}, u_{h_n}\rangle_{\mathscr{V}_2} \\ &\le \|f\|_{\mathscr{V}}' \sup_n \|u_{h_n}\|_{\mathscr{V}} + \sup_n \|u_{h_n}\|_{\mathscr{V}} \sup_n \|\mathscr{A}(u_{h_n})\|_+ < +\infty.\end{aligned}$$

Hence, due to Condition (Π) for $\mathscr{B}$ estimate (2.156) follows.

From equality (2.153), estimates (2.154)–(2.156), using the Banach–Alaoglu Theorem, we obtain the existence of subsequences $\{u_{h_n}\}_{n\ge1} \subset \{u_h\}_{h>0}$, $\{d_{h_n}'\}_{n\ge1} \subset \{d_h'\}_{h>0}$, $\{d_{h_n}''\}_{n\ge1} \subset \{d_h''\}_{h>0}$, $(0 < h_n \to 0)$, denoted again by $\{u_h\}_{h>0}$, $\{d_h'\}_{h>0}$, $\{d_h''\}_{h>0}$, and $u \in \mathscr{V}$, $d' \in \mathscr{V}_1$, $d'' \in \mathscr{V}_2$, such that

$$u_h \rightharpoonup u \text{ in } \mathscr{V}, \quad d_h' \rightharpoonup d' \text{ in } \mathscr{V}_1^*, \quad d_h'' \rightharpoonup d'' \text{ in } \mathscr{V}_2^*, \quad \Lambda_h u_h \rightharpoonup \Lambda u \text{ in } \mathscr{V}^*.$$

From here, in particular, it follows that

$$v_h := d_h' + d_h'' \rightharpoonup d' + d'' =: w \text{ in } \mathscr{V}^*. \tag{2.157}$$

Let us introduce the following map: $\mathscr{C}(v) = \mathscr{A}(v) + \mathscr{B}(v) : \mathscr{V} \to C_v(\mathscr{V}^*)$. We shall prove that this map satisfies Property (M). For this it is enough to show the λ-pseudomonotonicity of $\mathscr{C}$ on $\mathscr{V}$. Indeed, if $\mathscr{C}$ is λ-pseudomonotone on $\mathscr{V}$ and $\{y_n\}_{n\ge0} \subset \mathscr{V}$, $d_n \in C(y_n)$, $\forall n \ge 1$, be such that

$$y_n \rightharpoonup y_0 \text{ in } \mathscr{V}, \quad d_n \rightharpoonup d_0 \text{ in } \mathscr{V}^* \quad \text{and} \quad \varlimsup_{n\to\infty} \langle d_n, y_n\rangle_{\mathscr{V}} \le \langle d_0, y_0\rangle_{\mathscr{V}},$$

then

$$\overline{\lim_{n\to\infty}} \langle d_n, y_n - y_0\rangle_{\mathscr{V}} \le \overline{\lim_{n\to\infty}} \langle d_n, y_n\rangle_{\mathscr{V}} + \overline{\lim_{n\to\infty}} \langle d_n, -y_0\rangle_{\mathscr{V}}$$
$$\le \langle d_0, y_0\rangle_{\mathscr{V}} - \langle d_0, y_0\rangle_{\mathscr{V}} = 0.$$

Hence, thanks to the λ-pseudomonotocity of $\mathscr{C}$ it follows the existence of $\{y_{n_k}\}_{k\ge1} \subset \{y_n\}_{n\ge1}$, $\{d_{n_k}\}_{k\ge1} \subset \{d_n\}_{n\ge1}$, such that

$$\underline{\lim_{k\to\infty}} \left\langle d_{n_k}, y_{n_k} - w\right\rangle_{\mathscr{V}} \ge [\mathscr{C}(y_0), y_0 - w]_- = [\overline{\text{co}}^{*}\mathscr{C}(y_0), y_0 - w]_-, \, \forall w \in \mathscr{V}.$$

From here

$$[\mathscr{C}(y_0), y_0 - w]_- \le \underline{\lim_{k\to\infty}} \left\langle d_{n_k}, y_{n_k} - w\right\rangle_{\mathscr{V}} \le \overline{\lim_{n\to\infty}} \langle d_n, y_n - w\rangle_{\mathscr{V}}$$
$$\le \langle d_0, y_0 - w\rangle_{\mathscr{V}}, \;\forall w \in \mathscr{V}.$$

Hence, Proposition 1 and $C(y) \in C_v(\mathscr{V}^*)$ imply $d_0 \in \mathscr{C}(y_0)$. Thus, $\mathscr{C}$ satisfies Property (M) on $\mathscr{V}$. Further, since $\mathscr{A}$ is a λ-pseudomonotone operator with bounded values on $\mathscr{V}_1$ and $\mathscr{B}$ is λ-pseudomonotone on $\mathscr{V}_2$, Lemma 1.15 implies that $\mathscr{C}$ is λ-pseudomonotone.

We use the fact that $\mathscr{C}$ satisfies Property (M) on $\mathscr{V}$. Take v from $\mathscr{V}\bigcap D(\Lambda^*;\mathscr{V}^*)$. From (2.153) and (2.157) it follows that

$$\left\langle u_h, \Lambda_h^* v\right\rangle_{\mathscr{V}} + \langle v_h, v\rangle_{\mathscr{V}} = (f, v). \tag{2.158}$$

But

$$\Lambda_h^* v = \frac{I - G(h)^*}{h} v + \frac{1-\theta_h}{h} G(h)^* v,$$

and, by (2.146), $\Lambda_h^* v \to \Lambda^* v$ in $\mathscr{V}^*$. Consequently, passing to the limit in (2.158) as $h \to 0$ we shall obtain that

$$\langle u, \Lambda^* v\rangle_{\mathscr{V}} + \langle w, v\rangle_{\mathscr{V}} = \langle f, v\rangle_{\mathscr{V}}, \quad \forall v \in \mathscr{V} \cap D(\Lambda^*;\mathscr{V}^*),$$

and then by (2.141), (2.142) we get $u \in D(\Lambda, \mathscr{V}, \mathscr{V}^*)$ and

$$\Lambda u + w = f,$$

The proof of Theorem 2.6 will be finished if we can show that

$$w \in \mathscr{C}(u). \tag{2.159}$$

From (2.153) and (2.157) for $v \in \mathscr{V} \cap D(\Lambda; \mathscr{V}^*) \subset \mathscr{H}$ we have

$$\begin{aligned}\langle v_h, u_h - v\rangle_{\mathscr{V}} &= \langle f, u_h - v\rangle_{\mathscr{V}} - \langle \Lambda_h v, u_h - v\rangle_{\mathscr{V}} - \langle \Lambda_h(u_h - v), u_h - v\rangle_{\mathscr{V}} \\ &\le \langle f, u_h - v\rangle_{\mathscr{V}} - \langle \Lambda_h v, u_h - v\rangle_{\mathscr{V}},\end{aligned}$$

as $\Lambda_h \ge 0$ in $\mathscr{L}(\mathscr{H}, \mathscr{H})$. From here

$$\overline{\lim}\ \langle v_h, u_h\rangle_{\mathscr{V}} \le \langle w, v\rangle_{\mathscr{V}} + \langle f, u - v\rangle_{\mathscr{V}} - \langle \Lambda v, u - v\rangle_{\mathscr{V}}, \quad \forall v \in \mathscr{V} \cap D(\Lambda; \mathscr{V}^*).$$

But, by (2.144), the same inequality is fulfilled for any $v \in D(\Lambda; \mathscr{V}, \mathscr{V}^*)$, and, putting $v = u$, we get

$$\overline{\lim}\ \langle v_h, u_h\rangle_{\mathscr{V}} \le \langle w, u\rangle_{\mathscr{V}},$$

and (2.159) follows, as $\mathscr{C}$ is an operator of type (M).

Theorem 2.6 is proved. □

In this section we shall apply our main Theorem to some particular equations.

Example 2.5. Let $\Omega \subset \mathbb{R}^n$ be a bounded region with regular boundary $\partial\Omega$, $S = [0, T]$ be a finite time interval, $Q = \Omega \times (0; T)$, $\Gamma_T = \partial\Omega \times (0; T)$. The operator $\mathscr{A}$ is defined by $(Au)(t) = A(u(t))$, where

$$\mathscr{A}(u) = -\sum_{i=1}^{n} \frac{\partial}{\partial x_i}\left(\left|\frac{\partial u}{\partial x_i}\right|^{p-2} \frac{\partial u}{\partial x_i}\right) + |u|^{p-2} u \tag{2.160}$$

(see [L69, Chap. 2.9.5]). Let V be a closed subspace in the Sobolev space $W^{1,p}(\Omega)$, $p > 1$, such that

$$W_0^{1,p}(\Omega) \subset V \subset W^{1,p}(\Omega). \tag{2.161}$$

We define the space

$$H := L^2(\Omega)$$

and

$$\mathscr{V}_1 = L^p(0, T; V), \quad \mathscr{H} = L^2(0, T; H), \quad \mathscr{V}_2 = L^2(0, T; H).$$

The operator $\mathscr{A} : \mathscr{V}_1 \to \mathscr{V}_1^*$ is bounded, +-coercive and pseudomonotone (see [L69, Chap. 2]). Thus, it is also λ-pseudomonotone.

Let us consider a convex lower semicontinuous functional $\psi : \mathbb{R} \to \mathbb{R}$. Assume the existence of constants $M, C > 0$ such that

$$\psi(s) \ge M s^2 + C, \forall u, \tag{2.162}$$

and also that $\psi(u) \in L^1((0, T) \times \Omega)$, for all $u \in \mathscr{H}$. Denote by $\Phi : \mathbb{R} \rightrightarrows \mathbb{R}$ its subdifferential. It is well known [B76, p. 61] that $\varphi : \mathscr{V}_2 \to \mathbb{R}$ defined by

$$\varphi(u) = \int_0^T \int_\Omega \psi(u(x))\, dx$$

is a convex, lower semicontinuous function in $\mathscr{V}_2$. Moreover, $w \in \partial\varphi(u)$ if and only if $w(x) \in \Phi(u(x))$, a.e. on $(0,T) \times \Omega$, and $w_i \in \mathscr{V}_2$. It follows easily from (2.162) that $\mathscr{B} = \partial\varphi$ is $+$-coercive.

Putting $\mathscr{V} = \mathscr{V}_1 \cap \mathscr{V}_2$ (and then $\mathscr{V}^* = L^q(0,T;V^*) + L^2(0,T;L^2(\Omega)))$, where $\frac{1}{p} + \frac{1}{q} = 1$, we have that (2.140) holds if $p \geq 2$. For $1 < p < 2$ we can take $\Phi = D(0,T;V)$ (see [L69]).

In our case $\Lambda = \dfrac{dy}{dt}$ is the derivative in the sense of scalar distributions $\mathscr{D}^*(0,T;V^*)$ and

$$D(\Lambda;\mathscr{V},\mathscr{V}^*) := W = \{y \in \mathscr{V} \cap \mathscr{H} \,:\, y' \in \mathscr{H} + \mathscr{V}^*,\ y(0) = \bar{0}\},$$
$$G(s)u(t) := \begin{cases} u(t-s) & \text{for } t \geq s, \\ 0 & \text{for } t \leq s. \end{cases}$$

The map Λ satisfies conditions (2.141)–(2.144) [L69, Sect. 2.9].

Then all conditions of Theorem 2.6 are satisfied, so that the problem

$$\begin{aligned}
&\int_Q \frac{dy(x,t)}{dt}\Big(v(x,t) - y(x,t)\Big)dxdt \\
&\quad + \sum_{i=1}^n \int_Q \left|\frac{\partial y(x,t)}{\partial x_i}\right|^{p-2} \frac{\partial y(x,t)}{\partial x_i}\left(\frac{\partial v(x,t)}{\partial x_i} - \frac{\partial y(x,t)}{\partial x_i}\right)dxdt \\
&\quad + \int_Q |y|^{p-2} y(v-y)dxdt + \int_Q \psi(v(x,t))dxdt - \int_Q \psi(y(x,t))dxdt \\
&\quad \geq \int_Q f(v-y)dxdt + \int_{\Gamma_T} g(v-y)dxdt, \quad \forall v \in \mathscr{V}, \qquad (2.163)
\end{aligned}$$

$$y(x,0) = 0 \qquad \text{a.e. on } \Omega, \qquad (2.164)$$

has a solution $y \in W$, obtained by the method of difference approximations. Note that in (2.163)–(2.164) $f \in L^2(Q)$, $g \in L^2(\Gamma_T)$ are fixed elements.

Example 2.6. Let $n \geq 1$, $k \geq 1$, $A \subset \mathbb{R}^k$ be a nonempty compact set, $\Omega \subset \mathbb{R}^n$ be a bounded region with boundary $\partial\Omega$. Let us also consider a family maps $U_\alpha : \mathbb{R}^n \to \mathbb{R}$, where $\alpha \in A$, that satisfies the following conditions:

1. The map $\mathbb{R}^n \times A \ni (\xi,\alpha) \to U_\alpha(\xi) \in \mathbb{R}$ is continuous.
2. $\mathbb{R}^n \ni \xi \to U_\alpha(\xi) \in \mathbb{R}$ is convex for all $\alpha \in A$.
3. there exist $a > 0, b > 0$ such, that $\|\partial U_\alpha(\xi)\|_+ \leq a + b\|\xi\|$ $\forall \xi \in \mathbb{R}^n$, $\forall \alpha \in A$.

Together with $\{U_\alpha(\xi)\}_{\alpha\in A}$ let us consider the function $U(\xi) := \max\limits_{\alpha\in A} U_\alpha(\xi) : \mathbb{R}^n \to \mathbb{R}$ and the multivalued map with compact values $A(\xi) := \{\alpha \in A \mid U_\alpha(\xi) = U(\xi)\}$, where $x \in \Omega$, $\xi \in \mathbb{R}^n$. Assume also the next coercitivity condition:

4. there exist constants $M, C > 0$ such that

$$U(\xi) \geq M\|\xi\|^2 + C, \forall \xi.$$

Let us consider the following problem:

$$y'(t,x) - \sum_{i=1}^{n} \frac{\partial}{\partial x_i}\left(\left|\frac{\partial y(t,x)}{\partial x_i}\right|^{p-2} \frac{\partial y(t,x)}{\partial x_i}\right) - \sum_{i=1}^{n} \frac{\partial}{\partial x_i} \overline{\text{co}}\left(\bigcup_{\alpha \in A(\nabla y(t,x))} \partial U_\alpha(\nabla y(t,x))\right) \ni f(t,x), \tag{2.165}$$

$$y(t,x)\big|_{\partial\Omega} = 0, \tag{2.166}$$

$$y(0,x) = 0. \tag{2.167}$$

From [Ps80, Theorem II.3.14] and conditions (1) and (2) it follows that inclusion (2.165) is equivalent to

$$y'(t,x) - \sum_{i=1}^{n} \frac{\partial}{\partial x_i}\left(\left|\frac{\partial y(t,x)}{\partial x_i}\right|^{p-2} \frac{\partial y(t,x)}{\partial x_i}\right) - \sum_{i=1}^{n} \frac{\partial}{\partial x_i} \partial U(\nabla y(t,x)) \ni f(t,x). \tag{2.168}$$

Integrating by parts we obtain the next differential-operator inclusion

$$y' + \mathscr{A}(y) + L^*\partial\varphi(Ly) \ni f, \qquad y(0) = \bar{0}, \tag{2.169}$$

where

$$\begin{aligned}
\mathscr{A} &: L^p(0,T;H_0^1(\Omega)) \to L^q(0,T;H^{-1}(\Omega)),\\
L &: H_0^1(\Omega) \to L^{2,n}(\Omega)\ (Lv = \nabla v,\ \forall v \in H_0^1(\Omega)),\\
L^* &: L^{2,n}(\Omega) \to H^{-1}(\Omega)\ (L^*v = -\text{div}\, v,\ \forall v \in L^{2,n}(\Omega)),\\
\varphi &: L^2(0,T;L^{2,n}(\Omega)) \to \mathbb{R},\\
\varphi(y) &= \int_Q U(y(t,x))dtdx,\ \forall y \in L^2(0,T;L^{2,n}(\Omega))),\\
f &\in L^2(0,T;H^{-1}(\Omega)) + L^q(0,T;H^{-1}(\Omega)),
\end{aligned}$$

and $\mathscr{V}_1 = L^p(0,T;H_0^1(\Omega))$, $\mathscr{H} = L^2(0,T;L^2(\Omega))$, $\mathscr{V}_2 = L^2(0,T;H_0^1(\Omega))$.

Hence, inclusion (2.169) is equivalent to

$$y' + \mathscr{A}(y) + \partial(\varphi \circ L)(y) \ni f, \qquad y(0) = \bar{0}$$

(see [ET99]). So, in a similar way as in the previous example, from conditions (1)–(4) we obtain that problem (2.165)–(2.167) has a solution $y \in L^2(0, T; H_0^1(\Omega)) \cap L^p(0, T; H_0^1(\Omega))$.

2.6 On Solvability for the Second Order Evolution Inclusions with the Volterra Type Operators

The progress in the investigation of nonlinear boundary problems for the partial differential equations became possible thanks to the intensive development of the nonlinear analysis methods which found their application in the different sections of mathematics. It has recently become natural to reduce these problems to the study of nonlinear operator and differential-operator equations and inclusions in functional spaces. At such approach the results for the concrete systems are obtained as the rather simple consequences of operator Theorems [L69, GGZ74].

The evolution differential equations and inclusions are studied rather strongly. To prove the properties of the resolving operator (nonemptyness, compactness, connectedness) the method of monotony, the method of compactness and their combinations are often used.

Here we are studing the solvability problem for the evolution inclusion of the sort

$$y'' + A(y') + B(y) \ni f.$$

with multivalued noncoercive maps, that is important for the applications.

The latest investigations, concern this direction enveloped the class of problems with strongly monotone operator A and multivalued operator B that can be presented as the sum of the single-valued linear self-adjoint monotone operator and the multivalued demiclosed bounded operator. These problems are coercive. They were considered for example by N.S. Papageorgiou, N. Yannakakis [Pap94, PY06]. More partial cases of evolution inclusions were considered by N.U. Ahmed., S. Kerbal [AK03], L. Gasinski, M. Smolka [GS02], A. Kartsatos, L. Markov [KM01], S. Migorski [M95] and others.

Our goal is to extend the given approach for wider class of problems, namely for problems with the multivalued noncoercive nonmonotone operator A and the multivalued operator B, that satisfies the similar conditions.

The idea on passing to the subsequences in the classical definition of single-valued pseudomonotone operator was presented by I.V. Scrypnik [S94]. It was developed for the first order differential-operator equations and inclusions in infinite dimensional spaces with $+$-coercive W_{λ_0}-pseudomonotone maps by V.S. Melnik, M.Z. Zgurovskiy, A.N. Novikov [ZMN04, ZM04] and P.O. Kasyanov [KK03a]–[KMY07]. This gave the possibility to investigate the substantially wider class of applied problems. Particularly, the given methodology combined with the noncoercive theory [GGZ74, ZMN04, KMY07], that we try to apply to the second order evolution inclusions, allows sufficiently extend the class of "noncoercive," "nonmonotone" problems with sufficiently multivalued maps, for which we can obtain the solvability. In virtue of the operators are multivalued, the extending faced with the principle difficulties, which are not typical for the differential-operator equations. Here the proof of the solvability is based on the method of singular perturbations [L69, KMY07]. This fact allows us to obtain important a priori estimations for the solutions. It gives the possibility to study the properties for the obtained solutions (for example, dynamics). As the example illustrating offered approach we consider the class of problems with nonlinear operators. The obtained results are new for the inclusions as well as for the equations.

We remark that such type second order evolution inclusions arises during the studying evolution problems which describe the dynamic contact of a viscoelastic body and a foundation (see for example [DM05, DMP03] and citations there). The contact is modeled by a general normal compliance condition and a friction law which are nonmonotone, possibly multivalued and of the subdifferential form while the damping operator is assumed to be coercive and generalized pseudomonotone. They derive a formulation of the model in the form of a multidimensional second order evolution inclusion. Then they establish the a priori estimates and the existence of weak solutions by using an abstract surjectivity result. Their approach has some disadvantages. At first, they claim rather strong uniformly —-coercivity condition for damping. The other summands complied to the nonlinear damping. At the second, they approach is not constructive. Here we try to develop the constructive scheme for investigation of such problem with nonlinear noncoercive (even not +-coercive) damping operator.

2.6.1 *Setting of the Problem*

Let H be the real Hilbert space with the inner product $(\cdot,\cdot)$, V_1 and V_2 are some real reflexive separable Banach spaces that continuously embedded in H and

$$V := V_1 \cap V_2 \text{ is dense in spaces } V_1,\ V_2 \text{ and } H.$$

We consider that one of the embeddings $V_i \subset H$ is compact. Let us note, that we identify topologically adjoint space with H (with regard to the bilinear form $(\cdot,\cdot)$) with H. Then we obtain

$$V_1 \subset H \subset V_1^*, \qquad V_2 \subset H \subset V_2^*$$

with continuous and dense embeddings, where $(V_i^*, \|\cdot\|_{V_i^*})$ is topologically adjoint space with V_i with regard to the canonical bilinear form

$$\langle\cdot,\cdot\rangle_{V_i} : V_i^* \times V_i \to \mathbb{R}$$

$(i = 1, 2)$, that coincides on $H \times V$ with the inner product $(\cdot, \cdot)$ in H. Let us consider reflexive functional spaces

$$X_i = L_{r_i}(S; H) \cap L_{p_i}(S; V_i), \quad Y = L_2(S; H),$$

where $S = [0, T]$, $1 < p_i \le r_i < +\infty$, $i = 1, 2$, $\max\{r_1; r_2\} \ge 2$.

Let us consider the reflexive Banach space $X = X_1 \cap X_2$ with norm $\|y\|_X = \|y\|_{X_1} + \|y\|_{X_2}$. Let us note, that the space X that continuously and densely embedded in Y, that is the norm $\|\cdot\|_Y$ is continuous with regard to $\|\cdot\|_X$ on X.

Let us identify spaces $L_{q_i}(S; V_i^*) + L_{r_i'}(S; H)$ and X_i^*. Similarly,

$$X^* = X_1^* + X_2^* \equiv L_{q_1}(S; V_1^*) + L_{q_2}(S; V_2^*) + L_{r_1'}(S; H) + L_{r_2'}(S; H), \quad Y^* \equiv Y,$$

where $r_i^{-1} + r_i'^{-1} = p_i^{-1} + q_i^{-1} = 1$.

Let us determine on $X^* \times X$ the form of duality:

$$\langle f, y \rangle = \int_S (f_{11}(\tau), y(\tau))_H \, d\tau + \int_S (f_{12}(\tau), y(\tau))_H \, d\tau + \int_S \langle f_{21}(\tau), y(\tau) \rangle_{V_1} \, d\tau$$
$$+ \int_S \langle f_{22}(\tau), y(\tau) \rangle_{V_2} \, d\tau = \int_S (f(\tau), y(\tau)) \, d\tau,$$

where $f = f_{11} + f_{12} + f_{21} + f_{22}$, $f_{1i} \in L_{r_i'}(S; H)$, $f_{2i} \in L_{q_i}(S; V_i^*)$. Let us consider

$$W = \{y \in X \mid y' \in X^*\},$$

where the derivative y' of the element $y \in X$ is considered in the sense of scalar distributions spaces $\mathscr{D}^*(S; V^*) = \mathscr{L}(\mathscr{D}(S); V_w^*)$, with $V = V_1 \cap V_2$, V_w^* is the space V^* with topology $\sigma(V^*, V)$. Let us introduce the graph norm on W:

$$\|y\|_W = \|y\|_X + \|y'\|_{X^*} \text{ for any } y \in W.$$

Let us note that the reflexive Banach space W is compactly embedded in Y, at that the norm $\|\cdot\|_Y$ is compact with regard to $\|\cdot\|_W$ on W.

Let $A, B : X \rightrightarrows X^*$ are the multivalued maps of the pseudomonotone type. Let us consider the Cauchy problem about the solvability for differential-operator inclusions with noncoercive multivalued maps of W_{λ_0}-pseudomonotone type.

$$\begin{cases} y'' + Ay' + By \ni f, \\ y(0) = a_0, \; y'(0) = \bar{0}, \; y \in C(S; V), \; y' \in W, \end{cases} \tag{2.170}$$

where $a_0 \in V$ and $f \in X^*$ are arbitrary fixed elements.

We consider more precise conditions on operators A and B in Theorem 2.7 and Corollary 2.6.

Remark 2.20. It is obviously, that the space W is continuously embedded in $C(S; V^*)$. So, initial conditions make sense.

Further we will suggest that the next condition is true for $A : X \rightrightarrows X^*$:

Definition 2.9. Multivalued map $A : X \rightrightarrows X^*$ satisfies *Condition* (H), if for any $y \in X$, $n \geq 1$, $\{d_i\}_{i=1}^n \subset A(y)$ and $E_j \subset S$, $j = \overline{1,n}$: $\forall j = \overline{1,n}$ E_j is measurable, $\cup_{j=1}^n E_j = S$, $E_i \cap E_j = \emptyset\ \forall i \neq j, i, j = \overline{1,n}$ element $d \in \overset{*}{\overline{\text{co}}}\, A(y)$, where

$$d(\tau) = \sum_{j=1}^{n} d_j(\tau)\chi_{E_j}(\tau), and \chi_{E_j}(\tau) = \begin{cases} 1, & \tau \in E_j, \\ 0, & \text{else.} \end{cases}$$

Let us consider multivalued duality map

$$J(y) = \{\xi \in X^* \mid \langle \xi, y\rangle_X = \|\xi\|_{X^*}^2 = \|y\|_X^2\} \in C_v(X^*) \quad \forall y \in X.$$

In consequence of [AE84, Theorem 4, p. 202 and Statement 8, p. 203] for any $f \in X^*$ map

$$\begin{aligned} J^{-1}(f) &= \{y \in X \mid f \in J(y)\} \\ &= \{y \in X \mid \langle f, y\rangle_X = \|f\|_{X^*}^2 = \|y\|_X^2\} \in C_v(X). \end{aligned}$$

is also defined on the whole space X and it is maximal monotone multivalued map.

We will approximate the inclusion from (2.170) by the next:

$$-\varepsilon \frac{d}{dt} J^{-1}\left(\frac{d}{dt}\left(\frac{d}{dt} y_\varepsilon\right)_\lambda\right) + \frac{d^2}{dt^2} y_\varepsilon + A\left(\frac{d}{dt} y_\varepsilon\right) + B(y_\varepsilon) \ni f. \qquad (2.171)$$

Definition 2.10. We will say,that $y \in X$ with $\frac{d}{dt} y \in W$ is the solution of problem (2.170) is obtained *by the method of singular perturbations*, if $\left\{y, \frac{d}{dt} y\right\}$ – the weak limit of some subsequence $\left\{y_{\varepsilon_{n_k}}, \frac{d}{dt} y_{\varepsilon_{n_k}}\right\}_{k \geq 1}$ of some sequence $\left\{y_{\varepsilon_n}, \frac{d}{dt} y_{\varepsilon_n}\right\}_{n \geq 1}$ ($\varepsilon_n \searrow 0+$ as $n \to \infty$) in the space $X \times W$, that for every $n \geq 1$ $y_{\varepsilon_n} \in X$ with $\frac{d}{dt} y_{\varepsilon_n} \in W$ is the solution of problem (2.171).

2.6.2 Results

Theorem 2.7. *Let* $\lambda_A \geq 0$ *is fixed,* $I : X \to X^*$ *is the identical motion,* $p_0 = \min\{p_1, p_2\}$, *the space* V *is compactly embedded in Banach space* V_0 *and the embedding* $V_0 \subset V^*$ *is continuous. Let us suggest, that* $A + \lambda_A I : X \to C_v(X^*)$ *is* $+$*-coercive, r.l.s.c. multivalued operator of the Volterra type with* $(X; W)$-s.c.v.

with $\|\cdot\|'_W = \|\cdot\|_{L_{p_0}(S;V_0)}$, *that satisfies Condition* (H); $B : Y \to C_v(Y^*)$ *is multivalued operator of the Volterra type, that satisfies Condition* (H), *the growth condition:*

$$\exists c_1, c_2 \geq 0 : \quad \|By\|_+ \leq c_1\|y\|_Y + c_2 \quad \forall y \in Y \tag{2.172}$$

the condition of continuity:

$$d_H(B(z), B(z_0)) \to 0, \text{ if } z \to z_0, \tag{2.173}$$

where $d_H(\cdot;\cdot)$ *is the Hausdorf metric in* $C_v(Y^*)$, *i.e.*

$$d_H(C; D) = \max\{\text{dist}(C; D), \text{dist}(D, C)\}, \ C, D \in C_v(Y^*).$$

Then for any $a_0 \in V$ *and* $f \in X^*$ *there exists at last one solution of problem (2.170)* $u \in X$, *that is obtained by the method of singular perturbations, moreover* $u' \in W$.

Remark 2.21. The inclusion $u'' + Au' + Bu \ni f$ we will consider as the inclusion in the space $\mathscr{D}^*(S; V^*)$. If $u \in C(S; V)$ with $u' \in X$ satisfies this inclusion, then $u'' \in f - Au' - Bu \subset X^*$. This means, that $u' \in W \subset C(S; H)$. From this it follows, that the conditions $u'(0) = \bar{0} \in H$ and $u' \in W$ are true.

Proof. Similarly to [GGZ74, Theorem VII.1.1] let us reduce the evolution inclusion from (2.170) to the first order inclusion. Let $R : X \to X$ ($Y \to Y$ respectively) be the operator of the Volterra type, that is defined by the relation

$$(Rv)(t) = a_0 + \int_0^t v(s)ds \quad \forall v \in X, \quad \forall t \in S.$$

R is Lipschitz-continuous operator from X to X (from Y to Y respectively). If u is the solution of problem (2.170): $u' \in W$, then $v = u'$ will be the solution of the problem

$$\begin{cases} v' + A(v) + B(Rv) \ni f, \\ v(0) = \bar{0}, \ v \in W. \end{cases} \tag{2.174}$$

Vice versa, if $v \in W$ is the solution of problem (2.174), then $u = Rv \in X$ is the solution of problem (2.170) such, that $u' \in W \subset X$.

Let us consider the multivalued operator

$$\mathscr{A} := A + B \circ R : X \to C_v(X^*)$$

and $I : X \to X \subset X^*$ is identical motion, $\lambda = \lambda_A + \lambda_B$, $\lambda_B = 1 + c_1 \cdot c_3$, c_3 is the Lipschitzean constant for the operator $R : Y \to Y$. For any $y \in X$ and a.a. $t \in S$ let us set

$$y_\lambda(t) = e^{-\lambda t}y(t), \quad f_\lambda(t) = e^{-\lambda t}f(t), \quad (A_\lambda y_\lambda)(t) = e^{-\lambda t}(\mathscr{A}y)(t) + \lambda y_\lambda(t).$$

i.e. $(d_\lambda \in A_\lambda(y_\lambda)) \Leftrightarrow (\forall w \in X \quad \langle d_\lambda, w\rangle_X \le [A(y) + \lambda y, w_\lambda]_+)$. Let us note, that $A_\lambda(y_\lambda)$ is nonempty, as $\exists d_\lambda \in A_\lambda(y_\lambda)$, that we can choose so:

$$d_\lambda(t) = e^{-\lambda t} d(t) + \lambda y_\lambda(t) \quad \text{for a.e.} \quad t \in S, d \in \mathscr{A}.$$

Then $A_\lambda : X \to C_v(X^*)$ and $v \in W$ is the solution of problem (2.174) if and only if $v_\lambda \in W$ satisfies the next:

$$v_\lambda' + A_\lambda v_\lambda \ni f_\lambda, \quad v_\lambda(0) = \bar{0}. \tag{2.175}$$

Let us check, that $A_\lambda : X \to C_v(X^*)$ satisfies such conditions:

(α_1) A_λ is $+$-coercive on X.
(α_2) A_λ is λ_0-pseudomonotone on W.
(α_3) A_λ is locally bounded on X.
(α_4) A_λ satisfies Condition (Π) on X.

Let us check (α_1). Let us fix $y \in X$. As $\|y_\lambda\|_X \le \|y\|_X$, then

$$\|y_\lambda\|_X^{-1}[A_\lambda y_\lambda, y_\lambda]_+ \ge \|y\|_X^{-1} \sup_{\zeta(y)\in A(y)} \int_S e^{-2\lambda t}\big(\zeta(y)(t) + \lambda_A y(t), y(t)\big)dt$$
$$+ \|y\|_X^{-1} \inf_{\zeta(y)\in B(Ry)} \int_S e^{-2\lambda t}\big(\zeta(y)(t) + \lambda_B y(t), y(t)\big)dt.$$

At first let us estimate the first term. Let us note, that

$$[(A + \lambda_A I)y, y]_+ \ge \gamma(\|y\|_X)\|y\|_X \quad \forall y \in X,$$

where $\gamma : \mathbb{R}_+ \to \mathbb{R}$ is bounded from bellow function on bounded in $\mathbb{R}_+$ sets such, that $\gamma(r) \to +\infty$ as $r \to \infty$ (it is obviously follows from $+$-coercivity $A + \lambda_A I$). From here, $\inf_{s\ge 0} \gamma(s) = a > -\infty$. For any $b > a$ let us consider nonempty bounded in $\mathbb{R}_+$ set $A_b = \{c \ge 0 \mid \gamma(c) \le b\}$. Let $c_b = \sup A_b$ for any $b > a$. Let us note, that $\forall b_1 > b_2 > a$, $+\infty > c_{b_1} \ge c_{b_2}$ and $c_b \to +\infty$ as $b \to +\infty$. Let us set

$$\hat{\gamma}(t) = \begin{cases} a, & t \in [0, c_{a+1}], \\ a + k, & t \in (c_{a+k}, c_{a+k+1}],\ k \ge 1. \end{cases}$$

Then, $\hat{\gamma} : \mathbb{R}_+ \to \mathbb{R}$ is bounded from bellow nondecreasing function on bounded in $\mathbb{R}_+$ sets such, that $\hat{\gamma}(r) \to +\infty$ as $r \to \infty$ and $\gamma(t) \ge \hat{\gamma}(t)$ $\forall t \ge 0$. As A is the operator of the Volterra type, then for fixed $y \in X$

$$\begin{aligned} \forall t \in S \ \sup_{\zeta\in A} \int_0^t \big(\zeta(y)(\tau) + \lambda_A y(\tau), y(\tau)\big)d\tau &= \sup_{\zeta\in A} \int_0^T \big(\zeta(y_t)(\tau) + \lambda_A y_t(\tau), y_t(\tau)\big)d\tau \\ &\ge \hat{\gamma}(\|y_t\|_X)\|y_t\|_X = \hat{\gamma}(\|y\|_{X_t})\|y\|_{X_t}, \end{aligned}$$

where $\|y\|_{X_t} = \|y_t\|_X$. Let

$$g_\zeta(\tau) = \big(\zeta(y)(\tau) + \lambda_A y(\tau), y(\tau)\big), \quad \zeta \in A, \tau \in S,$$
$$h(t) = \hat{\gamma}(\|y\|_{X_t})\|y\|_{X_t}, \quad t \in S.$$

For all $t \in S$ $h(t) \ge \min\{\hat{\gamma}(0), 0\}\|y\|_X$ and

$$\sup_{\zeta\in A} \int_0^t g_\zeta(\tau)d\tau \ge h(t) \quad \forall t \in S,$$

For any $y \in X$ in consequence of Condition (H)

$$\begin{aligned}
&\sup_{\zeta\in A} \int_0^T e^{-2\lambda\tau}\big(\zeta(y)(\tau) + \lambda_A y(\tau), y(\tau)\big)d\tau \\
&= e^{-2\lambda T} \sup_{\zeta\in A} \int_0^T g_\zeta(\tau)d\tau + \sup_{\zeta\in A} \int_0^T \left[e^{-2\lambda\tau} - e^{-2\lambda T}\right] g_\zeta(\tau)d\tau \\
&\ge e^{-2\lambda T} h(T) + 2\lambda \sup_{\zeta\in A} \int_0^T e^{-2\lambda s} \int_0^s g_\zeta(\tau)d\tau ds \\
&\ge e^{-2\lambda T} h(T) + 2\lambda T \sup_{\zeta\in A} \inf_{s\in S} e^{-2\lambda s} \int_0^s \big(\zeta(y)(\tau) + \lambda_A y(\tau), y(\tau)\big)d\tau.
\end{aligned}$$

Let us prove that for all $y \in X$

$$\sup_{\zeta\in A} \inf_{s\in S} e^{-2\lambda s} \int_0^s \big(\zeta(y)(\tau) + \lambda_A y(\tau), y(\tau)\big)d\tau \ge -C_1\|y\|_X,$$

where $C_1 = \max\{-\hat{\gamma}(0), 0\} \ge 0$ does not depend on $y \in X$.

Indeed, let $y \in X$ be fixed. Let us set

$$\varphi(s, d) = e^{-2\lambda s} \int_0^s \big(d(\tau) + \lambda_A y(\tau), y(\tau)\big)d\tau, \quad a = \sup_{d\in A(y)} \inf_{s\in S} \varphi(s, d),$$
$$A_d = \{s \in S \mid \varphi(s, d) \le a\}, \quad s \in S,\ d \in A(y).$$

From the continuity of $\varphi(\cdot, d)$ on S it follows, that A_d is nonempty closed set for any $d \in A(y)$. Indeed, for fixed $d \in A(y)$ there exists $s_d \in S$ such, that

$$\varphi(s_d, d) = \min_{\hat{s}\in S} \varphi(\hat{s}, d) \le a.$$

The closure of A_d follows from the continuity of $\varphi(\cdot, d)$ on S.

Now let us prove, that the system $\{A_d\}_{d\in A(y)}$ is centralized. For fixed $\{d_i\}_{i=1}^n \subset A(y)$, $n \geq 1$, let us set

$$\psi_i(\tau) = \big(d_i(\tau) + \lambda_A y(\tau), y(\tau)\big), \quad \psi(\tau) = \max_{i=1}^{n} \psi_i(\tau), \quad \tau \in S \quad \text{a.e.},$$
$$E_0 = \emptyset, \quad E_j = \left\{\tau \in S \setminus \left(\cup_{i=0}^{j-1} E_i\right) \mid \psi_j(\tau) = \psi(\tau)\right\}, \quad j = \overline{1,n},$$
$$d(\tau) = \sum_{j=1}^{n} d_j(\tau)\chi_{E_j}(\tau), \quad \chi_{E_j}(\tau) = \begin{cases} 1, & \tau \in E_j, \\ 0, & \text{else.} \end{cases}$$

Let us note, that $\forall j = \overline{1,n}$ E_j is measurable, $\cup_{j=1}^n E_j = S$, $E_i \cap E_j = \emptyset$ $\forall i \neq j$, $i, j = \overline{1,n}$, $d \in X^*$. Moreover,

$$\varphi(s, d_i) = e^{-2\lambda s} \int_0^s \psi_i(\tau) d\tau \leq e^{-2\lambda s} \int_0^s \psi(\tau) d\tau = \varphi(s, d), \quad s \in S,\ i = \overline{1,n}.$$

Therefore, thanks to Condition (H) for A, $d \in A(y)$ and for some $s_d \in S$

$$\varphi(s_d, d_i) \leq \varphi(s_d, d) = \min_{\hat{s}\in S} \varphi(\hat{s}, d) \leq a, \quad i = \overline{1,n}.$$

So, $s_d \in \cap_{i=1}^n A_{d_i} \neq \emptyset$.

As S is compact, and the system of closed sets $\{A_d\}_{d\in A(y)}$ is centralized, then $\exists s_0 \in S$: $s_0 \in \cap_{d\in A(y)} A_d$ [RS80]. This means, that

$$\begin{aligned}
&\sup_{\zeta\in A} \inf_{s\in S} e^{-2\lambda s} \int_0^s \big(\zeta(y)(\tau) + \lambda_A y(\tau), y(\tau)\big) d\tau \\
&\quad \geq \sup_{\zeta\in A} e^{-2\lambda s_0} \int_0^{s_0} \big(\zeta(y)(\tau) + \lambda_A y(\tau), y(\tau)\big) d\tau \\
&\quad = e^{-2\lambda s_0} \sup_{\zeta\in A} \int_0^{s_0} g_\zeta(\tau) d\tau \geq e^{-2\lambda s_0} h(s_0) \\
&\quad \geq e^{-2\lambda s_0} \min\{\hat{\gamma}(0), 0\} \|y\|_X \geq -\max\{-\hat{\gamma}(0), 0\} \|y\|_X = -C_1 \|y\|_X.
\end{aligned}$$

So,

$$\begin{aligned}
&\|y\|_X^{-1} \sup_{\zeta\in A} \int_0^T e^{-2\lambda\tau} \big(\zeta(y)(\tau) + \lambda_A y(\tau), y(\tau)\big) d\tau \\
&\quad \geq e^{-2\lambda T} \hat{\gamma}(\|y\|_X) - 2\lambda C_1 T, \qquad (2.176)
\end{aligned}$$

where $\hat{\gamma} : \mathbb{R}_+ \to \mathbb{R}$ is bounded from bellow nondecreasing function on bounded in $\mathbb{R}_+$ sets such, that $\hat{\gamma}(r) \to +\infty$ as $r \to \infty$, $C_1 \geq 0$ does not depend on $y \in X$.

Let us estimate the second term. Similarly to the first case, as $B \circ R$ is the operator of the Volterra type, then $\forall t \in S$

$$\begin{aligned}
&\inf_{\zeta \in B\circ R} \int_0^t \big(\zeta(y)(\tau) + \lambda_B y(\tau), y(\tau)\big) d\tau \\
&= \inf_{\zeta \in B\circ R} \int_0^T \big(\zeta(y_t)(\tau) + \lambda_B y_t(\tau), y_t(\tau)\big) d\tau \\
&\geq \lambda_B \|y_t\|_Y^2 - \|B(Ry_t)\|_+ \|y_t\|_Y \\
&\geq (1 + c_1 \cdot c_3)\|y_t\|_Y^2 - c_2\|y_t\|_Y - c_1\|Ry_t\|_Y\|y_t\|_Y \\
&\geq \|y_t\|_Y^2 - (c_2 + c_1\|R\bar{0}\|_Y)\|y_t\|_Y \geq -(c_2 + c_1\|R\bar{0}\|_Y)c_4\|y\|_X > -\infty,
\end{aligned}$$

where $c_4 > 0$ such, that $\|\cdot\|_Y \leq c_4\|\cdot\|_X$. Let

$$g_\zeta(\tau) = \big(\zeta(y)(\tau) + \lambda_B y(\tau), y(\tau)\big), \quad \zeta \in B \circ R, \tau \in S.$$

Then,

$$\begin{aligned}
&\inf_{\zeta \in B\circ R} \int_0^T e^{-2\lambda\tau} \big(\zeta(y)(\tau) + \lambda_B y(\tau), y(\tau)\big) d\tau \\
&\geq e^{-2\lambda T} h(T) + 2\lambda \inf_{\zeta \in B\circ R} \int_0^T e^{-2\lambda s} \int_0^s g_\zeta(\tau) d\tau ds \\
&\geq -e^{-2\lambda T}(c_2 + c_1\|R\bar{0}\|_Y)c_4\|y\|_X \\
&\quad + 2\lambda \int_0^T e^{-2\lambda s} \inf_{\zeta \in B\circ R} \int_0^s \big(\zeta(y)(\tau) + \lambda_B y(\tau), y(\tau)\big) d\tau ds \\
&\geq -(c_2 + c_1\|R\bar{0}\|_Y)c_4\|y\|_X \quad \forall y \in X.
\end{aligned}$$

So, with the help of (2.176), it follows, that

$$\begin{aligned}
\|y_\lambda\|_X^{-1}[A_\lambda y_\lambda, y_\lambda]_+ &\geq e^{-2\lambda T}\hat{\gamma}(\|y\|_X) - 2\lambda C_1 T - (c_2 + c_1\|R\bar{0}\|_Y)c_4 \\
&\geq e^{-2\lambda T}\hat{\gamma}(\|y_\lambda\|_X) - 2\lambda C_1 T - (c_2 + c_1\|R\bar{0}\|_Y)c_4 \quad \forall y \in X,
\end{aligned}$$

as $\|y\|_X \geq \|y_\lambda\|_X$. So, $A_\lambda : X \to C_v(X^*)$ is $+$-coercive.

Let us check (α_2). For any $y \in X$ and a.a. $t \in S$ let us set

$$(A^1_\lambda y_\lambda)(t) = e^{-\lambda t}(Ay)(t) + \lambda_A y_\lambda(t), \quad (A^2_\lambda y_\lambda)(t) = e^{-\lambda t}(B(Ry))(t) + \lambda_B y_\lambda(t).$$

Let us note, that $A^1_\lambda(y_\lambda) + A^2_\lambda(y_\lambda) = A_\lambda(y_\lambda)\ \forall y \in X$.

At first let us show, that A^1_λ is r.l.s.c. operator with $(X; W)$-s.b.v. It is to check the radial lower semicontinuity. Let us prove the semiboundness of variation. In view of the conditions of the Theorem for all $R > 0$, $y, \xi \in X$: $\|y\|_X \le R$, $\|\xi\|_X \le R$ the next inequality is true

$$[A(y) - A(\xi) + \lambda_A y - \lambda_A \xi, y - \xi]_- + C_A(R; \|y - \xi\|'_W) \ge 0.$$

Let us set $\hat{C}_A(R; \cdot) = \max\limits_{\tau \in [0,t]} C_A(R; \tau)$ for all $R, t \ge 0$ $(\hat{C}_A \in \Phi)$,

$$z_t(\tau) = \begin{cases} z(\tau), & 0 \le \tau \le t, \\ \bar{0}, & t < \tau \le T, \end{cases} \qquad t \in S,\ z \in X.$$

Let $\zeta, \eta \in A$ are fixed selectors. As A is the operator of the Volterra type, then $\forall t \in S$

$$\int_0^t \big(\zeta(y)(\tau) + \lambda_A y(\tau) - \eta(\xi)(\tau) - \lambda_A \xi(\tau), y(\tau) - \xi(\tau)\big) d\tau + \hat{C}_A\left(R; \|y - \xi\|'_{W_t}\right)$$
$$= \int_0^T \big(\zeta(y_t)(\tau) + \lambda_A y_t(\tau) - \eta(\xi_t)(\tau) - \lambda_A \xi_t(\tau), y_t(\tau) - \xi_t(\tau)\big) d\tau$$
$$+ \hat{C}_A\left(R; \|y_t - \xi_t\|'_W\right) \ge [(A + \lambda_A I)(y_t) - (A + \lambda_A I)(\xi_t), y_t - \xi_t]_-$$
$$+ \hat{C}_A(R; \|y_t - \xi_t\|'_W) \ge 0,$$

because $\|y_t\|_X \le \|y\|_X$ and $\|\xi_t\|_X \le \|\xi\|_X$. Here $\|\cdot\|'_{W_t} = \|\cdot\|_{L_{p_0}([0,t];V_0)}$.

Let us set

$$g(\tau) = \big(\zeta(y)(\tau) + \lambda_A y(\tau) - \eta(\xi)(\tau) - \lambda_A \xi(\tau), y(\tau) - \xi(\tau)\big), \quad \tau \in S,$$
$$h(t) = \hat{C}_A\left(R; \|y - \xi\|'_{W_t}\right), \quad t \in S.$$

Let us note, that function $S \ni t \to h(t)$ is monotone nondecreasing function and

$$\int_0^t g(\tau) d\tau \ge -h(t) \quad \forall t \in S.$$

So,

$$\int_0^T e^{-2\lambda\tau}\big(\zeta(y)(\tau)+\lambda_A y(\tau)-\eta(\xi)(\tau)-\lambda_A\xi(\tau), y(\tau)-\xi(\tau)\big)d\tau$$
$$=\int_0^T e^{-2\lambda\tau}g(\tau)d\tau = e^{-2\lambda T}\int_0^T g(\tau)d\tau + 2\lambda\int_0^T e^{-2\lambda\tau}\int_0^\tau g(s)ds d\tau$$
$$\geq -e^{-2\lambda T}h(T) - 2\lambda\int_0^T e^{-2\lambda\tau}h(\tau)d\tau \geq -h(T)\Big[e^{-2\lambda T}+2\lambda\int_0^T e^{-2\lambda\tau}d\tau\Big]$$
$$= -h(T) = -\hat{C}_A\left(R; \|y-\xi\|_W'\right).$$

Consequently,

$$\begin{aligned}
[A_\lambda^1 y_\lambda, y_\lambda-\xi_\lambda]_- &= \inf_{\zeta(y)\in A(y)}\int_S\big(e^{-\lambda t}\zeta(y)(t)+\lambda_A y(t)e^{-\lambda t}, e^{-\lambda t}(y(t)-\xi(t))\big)dt\\
&= \inf_{\zeta(y)\in A(y)}\int_S e^{-2\lambda t}\big(\zeta(y)(t)+\lambda_A y(t), y(t)-\xi(t)\big)dt\\
&\geq \sup_{\zeta(\xi)\in A(\xi)}\int_S e^{-2\lambda t}\big(\zeta(\xi)(t)+\lambda_A\xi(t), y(t)-\xi(t)\big)dt\\
&\quad -\hat{C}_A(R; \|y-\xi\|_{L_{p_0}(S;V_0)})\\
&= \sup_{\zeta(\xi)\in A(\xi)}\int_S\big(e^{-\lambda t}\zeta(\xi)(t)+\lambda_A\xi(t)e^{-\lambda t}, e^{-\lambda t}(y(t)-\xi(t))\big)dt\\
&\quad -\hat{C}_A(R; \|y-\xi\|_{L_{p_0}(S;V_0)})\\
&= [A_\lambda^1\xi_\lambda, y_\lambda-\xi_\lambda]_+ - \hat{C}_A(R; \|y-\xi\|_{L_{p_0}(S;V_0)}).
\end{aligned}\tag{2.177}$$

Let us consider weight space $L_{p_0,\lambda}(S;V_0)$, that consists of measurable functions $y_\lambda : S\to V_0$, for which integral $\int_S e^{\lambda t p_0}\|y_\lambda(t)\|_{V_0}^{p_0}dt$ is finite. Then

$$\|y-\xi\|_{L_{p_0}(S;V_0)} = \left(\int_S e^{\lambda t p_0}\|y_\lambda(t)-\xi_\lambda(t)\|_{V_0}^{p_0}dt\right)^{1/p_0} = \|y_\lambda-\xi_\lambda\|_{L_{p_0,\lambda}(S;V_0)}.$$

So from (2.177) we will have

$$[A_\lambda^1 y_\lambda, y_\lambda-\xi_\lambda]_- \geq [A_\lambda^1\xi_\lambda, y_\lambda-\xi_\lambda]_+ - \hat{C}_A(R; \|y_\lambda-\xi_\lambda\|_{L_{p_0,\lambda}(S;V_0)}).$$

The fact, that embedding $W \subset L_{p_0,\lambda}(S;V_0)$ is compact is finishes the proof $(X;W)$-s.b.v. for $A^1_\lambda : X \to C_v(X^*)$. It is the direct consequence of the continuity of embedding

$$W \subset \left\{ y \in L_{p_0}(S;V) \mid y' \in L_{\min\{r_1',r_2'\}}(S;V^*) \right\}$$

and Lemma about compactness with $B_0 = V$, $B = V_0$, $B_1 = V^*$, $p_0 = p_0$ and $p_1 = \min\{r_1', r_2'\}$.

As arbitrary r.l.s.c. multivalued operator with $(X;W)$-s.b.v. is λ_0-pseudomonotone on W [KMY07, Proposition 1.2.23], then we have required Proposition for A^1_λ.

Now let us consider A^2_λ. At first we will show, that A^2_λ is r.u.s.c. operator with N-s.b.v. on W. It is easy to check the radial upper semicontinuity. Let us prove the semiboundness of variation.

At first let us prove that if $B : Y \to C_v(Y^*)$ is continuous in the sense of (2.173) multivalued operator. Then $(B \circ R) : X \to C_v(X^*)$ is the operator with N-s.b.v. on W with $\|\cdot\|'_W = \|\cdot\|_Y$.

Let us build the function $C(R,t)$. For all $R \geq 0$ and $t \geq 0$ let

$$C(R,t) = t \cdot \sup_{\substack{z_1, z_2 \in X:\ \|z_i\|_X \leq R, \\ \|z_1 - z_2\|_Y \leq t,\ i=1,2}} d_H(B(Rz_1); B(Rz_2)) =: t \cdot c(R,t).$$

Let us check, that this map is defined correctly, i.e.

$$\forall R, t \geq 0 \quad c(R,t) < +\infty. \tag{2.178}$$

Let it be not so. Then for some $R > 0$ and $t > 0$ $C(R,t) = +\infty$. This means, that there exist sequences $\{z_i^n\}_{n\geq 1} \subset X$, $i = 1,2$ such, that $\|z_i^n\|_X \leq R$, $\|z_1^n - z_2^n\|_Y \leq t$ $\forall n \geq 1$ and

$$d_H(B(Rz_1^n); B(Rz_2^n)) \to +\infty \text{ as } n \to +\infty.$$

From the continuity of embedding $X \subset X^*$ and from global property of Lipschitz of the operator $R : X \to X$ it follows the boundness of the sequences $\{Rz_1^n\}_{n\geq 1}$ and $\{Rz_2^n\}_{n\geq 1}$ in W, and, so, from the compactness of embedding W in Y, and that fact, that the given sequences are precompact in Y.

So, within subsequences,

$$Rz_i^n \to \xi_i \text{ in } Y \quad \text{and} \quad d_H(B(Rz_i^n); B(\xi_i)) \to 0, \quad i = 1,2$$

for some $\xi_i \in Y$. So,

$$0 \leq \overline{\lim}\, d_H(B(Rz_1^n); B(Rz_2^n)) \leq d_H(B(\xi_1); B(\xi_2)) < +\infty.$$

The correctness of the definition of function C in the sense of (2.178) is checked. Let us note, that for all $R \geq 0$ and $t \geq 0$ $c(R,0) \equiv c(0,t) \equiv 0$.

Let us choose norm $\|\cdot\|'_W := \|\cdot\|_Y$ that is compact with regard to $\|\cdot\|_W$ on W and continuous with regard to $\|\cdot\|_X$ on X. Let us note, that for all $R \geq 0$, $y_1, y_2 \in X$ such, that $\|y_i\|_X \leq R$ and $d_i \in B(Ry_i)$, $i = 1, 2$

$$\langle d_1, y_2 - y_1 \rangle_X - \langle d_2, y_2 - y_1 \rangle_X \leq \|d_1 - d_2\|_{Y^*} \|y_1 - y_2\|_Y.$$

So,

$$\begin{aligned}
&[B(Ry_1), y_2 - y_1]_+ - [B(Ry_2), y_2 - y_1]_+ \\
&\quad \leq \operatorname{dist}(B(Ry_1); B(Ry_2))\|y_1 - y_2\|_Y \leq d_H(B(Ry_1); B(Ry_2))\|y_1 - y_2\|_Y, \\
&\quad [B(Ry_1), y_1 - y_2]_- - [B(Ry_2), y_1 - y_2]_- \\
&\quad \geq -d_H(B(Ry_1); B(Ry_2))\|y_1 - y_2\|_Y \geq -C(R; \|y_1 - y_2\|_Y).
\end{aligned}$$

So, to finish the proof of the given Proposition it remains to prove:

- $\frac{C(R,t)}{t} =: c(R,t) \to 0$ as $t \to 0+$;
- $\forall R \geq 0$ function $[0, +\infty) \ni t \to C(R,t)$ is continuous.

(1) Let us check, that for all $R > 0$ $c(R,t) \to 0$ as $t \to 0+$. In order to do that let us apply the method from the opposite. Let us suggest, that there exist $\varepsilon > 0$, $t_n \searrow 0+$, $z_1^n \in X$ and $z_2^n \in X$ such, that for all $i = 1, 2$ and $n \geq 1$

$$\|z_i^n\|_X \leq R, \quad \|z_1^n - z_2^n\|_Y \leq t_n, \quad d_H(B(Rz_1^n); B(Rz_2^n)) \geq \varepsilon.$$

From the continuity of embedding $X \subset X^*$ and from the global property of Lipschitz of operator $R : X \to X$ it follows the boundness of sequences $\{Rz_1^n\}_{n\geq 1}$ and $\{Rz_2^n\}_{n\geq 1}$ in W, and, so, from the compactness of embedding W in Y, it follows that fact, that the given sequences are precompact in Y. From here, within subsequences,

$$Rz_i^m \to \xi_i \text{ in } Y \quad \text{and} \quad d_H(B(Rz_i^m); B(\xi_i)) \to 0, \quad i = 1, 2$$

for some $\xi_i \in Y$, $i = 1, 2$. And from the global property of Lipschitz of the operator $R : Y \to Y$ with constant of Lipschitz $K > 0$ it follows, that:

$$\|\xi_1 - \xi_2\|_Y \leftarrow \|Rz_1^m - Rz_2^m\|_Y \leq K \cdot \|z_1^m - z_2^m\|_Y \leq K \cdot t_m \to 0$$

and $\xi_1 = \xi_2 =: \xi$. From the continuity of operator $B : Y \to Y^*$ we will have, that for big $m \geq 1$

$$0 = d_H(B(\xi_1); B(\xi_2)) \leftarrow d_H(B(Rz_1^m); B(Rz_2^m)) \geq \varepsilon > 0.$$

We have the contradiction.

(2) Now let us check, that $\forall R \geq 0$ function $[0, +\infty) \ni t \to C(R,t)$ is continuous. For this it is enough to check the continuity of $[0, +\infty) \ni t \to c(R,t)$.

The case, when $t = 0$ follows from the previous point. So, we will suggest, that $t > 0$. In order to check this we will apply the method from the opposite. Let there exist such $R > 0$, $t_0 > 0$, $t_n \to t_0$ as $n \to +\infty$ and $\varepsilon^* > 0$, that

$$|c(R,t_n) - c(R,t_0)| \geq \varepsilon^* \quad \forall n \geq 1. \tag{2.179}$$

Without losing generality, we will suggest, that or $t_n \to t_0$ from bellow, or $t_n \to t_0$ from above as $n \to +\infty$.

(2.1) Let $t_n \to t_0$ from above as $n \to \infty$. From the view of function $c(R,t)$ we can see, that for all $R \geq 0$ and $t_1 \geq t_2 > 0$ $c(R,t_1) \geq c(R,t_2)$. So, from (2.179) it follows

$$\forall n \geq 1 \quad c(R,t_n) \geq \varepsilon^* + c(R,t_0).$$

And this means, that for every $n \geq 1$ and $i = 1,2$ there exist $z_1^n \in X$ and $z_2^n \in X$ such, that $\|z_i^n\|_X \leq R$, $\|z_1^n - z_2^n\|_Y \leq t_n$ and

$$d_H(B(Rz_1^n); B(Rz_2^n)) \geq \frac{\varepsilon^*}{2} + c(R,t_0). \tag{2.180}$$

From the continuity of embedding $X \subset X^*$ and from the global property of Lipschitz of operator $R : X \to X$ it follows the boundness in W of sequences $\{Rz_1^n\}_{n\geq 1}$ and $\{Rz_2^n\}_{n\geq 1}$, and so, from the compactness of embedding W in Y it follows that fact, that the given sequences are precompact in Y. And so, within subsequences,

$$Rz_i^m \to \xi_i \text{ in} Y \quad \text{and} \quad d_H(B(Rz_i^m); B(\xi_i)) \to 0, \quad i = 1,2$$

for some $\xi_i \in Y$, $i = 1,2$. Moreover, as spaces X, X^*, Y, Y^* are reflexive and separable spaces, then, without losing generality, we can suggest:

$$z_i^m \to \zeta_i \text{ weakly in } X, \; z_i^m \to \zeta_i \text{ weakly in } Y,$$

$$\|\zeta_i\|_X \leq R, \; \|\zeta_1 - \zeta_2\|_Y \leq t_0 \quad \text{and } R\zeta_i = \xi_i.$$

From the continuity of operator $B : Y \to Y^*$ it is follows, that passing to the limit in (2.180) as $m \to +\infty$, we will obtain

$$c(R,t_0) \geq d_H(B(R\zeta_1); B(R\zeta_2)) \geq \frac{\varepsilon^*}{2} + c(R,t_0).$$

We obtained the contradiction.

(2.2) Now we consider the case, when $t_n \to t_0$ from bellow as $n \to \infty$. From (2.179) it follows, that

$$c(R,t_n) + \varepsilon^* \leq c(R,t_0) \leq d_H(B(Rz_1); B(Rz_2)) + \frac{\varepsilon^*}{2},$$

where $z_i \in X$ such, that $\|z_i\|_X \le R$ and $\|z_1 - z_2\|_Y \le t_0$. For arbitrary $n \ge 1$ and $i = 1, 2$ let $z_i^n := \frac{t_n}{t_0} z_i \to z_i$ in X (in Y, in W respectively) as $n \to \infty$, $\|z_i^n\|_X \le R$, $\|z_1^n - z_2^n\|_Y \le t_n$ and $Rz_i^n \to Rz_i$ in Y as $n \to \infty$, $i = 1, 2$. And so,

$$\begin{aligned} d_H\,(B(Rz_1); B(Rz_2)) + \tfrac{\varepsilon^*}{2} &\leftarrow d_H(B(Rz_1^n); B(Rz_2^n)) + \tfrac{\varepsilon^*}{2} \\ &\le c(R, t_n) + \tfrac{\varepsilon^*}{2} \le c(R, t_0) - \tfrac{\varepsilon^*}{2} \le d_H(B(Rz_1); B(Rz_2)). \end{aligned}$$

We obtained the contradiction again. So, function $\mathbb{R}_+ \ni t \to C(R,t) = t \cdot c(R,t)$ is continuous for every $R \ge 0$.

For some $C_B \in \Phi$ and for all $R > 0$, $y, \xi \in X$: $\|y\|_X \le R$, $\|\xi\|_X \le R$ it is follows:

$$[B(Ry) + \lambda_B y, y - \xi]_- - [B(R\xi) + \lambda_B \xi, y - \xi]_- + C_B(R; \|y - \xi\|_Y) \ge 0.$$

Let us set $\hat{C_B}(R; \cdot) = \max\limits_{\tau \in [0,t]} C_B(R; \tau)$ for all $R, t \ge 0$ ($\hat{C_B} \in \Phi$),

$$z_t(\tau) = \begin{cases} z(\tau), & 0 \le \tau \le t, \\ \bar{0}, & t < \tau \le T, \end{cases} \qquad t \in S,\ z \in X.$$

As $B \circ R$ is the operator of the Volterra type, then $\forall t \in S$

$$\begin{aligned} &\inf_{\zeta \in B \circ R} \int_0^t \big(\zeta(y)(\tau) + \lambda_B y(\tau), y(\tau) - \xi(\tau)\big) d\tau \\ &\quad - \inf_{\eta \in B \circ R} \int_0^t \big(\eta(\xi)(\tau) - \lambda_B \xi(\tau), y(\tau) - \xi(\tau)\big) d\tau + \hat{C_B}\,(R; \|y - \xi\|_{Y_t}) \\ &= \inf_{\zeta \in B \circ R} \int_0^T \big(\zeta(y_t)(\tau) + \lambda_B y_t(\tau), y_t(\tau) - \xi_t(\tau)\big) d\tau \\ &\quad - \inf_{\eta \in B \circ R} \int_0^T \big(\eta(\xi_t)(\tau) + \lambda_B \xi_t(\tau), y_t(\tau) - \xi_t(\tau)\big) d\tau + \hat{C_B}\,(R; \|y - \xi\|_{Y_t}) \\ &= [B(Ry_t) + \lambda_B y_t, y_t - \xi_t]_- - [B(R\xi_t) + \lambda_B \xi_t, y_t - \xi_t]_- \\ &\quad + \hat{C_B}(R; \|y_t - \xi_t\|_Y) \ge 0, \end{aligned}$$

because $\|y_t\|_X \le \|y\|_X$ and $\|\xi_t\|_X \le \|\xi\|_X$. Here $\|\cdot\|_{Y_t} = \|\cdot\|_{L_2([0,t];H)}$.

Let us set

$$\begin{aligned} g_\zeta(\tau) &= \big(\zeta(\tau), y(\tau) - \xi(\tau)\big), \quad \zeta \in X^*,\ \tau \in S, \\ h(t) &= \hat{C_B}\,(R; \|y - \xi\|_{Y_t}), \quad t \in S. \end{aligned}$$

Let us note, that function $S \ni t \to h(t)$ is monotone nondecreasing and

$$\inf_{\zeta \in B(Ry)+\lambda_B y} \int_0^t g_\zeta(\tau)d\tau - \inf_{\eta \in B(R\xi)+\lambda_B \xi} \int_0^t g_\eta(\tau)d\tau \geq -h(t) \quad \forall t \in S.$$

So, for fixed $\zeta \in B(Ry) + \lambda_B y$

$$\int_0^T e^{-2\lambda\tau}\big(\zeta(\tau), y(\tau) - \xi(\tau)\big)d\tau - \inf_{\eta \in B(R\xi)+\lambda_B \xi} \int_0^T e^{-2\lambda\tau}\big(\eta(\tau), y(\tau) - \xi(\tau)\big)d\tau$$
$$= \sup_{\eta \in B(R\xi)+\lambda_B \xi} \int_0^T e^{-2\lambda\tau}\big(\zeta(\tau) - \eta(\tau), y(\tau) - \xi(\tau)\big)d\tau.$$

For arbitrary $y \in X, t \in S$ as the consequence of condition (H) we have

$$\sup_{\eta \in B(R\xi)+\lambda_B \xi} \int_0^T e^{-2\lambda\tau}\big(\zeta(\tau) - \eta(\tau), y(\tau) - \xi(\tau)\big)d\tau$$
$$= e^{-2\lambda T} \sup_{\eta \in B(R\xi)+\lambda_B \xi} \int_0^T \big(\zeta(\tau) - \eta(\tau), y(\tau) - \xi(\tau)\big)d\tau$$
$$+ \sup_{\eta \in B(R\xi)+\lambda_B \xi} \int_0^T \left[e^{-2\lambda\tau} - e^{-2\lambda T}\right]\big(\zeta(\tau) - \eta(\tau), y(\tau) - \xi(\tau)\big)d\tau$$
$$\geq -e^{-2\lambda T}h(T) + 2\lambda \sup_{\eta \in B(R\xi)+\lambda_B \xi} \int_0^T e^{-2\lambda s} \int_0^s \big(\zeta(\tau) - \eta(\tau), y(\tau) - \xi(\tau)\big)d\tau ds$$
$$\geq -e^{-2\lambda T}h(T) + 2\lambda T \sup_{\eta \in B(R\xi)+\lambda_B \xi} \inf_{s \in S} e^{-2\lambda s} \int_0^s \big(\zeta(\tau) - \eta(\tau), y(\tau) - \xi(\tau)\big)d\tau.$$

Let us prove that for all $y \in X$

$$\sup_{\eta \in B(R\xi)+\lambda_B \xi} \inf_{s \in S} e^{-2\lambda s} \int_0^s \big(\zeta(\tau) - \eta(\tau), y(\tau) - \xi(\tau)\big)d\tau \geq -h(T).$$

Let us set

$$\varphi(s,d) = e^{-2\lambda s} \int_0^s \big(d(\tau), y(\tau) - \xi(\tau)\big)d\tau, \quad a = \sup_{d \in E} \inf_{s \in S} \varphi(s,d),$$
$$A_d = \{s \in S \mid \varphi(s,d) \leq a\}, \quad s \in S,\ d \in E := \zeta(\tau) - B(R\xi) - \lambda_B \xi.$$

From the continuity of $\varphi(\cdot, d)$ on S it is follows, that A_d is nonempty closed set for arbitrary $d \in A(y)$. Indeed, for fixed $d \in A(y)$ there exists $s_d \in S$ such, that

$$\varphi(s_d, d) = \min_{\hat{s} \in S} \varphi(\hat{s}, d) \le a.$$

The closure of A_d follows from continuity $\varphi(\cdot, d)$ on S.

Let us prove now, that the system $\{A_d\}_{d \in A(y)}$ is centralized. For fixed $\{d_i\}_{i=1}^n \subset E, n \ge 1$, let us set

$$\begin{aligned} &\psi_i(\tau) = \big(d_i(\tau), y(\tau) - \xi(\tau)\big), \quad \psi(\tau) = \max_{i=1}^{n} \psi_i(\tau), \quad \tau \in S \quad \text{a.e.,} \\ &E_0 = \emptyset, \quad E_j = \left\{\tau \in S \setminus \left(\cup_{i=0}^{j-1} E_i\right) \mid \psi_j(\tau) = \psi(\tau)\right\}, \quad j = \overline{1,n}, \\ &d(\tau) = \textstyle\sum_{j=1}^n d_j(\tau)\chi_{E_j}(\tau), \quad \chi_{E_j}(\tau) = \begin{cases} 1, \ \tau \in E_j, \\ 0, \text{ else.} \end{cases} \end{aligned}$$

Let us note, that $\forall j = \overline{1,n}$ E_j is measurable, $\cup_{j=1}^n E_j = S$, $E_i \cap E_j = \emptyset$ $\forall i \ne j$, $i, j = \overline{1,n}$, $d \in X^*$. Moreover,

$$\varphi(s, d_i) = e^{-2\lambda s} \int_0^s \psi_i(\tau) d\tau \le e^{-2\lambda s} \int_0^s \psi(\tau) d\tau = \varphi(s, d), \quad s \in S,\ i = \overline{1,n}.$$

So, thanks to Condition (H) for B, $d \in E$ and for some $s_d \in S$

$$\varphi(s_d, d_i) \le \varphi(s_d, d) = \min_{\hat{s} \in S} \varphi(\hat{s}, d) \le a, \quad i = \overline{1,n}.$$

Thus, $s_d \in \cap_{i=1}^n A_{d_i} \ne \emptyset$.

As S is compact, and the system of closed sets $\{A_d\}_{d \in E}$ is centralized, then $\exists s_0 \in S$: $s_0 \in \cap_{d \in E} A_d$ [RS80]. This means, that

$$\begin{aligned} &\sup_{\eta \in B(R\xi) + \lambda_B \xi} \inf_{s \in S} e^{-2\lambda s} \int_0^s \big(\zeta(\tau) - \eta(\tau), y(\tau) - \xi(\tau)\big) d\tau \\ &\ge \sup_{\eta \in B(R\xi) + \lambda_B \xi} e^{-2\lambda s_0} \int_0^{s_0} \big(\zeta(\tau) - \eta(\tau), y(\tau) - \xi(\tau)\big) d\tau \\ &= e^{-2\lambda s_0} \sup_{\eta \in B(R\xi) + \lambda_B \xi} \int_0^{s_0} \big(\zeta(\tau) - \eta(\tau), y(\tau) - \xi(\tau)\big) d\tau \end{aligned}$$

$$= e^{-2\lambda s_0}\left[\int_0^{s_0}(\zeta(\tau), y(\tau)-\xi(\tau))d\tau - \inf_{\eta\in B(R\xi)+\lambda_B\xi}\int_0^{s_0}(\eta(\tau), y(\tau)-\xi(\tau))d\tau\right]$$

$$\geq -e^{-2\lambda s_0}h(s_0) \geq -h(T).$$

Therefore,

$$\inf_{\zeta\in B(Ry)+\lambda_B y}\int_0^T e^{-2\lambda\tau}(\zeta(\tau), y(\tau)-\xi(\tau))d\tau$$

$$-\inf_{\eta\in B(R\xi)+\lambda_B\xi}\int_0^T e^{-2\lambda\tau}(\eta(\tau), y(\tau)-\xi(\tau))d\tau$$

$$\geq -e^{-2\lambda T}h(T) - 2\lambda T h(T) = -\left[e^{-2\lambda T}+2\lambda T\right]\cdot \hat{C_B}\left(R; \|y-\xi\|_Y\right).$$

Let us set $\widetilde{C}_B(R;t) = \left[e^{-2\lambda T}+2\lambda T\right]\cdot \hat{C_B}(R;t)$, $R, t \geq 0$ ($\widetilde{C}_B \in \Phi$). Then,

$$[A_\lambda^2 y_\lambda, y_\lambda - \xi_\lambda]_- - [A_\lambda^2 \xi_\lambda, y_\lambda - \xi_\lambda]_-$$

$$= \inf_{\zeta\in B(Ry)+\lambda_B y}\int_0^T e^{-2\lambda\tau}(\zeta(\tau), y(\tau)-\xi(\tau))d\tau$$

$$-\inf_{\eta\in B(R\xi)+\lambda_B\xi}\int_0^T e^{-2\lambda\tau}(\eta(\tau), y(\tau)-\xi(\tau))d\tau$$

$$\geq -\widetilde{C}_B\left(R; \|y-\xi\|_Y\right) = -\widetilde{C}_B(R; \|y-\xi\|_{L_2(S;H)}). \tag{2.181}$$

Let us consider the weight space $L_{2,\lambda}(S;H)$, that consists of measurable functions $y_\lambda : S \to H$, for which the integral $\int_S e^{2\lambda t}\|y_\lambda(t)\|_H^2 dt$ is finite. Then

$$\|y-\xi\|_{L_2(S;H)} = \left(\int_S e^{2\lambda t}\|y_\lambda(t)-\xi_\lambda(t)\|_H^2 dt\right)^{1/2} = \|y_\lambda - \xi_\lambda\|_{L_{2,\lambda}(S;H)}.$$

Therefore from (2.181) we will obtain

$$[A_\lambda^2 y_\lambda, y_\lambda - \xi_\lambda]_- \geq [A_\lambda^2 \xi_\lambda, y_\lambda - \xi_\lambda]_- - \widetilde{C}_B(R; \|y_\lambda - \xi_\lambda\|_{L_{2,\lambda}(S;H)}).$$

The proof of N-s.b.v. for $A_\lambda^2 : X \to C_v(X^*)$ finishes that fact, that the embedding $W \subset L_{2,\lambda}(S;H)$ is compact. It is the direct sequence of the compactness of embedding $W \subset Y$.

Let us check now λ_0-pseudomonotony A^2_λ on W. Let us $y_{\lambda,n} \to y_\lambda$ weakly in W (therefore $y_{\lambda,n} \to y_\lambda$ in $Y_\lambda := L_{2,\lambda}(S;H)$), $A^2_\lambda(y_{\lambda,n}) \ni d_{\lambda,n} \to d_\lambda \in X^*$ weakly in X^* and

$$\overline{\lim_{n\to\infty}} \langle d_{\lambda,n}, y_{\lambda,n} - y_\lambda\rangle \le 0.$$

Using the property of N-semibounded variation on W for the operator A^2_λ, we conclude, that for every $v_\lambda \in X$

$$\begin{aligned}\underline{\lim_{n\to\infty}} \langle d_{\lambda,n}, y_{\lambda,n} - v_\lambda\rangle &\ge \underline{\lim_{n\to\infty}} [A^2_\lambda(y_{\lambda,n}), y_{\lambda,n} - v_\lambda]_- \\ &\ge \underline{\lim_{n\to\infty}} [A^2_\lambda(v_\lambda), y_{\lambda,n} - v_\lambda]_- - \widetilde{C}_B(R; \|y_\lambda - v_\lambda\|_{y_\lambda}).\end{aligned} \tag{2.182}$$

At first let us estimate the first term of the right part in (2.182). Let us prove, that function $Y_\lambda \ni h_\lambda \mapsto [A^2_\lambda(v_\lambda), h_\lambda]_-$ is continuous $\forall v_\lambda \in X \subset Y_\lambda$. Let $z_{\lambda,n} \to z_\lambda$ in Y_λ, then for any $n \ge 1$ $\exists \xi_{\lambda,n} \in A^2_\lambda(v_\lambda)$ such, that

$$[A^2_\lambda(V_\lambda), z_{\lambda,n}]_- = \langle \xi_{\lambda,n}, z_{\lambda,n}\rangle.$$

From the sequence $\{\xi_{\lambda,n}; z_{\lambda,n}\}$ let us choose the subsequence $\{\xi_{\lambda,m}; z_{\lambda,m}\}$ such, that

$$\underline{\lim_{n\to\infty}} [A^2_\lambda(v_\lambda), z_{\lambda,n}]_- = \underline{\lim_{n\to\infty}} \langle \xi_{\lambda,n}, z_{\lambda,n}\rangle = \lim_{m\to\infty} \langle \xi_{\lambda,m}, z_{\lambda,m}\rangle$$

and in view of the weak compactness of the set $A^2_\lambda(v_\lambda)$ in Y_λ we find, that $\xi_{\lambda,m} \to \xi_\lambda$ weakly in Y^*_λ with $\xi_\lambda \in A^2_\lambda(v_\lambda)$. Therefore

$$\underline{\lim_{n\to\infty}} [A^2_\lambda(v_\lambda), z_{\lambda,n}]_- = \lim_{n\to\infty} \langle \xi_{\lambda,m}, z_{\lambda,m}\rangle = \langle \xi_\lambda, z_\lambda\rangle = [A^2_\lambda(v_\lambda), z_\lambda]_-,$$

and this proves the continuity of function $Y_\lambda \ni h_\lambda \mapsto [A^2_\lambda(v_\lambda), h_\lambda]_-$.

Therefore from (2.182) we obtain

$$\underline{\lim_{n\to\infty}} \langle d_{\lambda,n}, y_{\lambda,n} - v_\lambda\rangle \ge [A^2_\lambda(v_\lambda), y_\lambda - v_\lambda]_- - \widetilde{C}_B(R; \|y_\lambda - v_\lambda\|_{Y_\lambda}).$$

Then, substituting v_λ for y_λ in the last inequality, we have

$$\langle d_{\lambda,n}, y_{\lambda,n} - y_\lambda\rangle \to 0,$$

and therefore

$$\underline{\lim_{n\to\infty}} \langle d_{\lambda,n}, y_{\lambda,n} - v_\lambda\rangle \ge [A^2_\lambda(v_\lambda), y_\lambda - v_\lambda]_- - \widetilde{C}_B(R; \|y_\lambda - v_\lambda\|_{Y_\lambda}) \quad \forall v_\lambda \in X.$$

Substituting in the last inequality v_λ for $tw_\lambda + (1-t)y_\lambda$, where $w_\lambda \in X, t \in [0, 1]$, then dividing the result by t and passing to the limit as $t \to 0+$, thanks to the radial semicontinuity from above for A_λ^2, we obtain

$$\varliminf_{n\to\infty} \langle d_{\lambda,n}, y_{\lambda,n} - w_\lambda\rangle \geq [A_\lambda^2(y_\lambda), y_\lambda - w_\lambda]_- \quad \forall w_\lambda \in X.$$

Therefore, λ_0-pseudomonotony A_λ^2 on W is proved.

In order to prove the λ_0-pseudomonotony of A_λ on W we note, that the pair of maps $(A_\lambda^1, A_\lambda^2)$ is s-mutually bounded, as A_λ^2 is bounded as the consequence of the (2.172) and the boundness of identical motion.

Conditions (α_3) and (α_4) are clear.

In order to prove the solvability for problem (2.174) we will use Theorem 2.2. Let $L : W_{\bar{0}} \subset X \to X^*$ is linear densely defined operator ($Ly_\lambda = y_\lambda'$, $D(L) = W_{\bar{0}} = \{y \in W \mid y(0) = \bar{0}\}$), $A_\lambda : X \to C_v(X^*)$ is multivalued map. The next problem is being considered:

$$Ly_\lambda + A_\lambda(y_\lambda) \ni f_\lambda, \quad y_\lambda \in W_{\bar{0}}. \tag{2.183}$$

Remark 2.22. Let us note, that $D(L) = W_{\bar{0}}$ is reflexive Banach space with regard to the graph norm of the derivative. The closure of the graph of derivative in the sense of distributions on the space $W_{\bar{0}}$ provides the given condition.

Therefore, there exists at last one solution of problem (2.174) $v \in W$, that is obtained by the method of singular perturbations.

The Theorem is proved. □

Corollary 2.6. *Let $p_2 \geq 2$. Let $\lambda_A \geq 0$ is fixed, $I : X \to X^*$ is identical motion, $p_0 = \min\{p_1, p_2\}$, space V is compactly embedded in Banach space V_0 and embedding $V_0 \subset V^*$ is continuous. Let us suggest, that $A + \lambda_A I : X \to C_v(X^*)$ is $+$-coercive, r.l.s.c. multivalued operator of the Volterra type with $(X; W)$-s.b.v. with $\|\cdot\|_W' = \|\cdot\|_{L_{p_0}(S;V_0)}$, that satisfies Condition (H); $B : Y \to C_v(Y^*)$ is multivalued operator of the Volterra type, that satisfies Condition (H), the growth condition (2.172) and the continuity condition (2.173); $C : X \to X^*$ is the operator with such property:*

$$(Cu)(t) = C_0 u(t) \quad \forall u \in X, \quad \forall t \in S,$$

where $C_0 : V_2 \to V_2^*$ is linear, bounded, self-adjoint, monotone operator.

Then for arbitrary $a_0 \in V$ and $f \in X^*$ there exists at last one solution of the problem

$$\begin{cases} y'' + Ay' + By + Cy \ni f, \\ y(0) = a_0, \ y'(0) = \bar{0}, \ y \in C(S; V), \ y' \in W, \end{cases} \tag{2.184}$$

$y \in X$, that is obtained by the method of singular perturbations, at that $y' \in W$.

Proof. Let again R is the operator of the Volterra type, that is defined by relation

$$(Rv)(t) = a_0 + \int_0^t v(s)ds \quad \forall v \in X, \quad \forall t \in S.$$

Let as set $\hat{A} = A + C \circ R$, B is the operator from condition. Let us note, that in view of the properties of operator C_0 we have, that the next conditions are true:

(1) For $v \in X$ in view of that $p_2 \ge 2 \ge q_2$ and boundness C_0 we have

$$\|Cv\|_{X^*} = \left(\int_S \|C_0 v(t)\|_{V_2^*}^{q_2} dt \right)^{\frac{1}{q_2}} \le K_1 \left(\int_S \|v(t)\|_{V_2}^{q_2} dt \right)^{\frac{1}{q_2}}$$

$$\le K_2 \left(\int_S \|v(t)\|_{V_2}^{p_2} dt \right)^{\frac{1}{p_2}} = K_2 \|v\|_X,$$

where K_1 and K_2 are the constants, which don't depend on v. From here and from Lipschitz-continuity of $R : X \to X$ it follows the continuity of the operator $C \circ R : X \to X^*$;

(2) For $v, w \in X$ we have

$$\langle CRv - CRw, v - w \rangle = \int_S (C_0(Rv - Rw)(t), (Rv - Rw)'(t)) dt$$

$$= \frac{1}{2}(C_0(Rv - Rw)(T), (Rv - Rw)(T)) \ge 0$$

i.e. the operator $CR : X \to X^*$ is monotone;

(3) From the definition of operator R we have, that

$$\langle CRv, v \rangle \ge -\frac{1}{2}(C_0 a_0, a_0).$$

Therefore

$$\varliminf_{\|v\|_X \to \infty} \frac{1}{\|v\|_X} \langle CRv, v \rangle > -\infty.$$

Thus, in view of these conditions and conditions on the operator $A + \lambda_A I$ we have, that the operator $\hat{A} + \lambda_A I$ satisfies the conditions of Theorem 2.7. Therefore the problem

$$\begin{cases} y'' + \hat{A} y' + By \ni f, \\ y(0) = a_0, \ y'(0) = \bar{0}, \ y \in C(S; V), \ y' \in W, \end{cases}$$

has a solution. But, as $(Ry)' = y$, then the same is true for problem (2.184) too. □

2.6.3 Examples

Example 2.7. Let Ω from $\mathbb{R}^n$ is bounded region with regular boundary $\partial\Omega$, $S = [0; T]$, $Q = \Omega \times S$, $\Gamma_T = \partial\Omega \times S$, $1 < p = p_1 = p_2$, $\Phi : \mathbb{R} \to \mathbb{R}$ is continuous function, that satisfies the growth condition:

$$\text{for some } c_1, c_2 \in \mathbb{R} \quad |\Phi(t)| \leq c_1|t| + c_2 \quad \forall t \in \mathbb{R} \tag{2.185}$$

and the sign condition:

$$\exists c_3 > 0: \quad (\Phi(t) - \Phi(s))(t - s) \geq -c_3(s - t)^2 \quad \forall t, s \in \mathbb{R}; \tag{2.186}$$

$S \times \mathbb{R} \ni (t, y) \to \theta_i(t, y) \in \mathbb{R}_+, i = 1, 2$ are single-valued continuous functions, which satisfy such condition:

$$\exists c_1, c_2 \geq 0: \quad -c_2(1 + |x|) \leq \theta_1(t, x)$$

$$\leq \theta_2(t, x) \leq c_1(1 + |x|) \quad \forall t \in S, x \in \mathbb{R} \tag{2.187}$$

For arbitrary $f \in X^* = L_2(S; L_2(\Omega)) + L_q(S; W^{-1,q}(\Omega) + L_2(\Omega))$ let us consider the problem:

$$\frac{\partial^2 y(x,t)}{\partial t^2} - \sum_{i=1}^{n} \frac{\partial}{\partial x_i}\left(\left|\frac{\partial^2 y(x,t)}{\partial x_i\, \partial t}\right|^{p-2} \frac{\partial^2 y(x,t)}{\partial x_i\, \partial t}\right)$$

$$+ \left|\frac{\partial y(x,t)}{\partial t}\right|^{p-2} \frac{\partial y(x,t)}{\partial t} + \Phi\left(\frac{\partial y(x,t)}{\partial t}\right)$$

$$+ [\theta_1(t, y(x,t)); \theta_2(t, y(x,t))] \ni f(x,t) \qquad \text{a.e. on } Q, \tag{2.188}$$

$$y(x,0) = 0, \qquad \left.\frac{\partial y(x,t)}{\partial t}\right|_{t=0} = 0 \qquad \text{a.e. on } \Omega,$$

$$y(x,t) = 0 \qquad \text{a. e. on } \partial\Omega.$$

As the operator $A : L_p(S; W_0^{1,p}(\Omega) \cap L_2(\Omega)) \to L_q(S; W^{-1,q}(\Omega) + L_2(\Omega))$ let us take $(Au)(t) = A(u(t))$ [ZK07, Z06], where

$$A(\varphi) = A_1(\varphi) + A_2(\varphi) \qquad \forall \varphi \in C_0^2(\bar{\Omega}),$$

$$A_1(\varphi) = -\sum_{i=1}^{n} \frac{\partial}{\partial x_i}\left(\left|\frac{\partial \varphi}{\partial x_i}\right|^{p-2} \frac{\partial \varphi}{\partial x_i}\right) + |\varphi|^{p-2}\varphi, \quad A_2(\varphi) = \Phi(\varphi),$$

and as the operator $B : L_2(Q) \to L_2(Q)$ let us take

$$B(u) = \{v \in L_2(Q) \mid \theta_1(t, u(x,t)) \leq v(x,t) \leq \theta_2(t, u(x,t)) \text{ for a.e. } (x,t) \in Q\}.$$

Let us set $H = L_2(\Omega)$, $V_1 = V_2 = V = W_0^{1,p}(\Omega) \cap L_2(\Omega)$ and consider

$$X = L_p(S; V) \cap L_2(S; H), \quad X^* = L_q(S; V^*) + L_2(S; H),$$
$$Y = L_2(S; H) = L_2(Q).$$

Then problem (2.188) has the solution $y \in X, y' \in C(S;V), y'' \in X^*$, that is obtained by the method of singular perturbations.

Example 2.8. Let Ω from $\mathbb{R}^n$ is bounded region with regular boundary $\partial\Omega$, $S = [0;T]$, $Q = \Omega \times S$, $\Gamma_T = \partial\Omega \times S$, $p = p_1 = p_2 \geq 2$, $\Phi : \mathbb{R} \to \mathbb{R}$ is continuous function, that satisfies the growth condition (2.185) and the "sign condition" (2.186); $S \times \mathbb{R} \ni (t,y) \to \theta_i(t,y) \in \mathbb{R}_+, i = 1,2$ are single-valued continuous functions, which satisfy the condition (2.187).

For arbitrary $f \in X^* = L_2(S;L_2(\Omega)) + L_q(S;W^{-1,q}(\Omega))$ let us consider the problem:

$$\begin{aligned}
&\frac{\partial^2 y(x,t)}{\partial t^2} - \sum_{i=1}^{n} \frac{\partial}{\partial x_i}\left(\left|\frac{\partial^2 y(x,t)}{\partial x_i\,\partial t}\right|^{p-2} \frac{\partial^2 y(x,t)}{\partial x_i\,\partial t}\right) + \left|\frac{\partial y(x,t)}{\partial t}\right|^{p-2} \frac{\partial y(x,t)}{\partial t} \\
&\quad + \Phi\left(\frac{\partial y(x,t)}{\partial t}\right) - \Delta y(x,t) \\
&\quad + [\theta_1(t,y(x,t)); \theta_2(t,y(x,t))] \ni f(x,t) \qquad \text{a.e. on } Q, \qquad (2.189) \\
&\quad y(x,0) = 0, \quad \left.\frac{\partial y(x,t)}{\partial t}\right|_{t=0} = 0 \qquad \text{a.e. on } \Omega, \\
&\quad y(x,t) = 0 \qquad \text{a. e. on } \partial\Omega.
\end{aligned}$$

As the operator $A{:}L_p(S;W_0^{1,p}(\Omega)) \to L_q(S;W^{-1,q}(\Omega))$ let us take $(Au)(t) = A(u(t))$ [ZK07, Z06], where

$$\begin{aligned}
&A(\varphi) = A_1(\varphi) + A_2(\varphi) \qquad \forall \varphi \in C_0^2(\bar{\Omega}), \\
&A_1(\varphi) = -\sum_{i=1}^{n} \frac{\partial}{\partial x_i}\left(\left|\frac{\partial\varphi}{\partial x_i}\right|^{p-2} \frac{\partial\varphi}{\partial x_i}\right) + |\varphi|^{p-2}\varphi, \quad A_2(\varphi) = \Phi(\varphi);
\end{aligned}$$

as the operator $B : L_2(Q) \to L_2(Q)$ let us take

$$B(u) = \{v \in L_2(Q) \mid \theta_1(t,u(x,t)) \leq v(x,t) \leq \theta_2(t,u(x,t)) \text{ for a.e. } (x,t) \in Q\},$$

and as the operator $C : L_2(S;H_0^1(\Omega)) \to L_2(S;H^{-1}(\Omega))$ let us take the operator with such property

$$(Cu)(t) = C_0u(t), \quad \text{where} \quad C_0(v) = -\Delta v, \quad v \in H_0^1(\Omega).$$

Let us set $H = L_2(\Omega)$, $V_1 = W_0^{1,p}(\Omega)$, $V_2 = H_0^1(\Omega)$ and consider

$$X = L_p(S;V_1) \cap L_2(S;H) \cap L_2(S;V_2), \quad X^* = L_q(S;V_1^*) + L_2(S;H),$$

$$Y = L_2(S;H) = L_2(Q).$$

Then problem (2.189) has the solution $y \in X, y' \in X, y'' \in X^*$, that is obtained by the method of singular perturbations.

2.7 The Practical Experience of the Modelling of Harmful Impurities Expansion Process in the Atmosphere Taking into Account the "Heat Island" Effect

The elaboration of mathematical methods for the processes of harmful impurities expansion in the atmosphere is the necessary stage for the mathematical formalization of physical processes of the environment pollution. The use of the method of mathematical physics is the natural approach to the description of physical processes of harmful ingredients expansion. The models of atmospheric diffusion and transfer of impurities [M82] based on these methods were worked up and how they are widespread.

According to the approach of mathematical physics, the problems of atmospheric diffusion and transfer are formulated as variation principles which are reduced to boundary value problems for partial differential equations if there are no additional requirements for the solution [M82]. If it is necessary to take into account the additional conditions for the solutions of problem then, according to practice, the models of processes, which are under consideration, are being expanded at the expense of relations, which describe these conditions or correct the original model [M82]. That is why the mentioned additional relation often have empiric character.

Let us take into account of additional conditions for the solution of the problem on the stage of its variation definition. The more substantial way of the working up of models of processes, which are under consideration. Such approach leads to variation inequalities, which are introduced and developed in [DL76,P85,F64,S33].

Let us characterize the problem that is being considered. Meteorological factors as well as physical–chemical properties of impurities have the substantial influence on the processes of harmful impurity transfer in the atmosphere. The influence of many factors as, for example, wind, temperature and others, are taken into account in the context of traditional variation definitions and can be formalized in the context of boundary value problems for the partial differential equations. At the same time the influence of some meteorological factors (inverse stratification – inversion, temperature anomalies of atmosphere), physical–chemical properties of impurities and other factors leads to the appearance of specific effects of unidirectional conduction of boundary and effects of obstacle inside the boundary. The equations of diffusion and transfer in this case are not the best form of introduction for the processes which are under consideration [ANW67]. Let us consider several varieties of the process of harmful impurities expansion in the atmosphere which appear under the influence of mentioned factors. Let us formulate these processes in the form of variation inequalities, propose algorithms of their numerical realization and give the results of calculations.

Let us consider the problem of modeling the process of harmful impurities expansion in the atmosphere taking into account the "heat island" effect. As the example let us consider the problem of short-term prediction of processes of impurities expansion over some city under the condition of mentioned anomaly.

Heat anomalies over big settlements which appear under the action of meteorological and anthropogenic factors lie in systematical excess of the atmosphere air temperature in the centers of mentioned settlements over the temperature of the neighbourings ones. The sequence of this anomaly is the ascending air movement and the convergence of air current in the horizontal plane. We can point out the situation when as a result of action, for example, of elevated inversion the ascending air remains in the inversion layer, at that the horizontal convergence is being going on. As investigations showen, the wind velocity, caused by the "heat island" is small (within 1 m/s) [VLM91]. The spatio-temporal characteristics of the action of boundary conditions are unknown as in the previous example.

The influence of "heat island" on the processes of impurity expansion in the atmosphere leads to different anomalies, particularly to the accumulation of harmful impurity in the area of "heat island" activity. Let us work up the mathematical model of the processes of harmful impurity expansion in the atmosphere taking into account the "heat island" effect and propose the calculating procedure of its realization.

In order to solve this problem we will use the variation inequalities theory [VLM91] that takes into account the nonlinear effects of boundary unidirectional conduction.

Let $y(t,x)$ be the concentration of harmful impurity in the atmosphere defined on the bounded open set Ω of the space $\mathbb{R}^3$ with the smooth bound $\Gamma = \Gamma_L \cup \Gamma_R \cup \Gamma_U \cup \Gamma_D \cup \Gamma_F \cup \Gamma_T$ and on the interval of time $(0, t_k)$ for $t_k < \infty$, $Q = \Omega \times (0, t_k)$, $\Sigma = \Gamma \times (0, t_k)$ and it is the solution of the variation inequality [DL76]:

$$\left(\frac{\partial y}{\partial t}, v - y\right) + (A(\lambda)y, v) + \psi_i(v) - \psi_i(y) \geq (f, v - y) \text{ in } Q,$$

$$i = 1, 2 \quad \forall v \in H^1(\Omega) = V \tag{2.190}$$

with the initial condition

$$y|_{t=0} = y_0 \text{ in } Q, \tag{2.191}$$

where (f, g) is the action of functional $f \in (H^1(\Omega))^*$ on the element $g \in (H^1(\Omega))^*$ and the operator $A(\lambda) : V \to V^*$ is defined by bilinear form

$$(A(\lambda)y, \xi) = -\sum_{i=1}^{3} \int_{\Omega} \left(k(x)\frac{\partial y}{\partial x_i}\frac{\partial \xi}{\partial x_i} - c_i(x)\frac{\partial y}{\partial x_i}\xi\right) dx$$

$$+ \int_{\Omega} d(x)y\xi dx \quad \forall\, \xi \in H^1(\Omega) \tag{2.192}$$

if $f, g \in L^2(\Omega)$, the operation (f, g) coincides with the inner product in $L^2(\Omega)$, i.e. $(f, g) = \int_{\Omega} f(x)g(x)dx$; $k(x)$ is the coefficient of turbulent diffusion, $x = (x_1, x_2, x_3)$; $c_i(x)$ for $i = 1, 2, 3$ the wind axillary parameters on x_1, x_2 and x_3 respectively; $d(x)$ is the impurity absorption coefficient. The variable $f(t, x) = \sum_{j=1}^{k} q_j(t)\delta(x - x^j)$ is the force function of the process; $q_j(t)$ is the sources function operated in subspaces $\Omega_j \in \Omega$, $j = 1, \ldots, K$; K is the number of external action points: $\delta(x - x_j)$ is the characteristic function. Let us set up that parameters $\lambda = \{k(x), c_j(x), d(x)\}$ and the force function of the process f have the technological restrictions.

We will take the system (2.190), (2.191) as the mathematical description of the processes of harmful impurities expansion in the atmosphere which progress under the "heat island" influence. Physical conditions provide the fulfilment of the next conditions on the lateral surface $\Gamma_s = \Gamma_l \cup \Gamma_r \cup \Gamma_f \cup \Gamma_t$ of the considering spatial region Ω: if $y_{ext}(s) > y(t, x)|_{\Gamma_s}$, where $y(t, x)|_{\Gamma_s}$ and $y_{ext}(s)$ are the concentrations of harmful impurity on the bound Γ_s and its external side then the bound is open and the impurity penetrates from outside into considering region. When $y(t, x)|_{\Gamma_s} \geq y_{ext}(s)$ then the capacity of the bound Γ_s has zero values and the bound is being closed. The formulated physical conditions are satisfied by the system (2.190), (2.191), if the functional ψ_1 is defined by

$$\psi_1(y) = \begin{cases} \frac{1}{2}\tau(t, s)\left(y(t, x)|_{\Gamma_s} - y_{ext}(s)\right)^2, & y(t, x)|_{\Gamma_s} < y_{ext}(s), \\ 0, & y(t, x)|_{\Gamma_s} \geq y_{ext}(s), \end{cases} \tag{2.193}$$

where $\tau(t, s)$ is the capacity of the bound Γ_s. Let us define τ by $\tau \in L^\infty(Q)$, then the space of parameters τ is defined by $T = L^\infty(Q)$ with the norm $\|\tau\|_T = \|\tau\|_{L^\infty(Q)}$. The admissible parameters set T_{adm} has the form $T_{adm} = \{\tau \in T | \tau_{max} \geq \tau \geq 0 \text{ a.e.}\}$

Taking into account (2.193) the function $\varphi_1(y)$ satisfying the semiimpenetrability conditions (thick side) will have the form:

$$\varphi_1(y) = \begin{cases} \tau(t, s)[y(t, x)|_{\Gamma_s} - y_{ext}(s)], & y(t, x)|_{\Gamma_s} < y_{ext}(s), \\ 0 & y(t, x)|_{\Gamma_s} \geq y_{ext}(s) \end{cases}$$

or $\varphi_1(y) = \tau(y; t, s)y(t, x)|_{\Gamma_s}$

$$\tau(y; t, s) = \begin{cases} \tau(t, s), & y(t, x)_{\Gamma_s} < y_{ext}(s), \\ 0, & y(t, x)_{\Gamma_s} \geq y_{ext}(s), \end{cases} \tag{2.194}$$

where $\tau(t, s)$ is known. So, the relation (2.194) defines the coefficient $\tau(y; t, s)$ of known structure, the spatio-temporal characteristics for which are unknown.

The problem of variation inequalities (2.190), (2.191) for the functional ψ_1 defined by the relation (2.193), can be reduced to nonlinear problem with unknown bounds:

$$\frac{\partial y}{\partial t} + A(\lambda)y = f \text{ in } Q; \tag{2.195}$$

$$\left.\frac{\partial y}{\partial n}\right|_{\Gamma_s} = -\varphi_1(y); \tag{2.196}$$

$$\left.\frac{\partial y}{\partial n}\right|_{\Gamma-\Gamma_s} = v, \quad v = \text{const}; \tag{2.197}$$

$$y_{|t=0} = y_0 \text{ in } \Omega.$$

The solution of this system are defined by the pair $\{\hat{y}(t,x), \hat{\tau}(y;t,s)\}$. If unknown $\hat{y}(t,x)$ is the solution of the system (2.195)–(2.197), then the search of spatio-temporal characteristics for the coefficient $\hat{\tau}(y;t,s)$ is individual problem.

In order to solve the mentioned problem let us suppose that $\tau(t,s)$ in the relation (2.194) is unknown. We will replace the search of unknown spatio-temporal characteristics for coefficients $\hat{\tau}(y;t,s)$ by the search of unknown $\widehat{\tau}(t,s)$. Taking into account the influence on the capacity of bound of physical effects of inversion, when $y(t,x) \geq y_{\max}(x)$ might have been equal and when $y_{\max}(x) > y(t,x)$, $\hat{\tau}(t,s)$ might reach the maximal value.

Taking into account mentioned facts the problem of solving of variation inequality (2.190), (2.191) is being reduced to the optimization problem of search of unknown parameter $\widehat{\tau}(t,s)$ that satisfies the system (2.195)–(2.197) and provides the minimum of functional

$$J(\tau) = \int_0^{t_k}\int_{\Gamma_s} \exp\left(y(t,s) - y_{\text{ext}}(s)\right) ds dt, \begin{cases} y|_{\Gamma_s} \geq y_{\text{ext}}, \\ y|_{\Gamma_s} < y_{\text{ext}}, \end{cases} \to \inf_{\tau\in T_{\text{adm}}}, \tag{2.198}$$

where the upper branch of the functional $J(\cdot)$ minimizes $\tau(t,s)$ in the case $y(t,x)|_{\Gamma_s} \geq y_{\text{ext}}(s)$ otherwise the lower branch of this functional maximizes $\tau(t,s)$. Taking into account just mentioned facts as the solution of sought problem we will understand the pair $\{\hat{y}(t,x), \tau(t,s)\}$. Let us note that the functional $J(\cdot)$ is continuously differentiable by y and τ.

Let us solve the formulated minimization problem by the Lagrange method [161]. In this case the system (2.195)–(2.198) will have the form

$$L(\tau,y,p) = J(\cdot) + \left.\left(\frac{\partial y}{\partial t} + A(\lambda)y - f, p\right)\right|_{\Gamma_s} \to \inf_{\tau\in T_{\text{adm}}} \tag{2.199}$$

with boundary and initial conditions (2.196), (2.197), where $p(t,x)$ is unknown variable that will be defined later.

The necessary conditions of optimum for the formulated problem with regard to unknown parameter $\tau \in T_{\text{adm}}$ will have the form

$$\delta L(\tau) = \frac{\partial L}{\partial \tau}\delta\tau = 0 \quad \forall \tau \in T_{\text{adm}} \tag{2.200}$$

For $\tau \notin T_{\text{adm}}$ it is necessary to complete 2.200 with conditions

$$\frac{\partial L}{\partial \tau}(\cdot) = 0 \quad \forall \tau \notin T_{\text{adm}} \tag{2.201}$$

Varying the functional (2.199), we can show that in (2.201)

$$\frac{\partial L}{\partial \tau}(\cdot) = -\left(y(t,x)|_{\Gamma_s} - y_{\text{ext}}(s)\right) p(t,x)|_{\Gamma_s} \tag{2.202}$$

where $p(t,x)$ is the adjoint function that can be found from the solution of nonlinear adjoint system

$$-\frac{\partial p}{\partial t} + A^*(\lambda)p = 0, \tag{2.203}$$

with boundary and terminal conditions

$$\left.\frac{\partial p}{\partial n}\right|_{\Gamma_s} = \tau(t,s)p(t,x)|_{\Gamma_s} - \exp\left(y(t,s) - y_{\text{ext}}(s)\right), \quad \begin{cases} y|_{\Gamma_s} \geq y_{\text{ext}}, \\ y|_{\Gamma_s} < y_{\text{ext}}. \end{cases} \tag{2.204}$$

$$\left.\frac{\partial y}{\partial n}\right|_{\Gamma - \Gamma_s} = 0, \quad p|_{t=t_k} = 0, \tag{2.205}$$

$$A^*(\lambda)(\cdot) = -\left\{\sum_{i=1}^{3}\left[\frac{\partial}{\partial x_i}\left(k(x)\frac{\partial}{\partial x_i}\right) + c_i(x)\frac{\partial}{\partial x_i}\right] - d(x)\right\}(\cdot).$$

The procedure of search of unknown parameter τ is based on the relation of gradient in the form

$$\tau^{i+1} = \Pr\left\{\tau^i - \lambda_\tau\left(\frac{\partial L}{\partial \tau}\right)^i\right\}, \tag{2.206}$$

where i is the number of gradient cycle τ^0 and λ_τ are given.

The search of unknown variable based on the gradient relation (2.206), when the criterion of finishing is over:

$$\frac{\left|J^i - J^{i+1}\right|}{J^i} \leq \varepsilon, \tag{2.207}$$

and unknown variable will have the value $\hat{\tau}$.

Joining relations (2.195)–(2.197), (2.201)–(2.205) with (2.206), (2.207), we obtain the algorithm of realization of our problem:

1. To $i = 0$, where i is the index of current iteration we give the starting value τ^0.
2. For the step $i+1$, taking into account known τ^i in view of (2.201) and (2.202), we compute $\frac{\partial L}{\partial \tau}$, where y and p are defined by relations (2.195)–(2.197) and (2.203)–(2.205) respectively.
3. In view of (2.206) we define the value of τ^{i+1}.

4. We compute (2.198) and check the condition (2.207). If it is fulfilled then we have the finishing of algorithm else we pass to point 2.

The result of realization of un. 1–4 of algorithm is the totality $\{\hat{y}, \hat{\tau}\}$, that defines the solution of the problem of modelling the processes of harmful impurity expansion in the atmosphere taking into account the effects defined by the temperature inversion.

References

[ANW67] Alberg J, Nilson E, Walsh J (1967) The theory of splines and their applications. Academic, New York

[AK03] Ansari QH, Khan Z (2003) Relatively B-pseudomonotone variational inequalities over product of sets. J Inequal Pure Appl Math 4, Art. 6

[AE84] Aubin J-P, Ekeland I (1984) Applied nonlinear analysis. Wiley-Interscience, New York

[B76] Barbu V (1976) Nonlinear semigroups and differential equations in Banach spaces. Editura Academiei Noordhoff International, Bucuresti

[CVM04] Carl S, Le VyK, Motreanu D (2004) Existence and comparison results for quasilinear evolution hemivariation inequalities. Electron J Differ Equat 57:1–17 http://www.emis.de/journals/EJDE/2004/57/carl. Accessed 13 Apr 2004

[C90] Clarke FH (1990) Optimization and nonsmooth analysis. SIAM, Philadelphia

[DMP03] Denkowski Z, Migorski S, Papageorgiou NS (2003) An introduction to nonlinear analysis: applications. Kluwer, New York

[DM05] Denkowski Z, Migorski S (2005) Existence of solutions to evolution second order hemivariational inequalities with multivalued damping. System modeling and optimization. doi:10.1007/0-387-23467-5_14

[D65] Dubinskii Yu (1965) A weak convergence in nonlinear elliptic and parabolic equations. Mat Sb 67:609–642

[DL76] Duvaut G, Lions J-L (1976) Inequalities in mechanics and physics [in Russian], Springer, Berlin

[ET99] Ekeland I, Temam R (1999) Convex analysis and variational problems. SIAM, Philadelphia

[F64] Fichera G (1964) Problemi elastostatici con vincoli unilateral: il problema di Signorini con ambigue condizioni al contorno. Mem Accad Naz Lincei 8:91–140

[GGZ74] Gajewski H, Gröger K, Zacharias K (1974) Nichtlineare operatorgleichungen und operatordifferentialgleichungen. Akademie, Berlin

[GS02] Gasinski L, Smolka M (2002) An existence theorem for wave-type hyperbolic hemivariational inequalities. Math Nachr. doi:10.1002/1522-2616

[K94] Kamenskii MI (1994) Applications des operateurs condencents au problemes des solutions periodiques pour les equations paraboliques quasilineares. C R Acad Sci Paris Ser I Math 318:719–722

[KO91] Kamenskii MI, Obukhovskii VV (1991) On periodic solutions of differential inclusions with unbounded operators in Banach spaces. Zb Rad Prirod Mat Fak Ser Mat 21: 173–191

[KO93] Kamenskii MI, Obukhovskii VV (1993) Condensing multioperators and periodic solutions of parabolic functional-differential inclusions in Banach spaces. Nonlinear Anal Theor Meth Appl. doi:10.1016/0362-546X(93)90068-4

[KOZ96] Kamenskii MI, Obukhovskii VV, Zecca P (1996) Method of the solution sets for a quasilinear functional-differential inclusion in a Banach space. Diff Equat Dyn Syst 4:339–350

[KMV03] Kapustyan OV, Mel'nik VS, Valero J (2003) Attractors of multivalued dynamical processes generated by phase-field equations. Int J Bifurcat Chaos. doi:10.1142/S0218127403007801

[KK03a] Kapustyan AV, Kasyanov PO (2003a) Multivalued dinamics for nonautinomous inclusion with upper semicontinuous right-hand side. Rep NAS Ukr 5:13–16

[KK03b] Kapustyan AV, Kasyanov PO (2003b) Global attractor for nonautinomous inclusion with discontinuous right-hand side. Ukr Math J. doi:10.1023/B:UKMA.0000027041.12041.e8

[KMV07] Kapustyan OV, Mel'nik VS, Valero J (2007) A weak attractor and properties of solutions for the three-dimensional Benard problem. Discrete Contin Dyn Sys 18:449–481

[KKK08a] Kapustyan VO, Kasyanov PO, Kohut OP (2008a) On the solvability of one class of parameterized operator inclusions. Ukr Math J. doi:10.1007/s11253-009-0179-z

[KKK08b] Kapustyan VO, Kasyanov PO, Kohut OP (2008b) On properties of solutions for parameterized operator inclusions. Rep NTUU KPI 5:129–136

[K05a] Kasyanov PO (2005a) Galerkin method for one class of differential-operator inclusions with multivalued maps of pseudomonotone type. Naukovi visty NTUU "KPI" 2: 139–151

[K05b] Kasyanov PO (2005b) Faedo-Galerkin method for differential-operator inclusions. Rep NAS Ukr 9:20–24

[KMP08] Kasyanov PO, Mel'nik VS, Piccirillo AM (2008) Local subdifferentials and multivariational inequalities in Banach and Frechet spaces. Opusc matem 28:295–311

[KMV08] Kasyanov PO, Mel'nik VS, Valero J (2008) On the method of approximation for evolutionary inclusions of pseudomonotone type. Bull Aust Math Soc. doi:10.1017/S0004972708000130

[KM05a] Kasyanov PO, Mel'nik VS (2005a) Faedo-Galerkin method for differential-operator inclusions in Banach spaces with w_{λ_0}-pseudomonotone maps. Nat Acad Sci Ukr Kiev Inst Math Prepr Part 2 1:82–105

[KM05b] Kasyanov PO, Mel'nik VS (2005b) On some properties of subdifferential mappings in Frechet spaces. Ukr Math J. doi:10.1007/s11253-006-0017-5

[KM05c] Kasyanov PO, Mel'nik VS (2005c) On subdifferential maps in Banach and Frechet spaces. Rep NAS Ukr 12:23–27

[KM06] Kasyanov PO, Mel'nik VS (2006) Differential-operator inclusions in Banach spaces with W_λ-pseudomonotone maps. Nonlinear Bound Value Probl 16:46–68

[KMT05] Kasyanov PO, Mel'nik VS, Toscano L (2005) Method of approximation of evolutionary inclusions and variation inequalities by stationary. Sys Res Inf Tech 4:106–119

[KMT06] Kasyanov PO, Mel'nik VS (2006) Toscano S periodic solutions for nonlinear evolution equations with W_{λ_0}-pseudomonotone maps. Nonlinear Oscil. doi:10.1007/s11072-006-0037-y

[KMP06a] Kasyanov PO, Mel'nik VS, Piccirillo AM (2006a) On shauder basis in some Banach spaces. Rep NAS Ukr 4:23–30

[KMP06b] Kasyanov PO, Mel'nik VS, Piccirillo AM (2006b) On some theorems of compact embedding for Banach and locally convex spaces. Rep NAS Ukr 5:34–40

[KMY07] Kasyanov PO, Mel'nik VS, Yasinsky VV (2007) Evolution inclusions and inequalities in Banach spaces with w_λ-pseudomonotone maps. Naukova Dumka, Kiev

[KZY09] Kasyanov PO, Zadoyanchuk NV, Yasinsky VV (2009) Periodic solutions for a class of nonlinear hyperbolic evolution equations. Cybern Syst Anal. doi:10.1007/s10559-009-9149-6

[KM01] Kartsatos A, Markov L (2001) An L^2-approach to second-order nonlinear functional evolutions involving m-accretive operators in Banach spaces. Diff Integral Eqn 14:833–866

[K91] Kisielewicz M (1991) Differential inclusions and optimal control. Polish Scientific Publishers, Warszawa; Kluwer, Dordrecht

[K96] Kogut PI(1996) Asymptotic analysis of controllable parabolic systems with the boundary evolution. Rep NAS Ukr 10:99–103

[K99] Kogut PI (1999) On the topological convergence of epigraphs in partially ordered vector spaces. Math Stud 12:161–170
[K01a] Kogut PI (2001a) On the variational S-compactness of conditional minimization problems. J Math Sci. doi:10.1023/A:1012302232380
[K01b] Kogut PI (2001b) On the topological convergence of sets on parially ordered spaces. Nonlinear Bound Value Probl 11:94–97
[KM01] Kogut PI, Musejko OV (2001) On the S-homogenization of variation inequalities in Banach spaces. Rep NAS Ukr 5:73–76
[K01] Konnov IV (2001) Relatively monotone variation inequalities over product sets. Oper Res Lett. doi:10.1016/S0167-6377(00)00063-8
[KN97] Kovalevsky A, Nicolosi Fr (1997) Boundedness of solutions of variational inequalities with nonlinear degenerated elliptic operators of high order. Appl Anal. doi:10.1080/00036819708840560
[KN99a] Kovalevsky A, Nicolosi Fr (1999a) Boundedness of solutions of degenerate nonlinear elliptic variational inequalities. Nonlinear Anal theor Meth Appl. doi:10.1016/S0362-546X(98)00110-2
[KN99b] Kovalevsky A, Nicolosi Fr (1999b) Integral estimates for solutions of some degenerate local variational inequalities. Appl Anal. doi:10.1080/00036819908840789
[K66] Kuratowski K (1966) Topology. Academic, New York
[LSU68] Ladyzhenskaya OA, Solonnikov VA, Uraltseva NN (1968) Linear and quasilinear equations of parabolic type. American Mathematical Society, Providence
[LO71] Lasota A, Opial Z (1971) An approximation theorem for multi-valued mappings. Podst Starow 1:71–75
[L69] Lions JL (1969) Quelques methodes de resolution des problemes aux limites non lineaires. Dunod Gauthier-Villars, Paris
[LK83] Lyashko AD, Korchevsky MM (1983) Difference methods of resolution of nonlinear problems of filtration theory. Izv Vuzov Matematika 7:28–45
[MNS96] Makler-Scheimberg S, Nguyen VH, Strodiot JJ (1996) Family of perturbation methods for variational inequalities. J Optim Theor Appl. doi:10.1007/BF02192537
[M82] Marchouk G (1982) Mathematical modeling in the environment problem [in Russian]. Nauka, Moscow
[M84] Mel'nik VS (1984) Method of monotone operators in theory of optimal systems with restrictions. Docl USSR A 7:64–67
[M97] Mel'nik VS (1997) Critical points of some classes of multivalued mappings. Cybern Syst Anal. doi:10.1007/BF02665895
[M98] Mel'nik VS (1998) Estimates of the fractal and hausdorff dimensions of sets invariant under multimappings. Math Notes. doi:10.1007/BF02308758
[M00] Mel'nik VS (2000) Multivariational inequalities and operator inclusions in Banach spaces with mappings of the class $(S)_+$. Ukr Mat Zh 52. doi:10.1023/A:1010431221039
[M06] Mel'nik VS (2006) Topological methods in the theory of operator inclusions in Banach spaces. Int Ukr Math J. doi:10.1007/s11253-006-0062-0
[MK98] Mel'nik VS, Kogut PI (1998) On one class of extremum problems for nonlinear operator systems. Cybern Syst Anal. doi:10.1007/BF02667159
[MS97] Mel'nik VS, Solonoukha OV (1997) Stationary variational inequalities with multivalued operators. Cybern Syst Anal. doi:10.1007/BF02733070
[MV98] Mel'nik VS, Valero J (1998) On attractors of multivalued semi-flows and differential inclusions. Set-Valued Anal. doi:10.1023/A:1008608431399
[MV00] Mel'nik VS, Valero J (2000) On global attractors of multivalued semiprocesses and nonautonomous evolution inclusions. Set-Valued Anal. doi:10.1023/A:1026514727329
[M95] Migorski S (1995) Existence, variational and optimal problems for nonlinear second order evolution inclusions. Dynam Syst Appl 4:513–528
[M67] Mosko U (1967) A remark on a theorem of F.E. Browder. J Math Anal Appl. doi:10.1016/0022-247X(67)90108-4

[N99] Nagurney A (1999) Network economics: a variational inequality approach. Kluwer, Dordrecht

[NP95] Naniewicz Z, Panagiotopoulos PD (1995) Mathematical theory of hemivariational inequalities and applications. Marcel Dekker, New York

[NJJ98] Nieto JJ, Jiang Y, Jurang Y (1998) Monotone iterative method for functional differential equations. Nonlinear Anal. doi:10.1016/S0362-546X(97)00524-5

[NOZ94] Nistri P, Obukhovskii VV, Zecca P (1994) On the solvability of systems of inclusions involving noncompact operators. Trans Am Math Soc 342:543–562

[O92] Obukhovskii VV (1992) Semilinear functional-differential inclusions in a Banach space and controlled parabolic systems. Sov J Autom Inform Sci 24:71–79

[OS89] Omatu S, Seinfeld JH (1989) Distributed parameter systems: theory and applications. Clarendon, New York

[P87] Pachpatte BG (1987) Existense theorems for Hammerstein-type integral equation. An sti Univ Iasi Sec Ia 33:197–201

[P85] Panagiotopoulos PD (1985) Inequality problems in mechanics and applications: convex and nonconvex energy functions. Birkhäuser, Boston

[Pan85] Pang J-S (1985) Asymmetric variational inequality problems over product sets: applications and iterative methods. Math Prog. doi:10.1007/BF02591749

[Pap85] Papageorgiou NS (1985) On the theory of Banach space valued multifunctions. 1. Integration and conditional expectation. J Multivar Anal. doi:10.1016/0047-259X(85)90078-8

[Pap86] Papageorgiou NS (1986) Random differential inclusions in Banach spaces. J Differ Equ. doi:10.1016/0022-0396(86)90021-5

[Pap87a] Papageorgiou NS (1987a) On multivalued evolution equations and differential inclusions in Banach spaces. Comment Math Univ Sancti Pauli 36:21–39

[Pap87b] Papageorgiou NS (1987b) Convergence theorems for Banach space valued integrable multifunctions. Int J Math Math Sci. doi:10.1155/S0161171287000516

[Pap88a] Papageorgiou NS (1988a) Volterra integral inclusions in Banach spaces. J Integr Equ Appl. doi:10.1216/JIE-1988-1-1-65

[Pap88b] Papageorgiou NS (1988b) Existence and convergence results for integral inclusions in Banak spaces. J Integr Equ Appl. doi:10.1216/JIE-1988-1-2-265

[Pap94] Papageorgiou NS (1994) Existence of solutions for second-order evolution inclusions. J Appl Math Stoch Anal. doi:10.1155/S1048953394000407

[PY06] Papageorgiou NS, Yannakakis N (2006) Second order nonlinear evolution inclusions II: structure of the solution set. Acta Math Sin English Ser. doi: 10.1007/s10114-004-0509-x

[PKZ08] Perestyuk MO, Kasyanov PO, Zadoyanchuk NV (2008) On solvability for the second order evolution inclusions with the volterra type operators. Miskolc Math Notes 9: 119–135

[Pe67] Petunin Yu I (1967) The reflexivity criterium for a Banach space. RAS USSR 140: 56–58

[Ps80] Pshenichniy (1980) Convex analysis and extremal problems [in Russian]. Nauka, Moscow

[RS98] Repovs D, Semenov PV (1998) Continuous selections of multivalued mappings. Kluwer, Dordrecht

[RS80] Reed M, Simon B (1980) Methods of mathematical physics. In: Functional analysis, 1st edn. Academic, New York

[RR64] Robertson AP, Robertson WJ (1964) Topological vector soaces. Cambridge University Press, New York

[R73] Rudin W (1973) Functional analysis. McGraw-Hill, New York

[S71] Schaefer H (1971) Topological vector spaces. Springer, New York

[SY02] Sell GR, You Yu (2002) Dynamics of evolutionary equations. Springer, New York

[S97] Showalter R (1997) Monotone operators in Banach space and nonlinear partial differential equations. AMS, Providence

[S33] Signorini A (1933) Sopra alcune questioni di elastostatica. Atti Soc lal Progr Sci Roma

[S86] Simon J (1986) Compact sets in the space $L^p(0, T; B)$. J Annali di Matematica Pura ed Applicata. doi:10.1007/BF01762360

[S94] Skrypnik IV (1994) Methods for analysis of nonlinear elliptic boundary value problems. AMS, Providence

[SV06] Sliusarchuk V Yu (2006) Diffirential equations in Banach spaces. Nat Univ Vodn Gosp i Prirodokor, Rivne

[St64] Stampacchia G (1964) Formes bilineaires coercitives sur les ensembles convexes. C R Acad Sci Paris 258:4413–4416

[T75] Temam R (1975) A non-linear eigenvalue problem: the shape at equilibrium of a confined plasma. Avchd Rational Mech Anal. doi:10.1007/BF00281469

[T92] Tolstonogov AA (1992) Solutions of evolution inclusions. Int Siberian Math J 33. doi:10.1007/BF00970899

[TU92] Tolstonogov AA, Umanskii Yu A (1992) On solutions of evolution inclusions. II. Int Siberian Math J 33. doi:10.1007/BF00971135

[Tr78] Triebel H (1978) Interpolation theory, function spaces, differential operators. North-Holland Publishing, Amsterdam

[V73] Vainberg MM (1973) Variation methods and method of monotone operators. Wiley, New York

[VM98] Vakulenko AN, Mel'nik VS (1998) On a class of operator inclusions in Banach spaces. Rep NAS of Ukr 8:20–25

[VMS99] Vakulenko AN, Mel'nik VS (1999)Solvability and properties of solutions of one class of operator inclusions in Banach spaces. Naukovi visti NTUU "KPI" 3:105–112

[VM00] Vakulenko AN, Mel'nik VS (2000) On topological method in operator inclusions whith densely defined mappings in Banach spaces. Nonlinear Bound Value Probl 10:125–142

[VLM91] Vladimirov AM, Lyakhin Yu I, Matveev LT (1991) Environmental protection. Gidrometeoizdat, Leningrad

[Z06] Zadoyanchuk NV, Kasyanov PO (2006) On solvabbility for the second order nonlinear differential–operator equations with the noncoercive w_λ-pseudomonotone type operators. Rep NAS of Ukr 12:15–19

[ZK07] Zadoyanchuk NV, Kasyanov PO (2007) Faedo-Galerkin method for nonlinear second-order evolution equations with volterra operators. Nonlinear Oscill. doi:10.1007/s11072-007-0016-y

[ZM00] Zgurovskii MZ, Mel'nik VS (2000) Penalty method for variational inequalities with multivalued mappings. Part I Cybern Syst Anal. doi: 10.1007/BF02667060

[ZM02] Zgurovskii MZ, Mel'nik VS (2002) The Ki-Fan inequality and operator inclusions in Banach spaces. Cybern Syst Anal. doi: 10.1023/A:1016391328367

[ZM04] Zgurovsky MZ, Mel'nik VS (2004) Nonlinear analysis and control of physical processes and fields. Springer, Berlin

[ZN96] Zgurovsky MZ, Novikov AN (1996) Analysis and control of unilateral physical processes [in Russian]. Naukova Dumka, Kiev

[ZMN04] Zgurovsky MZ, Mel'nik VS, Novikov AN (2004) Applied methods of analysis and control by nonlinear processes and fields [in Russian]. Naukova Dumka, Kiev

Chapter 3
Evolution Variation Inequalities

Abstract Chapter 5 is devoted to developing of the multi-valued penalty method for the proof of solvability for evolutional multivariational inequalities with multi-valued maps of w_{λ_0}-pseudomonotone type. These objects describe new classes of nonlinear problems with unilateral constraint. At first we consider the equivalent representations for evolution variation inclusions with differential-operator inclusions. In Sect. 5.2 we consider strong solutions of evolutional multivariational inequalities with multivalued $+$-coercive w_{λ_0}-pseudomonotone maps. An example of the multi-variational inequality with differential operators of hydrodynamic type disturbed by subdifferential for a locally Lipschitzian functional is considered in this section. Then we consider weak solutions of evolutional multivariational inequalities with $+$-coercive maps. Examples of unilateral problems with differential operators of Leray–Lions are considered in this section. As application we consider some dynamical contact problem with multivalued damping. We also investigate unilateral diffusion processes. These examples demonstrate the obtained generalizations. The multi-valued penalty method allows us to consider the fundamentally wider class of approximative problems. By the help of such problems we search solutions of the given problem. The new a priori estimations for the time derivative of approximate solutions of the given problem are obtained.

3.1 Equivalent Representations for Evolution Variation Inequalities with Differential-Operator Inclusions

Let X be the reflexive Banach space, X^* be its topologically adjoint,

$$\langle \cdot, \cdot \rangle_X : X^* \times X \to \mathbb{R} \text{ is canonical paring.}$$

We assume that for some interpolation pair of reflexive Banach spaces X_1, X_2, $X = X_1 \cap X_2$. Then due to Theorem 1.3 $X^* = X_1^* + X_2^*$. Remark also that

$$\langle f, y \rangle_X = \langle f_1, y \rangle_{X_1} + \langle f_2, y \rangle_{X_2} \quad \forall f \in X^* \, \forall y \in X,$$

where $f = f_1 + f_2$, $f_i \in X_i^*$, $i = 1, 2$.

M.Z. Zgurovsky et al., *Evolution Inclusions and Variation Inequalities for Earth Data Processing II*, Advances in Mechanics and Mathematics 25,
DOI 10.1007/978-3-642-13878-2_3,

Further let $A : X \rightrightarrows X^*$ be strict multivalued map with bounded values.

Let $L : D(L) \subset X \to X^*$ be linear operator, with dense definitional domain, $A : X_1 \rightrightarrows X_1^*$ be the multivalued map, $\varphi : X_2 \to \mathbb{R}$ be a convex lower semicontinuous functional. We consider the following problem:

$$\begin{cases} \langle Ly, w-y\rangle_X + [A(y), w-y]_+ \\ \qquad +\varphi(w) - \varphi(y) \geq \langle f, w-y\rangle_X, \quad \forall w \in X, \\ y \in D(L), \end{cases} \tag{3.1}$$

where $f \in X^*$ is arbitrary fixed.

Remark 3.1. Further we will assume that $D(L)$ is reflexive Banach space with the norm

$$\|y\|_{D(L)} = \|y\|_X + \|Ly\|_{X^*} \quad \forall y \in D(L).$$

The given condition is true through the maximal monotony of L on $D(L)$ (Corollary 1.8).

Remark 3.2. Due to Theorem 3 and to Proposition 2, the evolution variation inequality from (3.1) is equivalent to the inclusion:

$$Ly + \overset{*}{\overline{\text{co}}}\, A(y) + \partial\varphi(y) \ni f, \quad y \in D(L), \tag{3.2}$$

where $\partial\varphi : X_2 \to C_v(X_2^*)$ is subdifferential of convex functional φ.

3.1.1 Singular Perturbations Method

Let us consider generally speaking the multivalued duality map

$$J(y) = \{\xi \in X^* \mid \langle \xi, y\rangle_X = \|\xi\|_{X^*}^2 = \|y\|_X^2\} \in C_v(X^*) \quad \forall y \in X,$$

namely

$$J(y) = \partial(\|\cdot\|_X^2/2)(y) \quad \forall y \in X,$$

which is the corollary of Proposition 8. From Theorem 2 it follows that this map is defined on the whole space X, and from [AE84] its maximal monotony follows. Besides, due to [AE84, Theorem 4, p. 202 and Proposition 8, p. 203] for each $f \in X^*$ the map

$$\begin{aligned} J^{-1}(f) &= \{y \in X \mid f \in J(y)\} \\ &= \{y \in X \mid \langle f, y\rangle_X = \|f\|_{X^*}^2 = \|y\|_X^2\} \in C_v(X). \end{aligned}$$

is also defined on the whole space X and it is the maximal monotone multivalued map.

We will approximate the inequality from (3.1) by the following:

$$\begin{cases} \varepsilon\left[L\omega - Ly, J^{-1}(Ly)\right]_+ + \langle Ly, \omega - y\rangle_X \\ \quad +[A(y), \omega - y]_+ + \varphi(\omega) - \varphi(y) \geq \langle f, \omega - y\rangle_X. \end{cases} \tag{3.3}$$

Definition 3.1. We will say that a solution $y \in D(L)$ of problem (3.1) turns out by *Singular perturbations method*, if y is a weak limit of a subsequence $\{y_{\varepsilon_{n_k}}\}_{k\geq 1}$ of the sequence $\{y_{\varepsilon_n}\}_{n\geq 1}$ ($\varepsilon_n \searrow 0+$ as $n \to \infty$) in the space $D(L)$ that for every $n \geq 1 \quad D(L) \ni y_{\varepsilon_n}$ is a solution of problem (3.3).

Remark 3.3. Due to Theorem 3 and to Proposition 2, the evolution variation inequality from (3.3) is equivalent to the inclusion:

$$\varepsilon L^* J^{-1}(Ly_\varepsilon) + Ly_\varepsilon + \overset{*}{\overline{\text{co}}}\, A(y_\varepsilon) + \partial\varphi(y_\varepsilon) \ni f, \tag{3.4}$$

where $\partial\varphi : X_2 \to C_v(X_2^*)$ is subdifferential map of convex functional φ.

Now let $(V_i; H; V_i^*)$ be evolutionary triples such that the space $V = V_1 \cap V_2$ is continuously and densely embedded in H, $\{h_i\}_{i\geq 1} \subset V$ – is complete in V countable system of vectors, $H_n = \text{span}\{h_i\}_{i=1}^n$, $n \geq 1$;

$$\begin{aligned} X &= L_{r_1}(S;H) \cap L_{r_2}(S;H) \cap L_{p_1}(S;V_1) \cap L_{p_2}(S;V_2), \\ X^* &= L_{q_1}(S;V_1^*) + L_{q_2}(S;V_2^*) + L_{r_2'}(S;H) + L_{r_1'}(S;H), \\ X_i &= L_{r_i}(S;H) \cap L_{p_i}(S;V_i), \quad X_i^* = L_{q_i}(S;V_i^*) + L_{r_i'}(S;H), \\ W &= \{y \in X \mid y' \in X^*\}, \quad W_i = \{y \in X_i \mid y' \in X^*\} i = 1, 2. \end{aligned}$$

with the norms corresponding, we assume that $p_0 := \max\{r_1, r_2\} < +\infty$ (see Remark 1.3).

For the multivalued map $C : X \rightrightarrows X^*$ and for a convex lower semicontinuous functional $\varphi : X \to \mathbb{R}$ we consider the problem:

$$\begin{cases} \langle y', w - y\rangle_X + [C(y), w - y]_+ \\ \quad + \varphi(w) - \varphi(y) \geq \langle f, w - y\rangle_X \quad \forall w \in X, \\ y(0) = y_0. \end{cases} \tag{3.5}$$

Here $f \in X^*$, $y_0 \in H$ is arbitrary fixed, y' is the derivative of an element $y \in X$ considered in the sense of scalar distributions space $\mathscr{D}^*(S;V^*)$, $S = [0;T]$.

Remark 3.4. Due to Theorem 3 and to Proposition 2, the evolution variation inequality from (3.5) is equivalent to the inclusion:

$$y' + \overset{*}{\overline{\text{co}}}\, C(y) + \partial\varphi(y) \ni f, \tag{3.6}$$

where $\partial\varphi : X \to C_v(X^*)$ is subdifferential of convex functional φ.

3.1.2 Faedo-Galerkin Method

3.1.2.1 Faedo-Galerkin Method I

For each $n \geq 1$ let us consider Banach spaces

$$X_n = L_{p_0}(S; H_n), \quad X_n^* = L_{q_0}(S; H_n), \quad W_n = \{y \in X_n \mid y' \in X_n^*\},$$

where $1/p_0 + 1/q_0 = 1$. Let us also remind, that for any $n \geq 1$ I_n be the canonical embedding of X_n in X, $I_n^* : X^* \to X_n^*$ is adjoint operator to I_n.

Let us introduce the following maps:

$$C_n := I_n^* C I_n : X_n \rightrightarrows X_n^*, \quad f_n := I_n^* f \in X_n^*.$$

Let us consider the sequence $\{y_{0n}\}_{n\geq 0} \subset H$:

$$\forall n \geq 1 \quad H_n \ni y_{0n} \to y_0 \text{ in } H \text{ as } n \to +\infty. \tag{3.7}$$

Together with problem (3.5) $\forall n \geq 1$ we consider the following class of problems:

$$\begin{cases} \langle y_n', w_n - y_n \rangle_{X_n} + [C_n(y_n), w_n - y_n]_+ \\ \qquad + \varphi(w) - \varphi(y_n) \geq \langle f_n, w_n - y_n \rangle_{X_n} \quad \forall w_n \in X_n, \\ y_n(0) = y_{0n}. \end{cases} \tag{3.8}$$

Definition 3.2. We will say, that the solution $y \in W$ of (3.5) turns out by *Faedo-Galerkin method*, if y is a weak limit of a subsequence $\{y_{n_k}\}_{k\geq 1}$ from $\{y_n\}_{n\geq 1}$ in W, which satisfies the following conditions:

(a) $\forall n \geq 1 \quad W_n \ni y_n$ is a solution of (3.8).
(b) $y_{0n} \to y_0$ in H as $n \to \infty$.

Remark 3.5. Due to Theorem 3 and to Proposition 2, the evolution variation inequality from (3.8) is equivalent to the inclusion:

$$y_n' + C_n(y_n) + I_n^* \partial\varphi(I_n y_n) \ni f_n, \tag{3.9}$$

where $\partial\varphi : X \to C_v(X^*)$ is subdifferential of convex functional φ.

For the multivalued map $A : X \rightrightarrows X^*$, for linear dense defined map $L : D(L) \subset X \to X^*$ and for a convex lower semicontinuous functional $\varphi : X \to \mathbb{R}$ we consider the problem:

$$\begin{cases} \langle Ly, w - y \rangle_X + [A(y), w - y]_+ \\ \qquad + \varphi(w) - \varphi(y) \geq \langle f, w - y \rangle_X \quad \forall w \in X, \\ y(0) \in D(L). \end{cases} \tag{3.10}$$

Here $f \in X^*$ is arbitrary fixed. On $D(L)$ we consider the graph norm

$$\|y\|_{D(L)} = \|y\|_X + \|Ly\|_{X^*} \quad \forall y \in D(L).$$

3.1.2.2 Faedo-Galerkin Method II

For each $n \geq 1$ let us consider Banach spaces

$$X_n = L_{p_0}(S; H_n), \quad X_n^* = L_{q_0}(S; H_n), \quad W_n = \{y \in X_n \mid y' \in X_n^*\},$$

where $1/p_0 + 1/q_0 = 1$. Let us also remind, that for any $n \geq 1$ I_n – the canonical embedding of X_n in X, $I_n^* : X^* \to X_n^*$ – is adjoint with I_n.

Let us introduce the following maps: $A_n := I_n^* A I_n : X_n \rightrightarrows X_n^*$,

$$L_n := I_n^* L I_n : D(L_n) = X_n \cap D(L) \subset X_n \rightrightarrows X_n^*, \quad f_n := I_n^* f \in X_n^*.$$

Together with problem (3.10) $\forall n \geq 1$ we consider the following class of problems:

$$\begin{cases} \langle L_n y_n, w_n - y_n \rangle_{X_n} + [A_n(y_n), w_n - y_n]+ \\ \qquad + \varphi(w) - \varphi(y_n) \geq \langle f_n, w_n - y_n \rangle_{X_n} \quad \forall w_n \in X_n, \\ y_n \in D(L_n). \end{cases} \tag{3.11}$$

Definition 3.3. We will say, that the solution $y \in W$ of (3.10) turns out by *Faedo-Galerkin method*, if y is the weak limit of a subsequence $\{y_{n_k}\}_{k \geq 1}$ from $\{y_n\}_{n \geq 1}$ in $D(L)$, where for each $n \geq 1$ y_n is a solution of problem (3.11).

Remark 3.6. Due to Theorem 3 and to Proposition 2, the evolution variation inequality from (3.11) is equivalent to the inclusion:

$$L_n y_n + A_n(y_n) + I_n^* \partial\varphi(I_n y_n) \ni f_n, \tag{3.12}$$

where $\partial\varphi : X \to C_v(X^*)$ is subdifferential of convex functional φ.

3.1.3 *The Method of Finite Differences*

Let Φ be a Hausdorff locally convex linear topological space, Φ^* be the topologically adjoint with Φ. By (f, ξ) we denote the canonical pairing of $f \in \Phi^*$ and $\xi \in \Phi$.

Let the spaces $\mathscr{V}$, $\mathscr{H}$ and $\mathscr{V}^*$ be given. Moreover

$$\Phi \subset \mathscr{V} \subset \Phi^*, \qquad \Phi \subset \mathscr{H} \subset \Phi^*, \qquad \Phi \subset \mathscr{V}^* \subset \Phi^*,$$

with continuous and dense embeddings. We assume that $\mathscr{H}$ is a Hilbert space with the scalar product $(h_1, h_2)_{\mathscr{H}}$ and norm $\|h\|_{\mathscr{H}}$, $\mathscr{V}$ is a reflexive separable Banach space with norm $\|v\|_{\mathscr{V}}$, $\mathscr{V}^*$ is the adjoint with $\mathscr{V}$ with the norm $\|f\|_{\mathscr{V}^*}$ associated with the bilinear form $(\cdot,\cdot)_{\mathscr{H}}$.

If $\xi, \psi \in \Phi$, then $(\xi, \psi) = (\xi, \psi)_{\mathscr{H}}$, i.e., it coincides with the scalar product in $\mathscr{H}$.

Let $\mathscr{V} = \mathscr{V}_1 \cap \mathscr{V}_2$ and $\|\cdot\|_{\mathscr{V}} = \|\cdot\|_{\mathscr{V}_1^*} + \|\cdot\|_{\mathscr{V}_2^*}$, where $(\mathscr{V}_i, \|\cdot\|_{\mathscr{V}_i})$, $i = \overline{1,2}$, are reflexive separable Banach spaces and the embeddings $\Phi \subset \mathscr{V}_i \subset \Phi^*$ and $\Phi \subset \mathscr{V}_i^* \subset \Phi^*$ are dense and continuous. The spaces $(\mathscr{V}_i^*, \|\cdot\|_{\mathscr{V}_i^*})$, $i = \overline{1,2}$, are the topologically adjoint with $(\mathscr{V}_i, \|\cdot\|_{\mathscr{V}_i})$. Then $\mathscr{V}^* = \mathscr{V}_1^* + \mathscr{V}_2^*$.

Let $\mathscr{A} : \mathscr{V}_1 \rightrightarrows \mathscr{V}_1^*$ be a multivalued map, $\varphi : \mathscr{V}_2 \to \mathbb{R}\mathscr{V}_2^*$ be a convex lower semicontinuous functional, $\Lambda : \mathscr{V} \to \mathscr{V}^*$ be an unbounded operator with domain $D(\Lambda; \mathscr{V}, \mathscr{V}^*)$. We consider the following problem

$$u \in D(\Lambda; \mathscr{V}, \mathscr{V}^*), \tag{3.13}$$

$$(\Lambda u, v-u) + [A(u), v-u]_+ + \varphi(v) - \varphi(u) \geq (f, v-u) \quad \forall v \in \mathscr{V}, \tag{3.14}$$

where $f \in V^*$ is a fixed element.

Remark 3.7. Due to Theorem 3 and to Proposition 2, the evolution variation inequality from (3.14) is equivalent to the inclusion:

$$\Lambda u + \overset{*}{\overline{\text{co}}}\, A(u) + \partial\varphi(u) \ni f, \tag{3.15}$$

where $\partial\varphi : \mathscr{V}_2 \to C_v(\mathscr{V}_2^*)$ is subdifferential of convex functional φ.

3.1.3.1 Method of Finite Differences

The natural approximation of inequality (3.14) is the inequality

$$\frac{(u_h, v-u_h) - (G(h)u_h, v-u_h)}{h} + [A(u_h), v-u_h]_+$$
$$+ \varphi(v) - \varphi(u_h) \geq (f, v-u_h) \quad \forall v \in \mathscr{V}, \quad (h > 0). \tag{3.16}$$

However, if $\mathscr{V}$ is not contained in $\mathscr{H}$, then (3.16), generally speaking, has no solutions, and it is necessary to modify the given inequality in an appropriate way. We shall choose a sequence $\theta_h \in (0,1)$ such that

$$\frac{1-\theta_h}{h} \to 0 \quad \text{as} \quad h \to 0. \tag{3.17}$$

We put $\theta_h = 1$ if $\mathscr{V} \subset \mathscr{H}$. Further, we define

$$\Lambda_h = \frac{I - \theta_h G(h)}{h} \tag{3.18}$$

and replace (3.16) by the inequality

$$(\Lambda_h u_h, v - u_h) + [A(u_h), v - u_h]_+ + \varphi(v) - \varphi(u_h) \geq (f, v - u_h) \quad \forall v \in \mathscr{V}. \tag{3.19}$$

Definition 3.4. We will say, that the solution u of (3.13)–(3.14) turns out by *finite differences method*, if u is the weak limit of a subsequence $\{u_{h_{n_k}}\}_{k\geq 1}$ from $\{u_{h_n}\}_{n\geq 1}$ ($h_n \searrow 0+$ as $n \to \infty$) in $\mathscr{V}$, where for each $n \geq 1$ $u_{h_n} \in \mathscr{V}$ is a solution of problem (3.19).

Remark 3.8. Due to Theorem 3 and to Proposition 2, the evolution variation inequality from (3.19) is equivalent to the inclusion:

$$\Lambda_h u_h + \overset{*}{\overline{\text{co}}}\, A(u_h) + \partial\varphi(u_h) \ni f, \tag{3.20}$$

where $\partial\varphi : \mathscr{V}_2 \to C_v(\mathscr{V}_2^*)$ is subdifferential of convex functional φ.

3.2 The Strong Solutions for Evolution Variation Inequalities

Let X be the reflexive Banach space, X^* be its topologically adjoint,

$$\langle\cdot,\cdot\rangle_X : X^* \times X \to \mathbb{R} \text{ is canonical paring.}$$

We assume that for some interpolation pair of reflexive Banach spaces X_1, X_2, $X = X_1 \cap X_2$. Then due to Theorem 1.3 $X^* = X_1^* + X_2^*$. Remark also that

$$\langle f, y\rangle_X = \langle f_1, y\rangle_{X_1} + \langle f_2, y\rangle_{X_2} \quad \forall f \in X^* \, \forall y \in X,$$

where $f = f_1 + f_2$, $f_i \in X_i^*$, $i = 1, 2$.

Further let $A : X \rightrightarrows X^*$ be strict multivalued map with bounded values.

Let $L : D(L) \subset X \to X^*$ be linear operator, with dense definitional domain, $A : X_1 \rightrightarrows X_1^*$ be the multivalued map, $\varphi : X_2 \to \mathbb{R}$ be a convex lower semicontinuous functional. We consider solutions of problem (3.1), that can be obtained by singular perturbations method.

Theorem 3.1. *Let X be a reflexive Banach space,*

$$L : D(L) \subset X \to X^*$$

be a linear, densely defined, maximal monotone on $D(L)$ operator, $A : X_1 \rightrightarrows X_1^*$ be finite-dimensionally locally bounded, λ_0-pseudomonotone on $D(L)$, +-coercive multivalued map, for which Condition (Π) is valid. Let also $\varphi : X_2 \to \mathbb{R}$ be a convex lower semicontinuous functional, that satisfy the next coercivity condition:

$$\frac{\varphi(y)}{\|y\|_{X_2}} \to +\infty \quad as \quad \|y\|_{X_2} \to +\infty. \tag{3.21}$$

Then there exists at least one solution $y \in D(L)$ of problem (3.1).

Proof. If we pass from problem (3.1) to the equivalent problem (3.2), and if we set as $B : X_2 \to C_v(X_2^*), -\partial\varphi$, we obtain that the given result is the direct corollary of Theorem 2.2. It follows from Proposition 7, from Remark 1, from Corollary 1, from Theorem 3 and from Lemma 1. □

Now let $(V_i; H; V_i^*)$ be evolutionary triples such that the space $V = V_1 \cap V_2$ is continuously and densely embedded in H, $\{h_i\}_{i\geq 1} \subset V$ – is complete in V countable vectors system, $H_n = \mathrm{span}\{h_i\}_{i=1}^n$, $n \geq 1$;

$$\begin{aligned}
X &= L_{r_1}(S;H) \cap L_{r_2}(S;H) \cap L_{p_1}(S;V_1) \cap L_{p_2}(S;V_2),\\
X^* &= L_{q_1}(S;V_1^*) + L_{q_2}(S;V_2^*) + L_{r_2'}(S;H) + L_{r_1'}(S;H),\\
X_i &= L_{r_i}(S;H) \cap L_{p_i}(S;V_i), \quad X_i^* = L_{q_i}(S;V_i^*) + L_{r_i'}(S;H),\\
W &= \{y \in X \mid y' \in X^*\}, \quad W_i = \{y \in X_i \mid y' \in X^*\} \quad i = 1,2.
\end{aligned}$$

with the norms corresponding, we assume that $p_0 := \max\{r_1, r_2\} < +\infty$ (see Remark 1.3).

For the multivalued map $C : X \rightrightarrows X^*$ and for a convex lower semicontinuous functional $\varphi : X \to \mathbb{R}$ we consider problem (3.5).

Theorem 3.2. *Let the multivalued map $C : X \to C_v(X^*)$ satisfies the following conditions:*

(1) C is λ_0-pseudomonotone on W.
(2) C is finite-dimensionally locally bounded.
(3) C satisfies Property (Π) on X.
(4) C satisfies the following coercive property:

$$\exists c > 0 : \quad \frac{[C(y), y]_+ - c\|C(y)\|_+}{\|y\|_X} \to +\infty \quad as \quad \|y\|_X \to \infty;$$

$\varphi : X \to \mathbb{R}$ be convex lower semicontinuous functional such that

$$\exists c > 0 : \quad |\varphi(y)| \leq c_1(1 + \|y\|_X) \quad \forall y \in X.$$

Besides, let the vectors system $\{h_j\}_{j\geq 1} \subset V_1 \cap V_2$ exists which is complete in V_1 and in V_2 and such that for $i = 1,2$ the triple $(\{h_j\}_{j\geq 1}; V_i; H)$ satisfies Condition (γ) with constant C_i. Then for arbitrary $f \in X^*$, $y_0 \in H$ the set

$$K_H(f) := \Big\{ y \in W \mid y \ \textit{is the solution of problem (3.5)}, \\ \textit{which turns out by Faedo-Galerkin method} \Big\}$$

is nonempty. Moreover the representation

$$K_H(f) = \bigcup_{\{y_{0n}\}_{n\geq 1} \subset H \ \textit{satisfies (3.7)}} \bigcap_{n\geq 1} \Big[\bigcup_{m\geq n} K_m(f_m)(y_{0m}) \Big]_{X_w}, \tag{3.22}$$

is true, where for each $n \geq 1$

$$K_n(f_n)(y_{0n}) = \Big\{ y_n \in W_n \mid y_n \ \textit{is a solution of problem (3.8)} \Big\},$$

$[\cdot]_{X_w}$ *is closuring operator in the space* $(X; \sigma(X; X^*))$.

The proof follows from Theorem 2.4.

Theorem 3.3. *Let* $L : D(L) \subset X \to X^*$ *be a linear operator such that* L *is maximal monotone on* $D(L)$ *and satisfies*

- *Condition* L_1*: for each* $n \geq 1$ *and* $x_n \in D(L_n)$ $Lx_n \in X_n^*$.
- *Condition* L_2*: for each* $n \geq 1$ *the set* $D(L_n)$ *is dense in* X_n.
- *Condition* L_3*: for each* $n \geq 1$ L_n *is maximal monotone on* $D(L)$.

Let also $A : X \rightrightarrows X^*$ *be* λ_0*-pseudomonotone on* $D(L)$*, +-coercive multivalued map that satisfies Condition* (Π)*;* $\varphi : X \to \mathbb{R}$ *be convex lower semicontinuous functional such that*

$$\varliminf_{\|y\|_X \to +\infty} \frac{\varphi(y)}{\|y\|_X} > -\infty. \tag{3.23}$$

Furthermore, let $\{h_j\}_{j\geq 1} \subset V$ *be a complete vectors system in* V_1*,* V_2*,* H *such that* $\forall i = 1, 2$ *the triple* $\big(\{h_j\}_{j\geq 1}; V_i; H\big)$ *satisfies Condition* (γ).

Then for each $f \in X^*$ *the set*

$$K_H(f) := \{ y \in D(L) \mid y \ \textit{is the solution of (3.10)}, \\ \textit{obtained by Faedo-Galerkin method} \}$$

is nonempty and the representation

$$K_H(f) = \bigcap_{n\geq 1} \Big[\bigcup_{m\geq n} K_m(f_m) \Big]_{X_w} \tag{3.24}$$

is true, where for each $n \geq 1$

$$K_n(f_n) = \{ y_n \in D(L_n) \mid y_n \ \textit{is the solution of (3.11)} \}$$

and $[\cdot]_{X_w}$ *is the closure operator in the space* X *with respect to the weak topology.*

The proof is similar to the proof of Theorem 3.1.

Let again Φ be a Hausdorff locally convex linear topological space, Φ^* be the topologically adjoint with Φ. By (f, ξ) we denote the canonical pairing of $f \in \Phi^*$ and $\xi \in \Phi$.

Let the spaces $\mathcal{V}$, $\mathcal{H}$ and $\mathcal{V}^*$ be given. Moreover

$$\Phi \subset \mathcal{V} \subset \Phi^*, \qquad \Phi \subset \mathcal{H} \subset \Phi^*, \qquad \Phi \subset \mathcal{V}^* \subset \Phi^*,$$

with continuous and dense embeddings. We assume that $\mathcal{H}$ is a Hilbert space with the scalar product $(h_1, h_2)_{\mathcal{H}}$ and norm $\|h\|_{\mathcal{H}}$, $\mathcal{V}$ is a reflexive separable Banach space with norm $\|v\|_{\mathcal{V}}$, $\mathcal{V}^*$ is the adjoint with $\mathcal{V}$ with the norm $\|f\|_{\mathcal{V}^*}$ associated with the bilinear form $(\cdot,\cdot)_{\mathcal{H}}$.

If $\xi, \psi \in \Phi$, then $(\xi, \psi) = (\xi, \psi)_{\mathcal{H}}$, i.e., it coincides with the scalar product in $\mathcal{H}$.

Let $\mathcal{V} = \mathcal{V}_1 \cap \mathcal{V}_2$ and $\|\cdot\|_{\mathcal{V}} = \|\cdot\|_{\mathcal{V}_1^*} + \|\cdot\|_{\mathcal{V}_2^*}$, where $(\mathcal{V}_i, \|\cdot\|_{\mathcal{V}_i})$, $i = \overline{1,2}$, are reflexive separable Banach spaces and the embeddings $\Phi \subset \mathcal{V}_i \subset \Phi^*$ and $\Phi \subset \mathcal{V}_i^* \subset \Phi^*$ are dense and continuous. The spaces $(\mathcal{V}_i^*, \|\cdot\|_{\mathcal{V}_i^*})$, $i = \overline{1,2}$, are the topologically adjoint with $(\mathcal{V}_i, \|\cdot\|_{\mathcal{V}_i})$. Then $\mathcal{V}^* = \mathcal{V}_1^* + \mathcal{V}_2^*$.

Theorem 3.4. *Assume the following conditions:*

1. *$A : \mathcal{V}_1 \rightrightarrows \mathcal{V}_1^*$ is a bounded, λ-pseudomonotone on $\mathcal{V}_1$ multivalued map, which satisfies the $+$-coerciveness condition on $\mathcal{V}_1$.*
2. *The functional $\varphi : \mathcal{V}_2 \to \mathbb{R}$ is convex, lower semicontinuous and satisfies the following coerciveness condition:*

$$\frac{\varphi(v)}{\|v\|_{\mathcal{V}_2}} \longrightarrow +\infty \qquad as \quad \|v\|_{\mathcal{V}_2} \longrightarrow +\infty;$$

3. *The operator Λ satisfies all the conditions given in (2.141)–(2.144).*

Then for any $f \in \mathcal{V}^*$ there exists $u \in \mathcal{V}$ satisfying (3.13)–(3.14).

3.3 The Multivalued Penalty Method for Evolution Variation Inequalities

Let again Φ be a Hausdorff locally convex linear topological space, Φ^* be the topologically adjoint with Φ. We denote the canonical pairing of $f \in \Phi^*$ and $\xi \in \Phi$ by (f, ξ).

Let the spaces $\mathcal{V}$, $\mathcal{H}$ and $\mathcal{V}^*$ be given. Moreover

$$\Phi \subset \mathcal{V} \subset \Phi^*, \qquad \Phi \subset \mathcal{H} \subset \Phi^*, \qquad \Phi \subset \mathcal{V}^* \subset \Phi^*,$$

with continuous and dense embeddings. We assume that $\mathcal{H}$ is a Hilbert space with the scalar product $(h_1, h_2)_{\mathcal{H}}$ and norm $\|h\|_{\mathcal{H}}$, $\mathcal{V}$ is a reflexive separable Banach space with norm $\|v\|_{\mathscr{V}}$, $\mathscr{V}^*$ is the adjoint with $\mathscr{V}$ with the norm $\|f\|_{\mathscr{V}^*}$ associated with the bilinear form $(\cdot,\cdot)_{\mathscr{H}}$.

If $\xi, \psi \in \Phi$, then $(\xi,\psi) = (\xi,\psi)_{\mathscr{H}}$, i.e., it coincides with the scalar product in $\mathscr{H}$.

Let us consider again the operators $\mathscr{A}$, Λ and the convex set K such that

The operator $\Lambda : D(\Lambda) = D(\Lambda; \mathscr{V}, \mathscr{V}^*) \subset \mathscr{V} \to \mathscr{V}^*$ satisfies all the mentioned in Sect. 2.5.2 conditions, in particular, the conditions given in (2.141)–(2.144); (3.25)

K is a convex closed subset from $\mathscr{V}$ such that for every $v \in K$ there exists a sequence $v_j \in K \cap D(\Lambda)$ such that $v_j \to v$ in $\mathscr{V}$ and $\overline{\lim\limits_{j\to\infty}} (\Lambda v_j, v_j - v) \le 0$ (see [LIO69, p. 396]); (3.26)

The multivalued map $A : \mathscr{V} \to C_v(\mathscr{V}^*)$ is λ_0 – pseudo monotone on $\mathscr{V}$, locally finite-dimensionally bounded, it satisfies Condition (Π) and for some $y_0 \in K \cap D(\Lambda)$ $\frac{[A(y), y - y_0]_+}{\|y\|_{\mathscr{V}}} \to +\infty$ as $\|y\|_{\mathscr{V}} \to \infty$; (3.27)

$\beta : \mathscr{V} \to C_v(\mathscr{V}^*)$ is the monotone, bounded, radially semicontinuous multivalued "penalty" operator, that corresponds to the set K, i.e. $K = \{y \in \mathscr{V} \mid \beta(y) \ni \bar{0}\}$. (3.28)

Remark 3.9. [LIO69, p. 284] The sufficient condition for (3.26) is

$$G(s)K \subset K \quad \forall s \ge 0.$$

If $\bar{0} \in K$, then this condition is fulfilled.

Theorem 3.5. *Let the conditions (3.25)–(3.28) hold, $f \in \mathscr{V}^*$ be arbitrary fixed. Then for each $\varepsilon > 0$ the problem*

$$\begin{aligned} \Lambda y_\varepsilon + A(y_\varepsilon) + \tfrac{1}{\varepsilon}\beta(y_\varepsilon) &\ni f, \\ y_\varepsilon &\in D(\Lambda) \end{aligned} \tag{3.29}$$

has a solution. Moreover, there is a sequence $\{y_\varepsilon\}_\varepsilon \subset D(\Lambda)$ such that

(a) For every $\varepsilon > 0$ y_ε is the solution of problem (3.29).
(b) There exists a subsequence $\{y_\tau\}_\tau \subset \{y_\varepsilon\}_\varepsilon$ such that for some $y \in \mathscr{V}$ $y_\tau \to y$ weakly in $\mathscr{V}$.
(c) y is the solution of the next problem:

$$\begin{aligned} (\Lambda v, v - y) + [A(y), v - y]_+ \ge (f, v - y) \quad &\forall v \in K \cap D(\Lambda), \\ y &\in K. \end{aligned} \tag{3.30}$$

Proof. By the analogy with [LIO69, p. 396] without loss of generality we may consider that $y_0 = \bar{0} \in K$. Otherwise, the maps $\tilde{A}(\cdot) = A(\cdot - y_0)$, $\tilde{f} = f - \Lambda y_0$, $\tilde{\Lambda} = \Lambda$, the set $\tilde{K} = K - y_0$ and $\tilde{y_0} = \bar{0}$ satisfy the conditions (3.25)–(3.28).

For every $\varepsilon > 0$ let us enter a new multivalued map:

$$A_\varepsilon(y) := A(y) + \frac{1}{\varepsilon}\beta(y), \quad y \in \mathscr{V}.$$

In virtue of Lemma 1.15, $A_\varepsilon : \mathscr{V} \to C_v(\mathscr{V}^*)$ is λ_0-pseudomonotone on $\mathscr{V}$. Due to the boundness of β, thanks to Condition (Π) and to the local finitedimensional boundness for A it follows that A_ε is locally finitedimensionally bounded and it satisfies Condition (Π).

Now, let us use the coercivity condition. From (3.27) it follows the existence of $R > 0$ such that

$$[A(y) - f, y]_+ \geq 0 \quad \forall y \in \mathscr{V} : \|y\|_{\mathscr{V}} = R.$$

Then, for every $\varepsilon > 0$

$$\begin{aligned}[A_\varepsilon(y) - f, y]_+ &\geq [A(y) - f, y]_+ + \frac{1}{\varepsilon}[\beta(y), y - \bar{0}]_- \\ &\geq [A(y) - f, y]_+ + \frac{1}{\varepsilon}[\beta(\bar{0}), y]_+ \\ &= [A(y) - f, y]_+ \geq 0 \quad \forall \|y\|_X = R.\end{aligned}$$

Hence, we can apply Theorem 2.6 for

$$\mathscr{V}_1 = \mathscr{V}_2 = \mathscr{V},\ D(\Lambda;\mathscr{V};\mathscr{V}^*) = D(\Lambda),\ \Lambda = \Lambda,\ \mathscr{A} \equiv \bar{0},\ \mathscr{B} = A_\varepsilon,\ f = f,\ R = R.$$

Then, we obtain that for every $\varepsilon > 0$ there exists $y_\varepsilon \in \mathscr{V}$ such that

$$y_\varepsilon \text{ is the solution of (3.29)}, \quad \|y_\varepsilon\|_{\mathscr{V}} \leq R. \tag{3.31}$$

We remark that the constant R does not depend on $\varepsilon > 0$.

From (3.31) it follows that there exist $d_\varepsilon \in A(y_\varepsilon)$, $b_\varepsilon \in \beta(y_\varepsilon)$ such that

$$\Lambda y_\varepsilon + d_\varepsilon + \frac{1}{\varepsilon} b_\varepsilon = f. \tag{3.32}$$

Due to $\bar{0} \in K \cap D(\Lambda)$ and to the monotony of L and β we have:

$$(d_\varepsilon, y_\varepsilon) \leq -(\Lambda y_\varepsilon, y_\varepsilon) + \frac{1}{\varepsilon}(b_\varepsilon, \bar{0} - y_\varepsilon) + (f, y_\varepsilon) \leq \|f\|_{\mathscr{V}^*} R < +\infty.$$

In virtue of Property (Π) for A and from (3.31) it follows that there exists $c_1 > 0$ such that

$$\|d_\varepsilon\|_{\mathscr{V}^*} \leq c_1 \quad \forall \varepsilon > 0. \tag{3.33}$$

Moreover, from (3.32) and (3.33) it follows that

$$\begin{aligned}0 &\le (b_\varepsilon, y_\varepsilon) = \varepsilon(f - d_\varepsilon - \Lambda y_\varepsilon, y_\varepsilon)\\ &\le \varepsilon(\|f\|_{\mathscr{V}^*} + c_1)R =: c_2 \cdot \varepsilon \to 0 \text{as} \quad \varepsilon \searrow 0+\,. \end{aligned} \tag{3.34}$$

From the monotony of β and from (3.34) it follows that for every $\omega \in \mathscr{V}$

$$\begin{aligned}(b_\varepsilon, \omega) &\le [\beta(y_\varepsilon), \omega - y_\varepsilon]_+ + (b_\varepsilon, y_\varepsilon) \le [\beta(\omega), \omega - y_\varepsilon]_+ + c_2\varepsilon\\ &\le \|\beta(\omega)\|_-(\|\omega\|_{\mathscr{V}} + R) + c_2\varepsilon.\end{aligned}$$

Hence, due to the Banach–Steinhauss Theorem there exists $c_3 > 0$ such that

$$\|b_\varepsilon\|_{\mathscr{V}^*} \le c_3 \quad \forall \varepsilon \in (0, \varepsilon_0), \tag{3.35}$$

for some $\varepsilon_0 > 0$.

The conditions (3.25) imply that for every $\omega \in D(\Lambda^*)$

$$(\Lambda y_\varepsilon, \omega) = (\Lambda^* \omega, y_\varepsilon) \le \|\omega\|_{D(\Lambda^*)} R \quad \forall \varepsilon > 0.$$

Hence, there exists $c_4 > 0$ such that

$$\|L y_\varepsilon\|_{D(\Lambda^*)^*} \le c_4 \quad \forall \varepsilon > 0.$$

From here, due to the equality (3.32) we obtain that

$$b_\varepsilon \to \bar{0} \quad \text{in} \quad D(\Lambda^*)^* \quad \text{as} \quad \varepsilon \searrow 0+\,. \tag{3.36}$$

From (3.32), (3.33) and from the monotony of Λ it follows that

$$\begin{aligned}(\Lambda y_\varepsilon, \omega) &= (\Lambda y_\varepsilon, \omega - y_\varepsilon) + (\Lambda y_\varepsilon, y_\varepsilon) \le (\Lambda \omega, \omega - y_\varepsilon) + (\|f\|_{\mathscr{V}^*} + c_1)R\\ &\le \|\Lambda \omega\|_{\mathscr{V}}(\|\omega\|_{\mathscr{V}} + R) + (\|f\|_{\mathscr{V}^*} + c_1)R \quad \forall \omega \in D(\Lambda).\end{aligned}$$

Therefore, there exists $c_5 > 0$ such that

$$\|\Lambda y_\varepsilon\|_{D(L)^*} \le c_5 \quad \forall \varepsilon > 0.$$

Passing to limit. From the estimations (3.31), (3.33), (3.35), from the convergence (3.36), due to the Banach–Alaoglu Theorem it follows that there exists a subsequence $\{y_\tau\}_\tau$ from $\{y_\varepsilon\}_\varepsilon$ such, that for some $y \in \mathscr{V}$, $d \in \mathscr{V}^*$

$$y_\tau \rightharpoonup y \text{ in } \mathscr{V}, \quad d_\tau \rightharpoonup d \text{ in } \mathscr{V}^*, \quad b_\tau \rightharpoonup \bar{0} \text{ in } \mathscr{V}^* \quad \text{as } \tau \searrow 0+\,. \tag{3.37}$$

The map β is λ_0-pseudomonotone on $\mathscr{V}$. Moreover, due to (3.34) and (3.37) we have:

$$\lim_{\tau \searrow 0+} (b_\tau, y_\tau - y) = \overline{\lim_{\tau \searrow 0+}} (b_\tau, y_\tau - y) \le 0.$$

Hence, due to a subsequence, for every $\omega \in \mathscr{V}$

$$0 = \varliminf_{\tau \searrow 0+} (b_\tau, y_\tau - \omega) \geq [\beta(y), y - \omega]_-.$$

The last relation is equivalent to $\bar{0} \in \beta(y)$. Hence, in virtue of (3.28), we obtain that

$$y \in K. \tag{3.38}$$

Now, let us show that

$$\varlimsup_{\tau \searrow 0+} (d_\tau, y_\tau - y) \leq 0. \tag{3.39}$$

Really, from (3.32) and from (3.28) it follows that for every $v \in D(\Lambda) \cap K$

$$\begin{aligned}(d_\tau, y_\tau - v) &= \frac{1}{\varepsilon}(b_\tau, v - y_\tau) + (f, y_\tau - v) + (\Lambda y_\tau, v - y_\tau) \\ &\leq \frac{1}{\varepsilon}[\beta(y_\tau), v - y_\tau]_+ + (f, y_\tau - v) + (\Lambda v, v - y_\tau) \\ &\leq \frac{1}{\varepsilon}[\beta(v), v - y_\tau]_- + (f, y_\tau - v) + (\Lambda v, v - y_\tau) \\ &\leq (f, y_\tau - v) + (\Lambda v, v - y_\tau),\end{aligned} \tag{3.40}$$

as $\bar{0} \in \beta(v)$. So,

$$\varlimsup_{\tau \searrow 0+} (d_\tau, y_\tau) \leq (d, v) + (f, y - v) + (\Lambda v, v - y) \quad \forall v \in D(\Lambda) \cap K.$$

But in virtue of (3.26) and (3.38) we can choose $v_j \in K \cap D(\Lambda)$ such that $v_j \to y$ in $\mathscr{V}$ and $\varlimsup_{j \to \infty} (\Lambda v_j, v_j - y) \leq 0$. If we put in the last relation $v = v_j$, we obtain:

$$\varlimsup_{\tau \searrow 0+} (d_\tau, y_\tau) \leq (d, y).$$

Therefore, due to (3.37), the inequality (3.39) is true.

Let us use the λ_0-pseudomonotony of A. From (3.37) and (3.39) it follows that there exist the subsequences $\{y_\nu\}_\nu \subset \{y_\tau\}_\tau$ and $\{d_\nu\}_\nu \subset \{d_\tau\}_\tau$ such that

$$\varliminf_{\nu \searrow 0+} (d_\nu, y_\nu - v) \geq [A(y), y - v]_- \quad \forall v \in \mathscr{V}, \tag{3.41}$$

in particular, from the inequality (3.39) it follows

$$\lim_{\nu \searrow 0+} (d_\nu, y_\nu - y) = 0.$$

Hence, due to (3.37), (3.40) and (3.41),

$$[A(y), y - v]_- \leq (f, y - v) + (\Lambda v, v - y) \quad \forall v \in K \cap D(\Lambda),$$

that is equivalent (due to Proposition 1) to (3.30).

The Theorem is proved. □

Now let $\mathscr{V} = \mathscr{V}_1 \cap \mathscr{V}_2$ and $\|\cdot\|_{\mathscr{V}} = \|\cdot\|_{\mathscr{V}_1^*} + \|\cdot\|_{\mathscr{V}_2^*}$, where $(\mathscr{V}_i, \|\cdot\|_{\mathscr{V}_i})$, $i = \overline{1,2}$, are reflexive separable Banach spaces and the embeddings $\Phi \subset \mathscr{V}_i \subset \Phi^*$ and $\Phi \subset \mathscr{V}_i^* \subset \Phi^*$ are dense and continuous. The spaces $(\mathscr{V}_i^*, \|\cdot\|_{\mathscr{V}_i^*})$, $i = \overline{1,2}$, are the topologically adjoint with $(\mathscr{V}_i, \|\cdot\|_{\mathscr{V}_i})$. Then $\mathscr{V}^* = \mathscr{V}_1^* + \mathscr{V}_2^*$.

For some multivalued map $\mathscr{A} : \mathscr{V}_1 \overset{\rightarrow}{\rightarrow} \mathscr{V}_1^*$ with nonempty convex closed (in the corresponding topology) bounded values, for some convex lower semicontinuous functional $\varphi : \mathscr{V}_2 \to \mathbb{R}$, for some linear dense defined operator $\Lambda : D(\Lambda; \mathscr{V}, \mathscr{V}^*) \subset \mathscr{V} \to \mathscr{V}^*$ and for some closed convex set $K \subset \mathscr{V}$, we consider the next problem of the solvability of the next evolution variation inequality in the space $\mathscr{V}$:

$$\begin{aligned}(\Lambda v, v - u) + [\mathscr{A}(u), v - u]_+ + \varphi(v) - \varphi(u) \\ \geq (f, v - u) \quad \forall v \in K \cap D(\Lambda; \mathscr{V}, \mathscr{V}^*),\end{aligned} \tag{3.42}$$

$$u \in K, \tag{3.43}$$

where $f \in \mathscr{V}^*$ be an arbitrary fixed element

Corollary 3.1. *Let the assumptions (3.25), (3.26) and (3.28) hold, $f \in \mathscr{V}^*$ be arbitrary fixed. Moreover, let*

The multivalued map $A : \mathscr{V}_1 \to C_v(\mathscr{V}_1^*)$ be λ_0 – pseudo monotone on $\mathscr{V}_1$, locally finite-dimensionally bounded, let A satisfy Condition (Π) and for some $y_0 \in K \cap D(\Lambda)$ $\frac{[A(y), y - y_0]_+}{\|y\|_{\mathscr{V}_1}} \to +\infty$ *as* $\|y\|_{\mathscr{V}_1} \to \infty$;

The functional $\varphi : \mathscr{V}_2 \to \mathbb{R}$ be convex, lower semicontinuous on $\mathscr{V}_2$ and satisfy the next coercivity condition: $\frac{\varphi(y)}{\|y\|_{\mathscr{V}_2}} \to +\infty$ *as* $\|y\|_{\mathscr{V}_2} \to \infty$;

Then for each $\varepsilon > 0$ the problem

$$\begin{aligned}(\Lambda y_\varepsilon, v - y_\varepsilon) + [A(y_\varepsilon), v - y_\varepsilon]_+ + \frac{1}{\varepsilon}[\beta(y_\varepsilon), v - y_\varepsilon]_+ + \varphi(v) - \varphi(y_\varepsilon) \\ \geq (f, v - y_\varepsilon) \quad \forall v \in \mathscr{V}, y_\varepsilon \in D(\Lambda)\end{aligned} \tag{3.44}$$

has a solution. Moreover, there is a sequence $\{y_\varepsilon\}_\varepsilon \subset D(\Lambda)$ such that

(a) For every $\varepsilon > 0$ y_ε is the solution of problem (3.44).
(b) There exists a subsequence $\{y_\tau\}_\tau \subset \{y_\varepsilon\}_\varepsilon$ such that for some $y \in \mathscr{V}$ $y_\tau \to y$ weakly in $\mathscr{V}$.
(c) y is the solution of problem (3.42)–(3.43).

Proof. At first let us consider the multivalued map

$$B(y) = \partial\varphi(y) \in C_v(\mathscr{V}_2^*) \quad \forall y \in \mathscr{V}_2.$$

Let us check, that the given map satisfies the next conditions:

(a) *Property* (Π). Let $y_0 \in \mathscr{V}_2$, $k > 0$ and the bounded set $B \subset \mathscr{V}_2$ be arbitrary fixed. Then $\forall y \in B$ and $\forall d(y) \in \partial\varphi(y)$ such that $(d(y), y - y_0) \le k$ is fulfilled. Let $u \in \mathscr{V}_2$ be arbitrary fixed, then

$$\begin{aligned}(d(y), u) &= (d(y), u + y_0 - y) + (d(y), y - y_0) \le \varphi(u + y_0) - \varphi(y) + k \\ &\le \varphi(u + y_0) - \inf_{y \in B} \varphi(y) + k \equiv const < +\infty,\end{aligned}$$

because every convex lower semicontinuous functional is bounded from below on every bounded set. Hence, by the Banach–Steinhaus Theorem, there exists $N = N(y_0, k, B)$ such that $\|d(y)\|_{\mathscr{V}_2} \le N$ for all $y \in B$.

(b) $+$*-coercivity on* $\mathscr{V}_2$. Let us put in the definition of subdifferential $v = y_0$. Then

$$\|y\|_{\mathscr{V}_2}^{-1} [\partial\varphi(y), y - y_0]_+ \ge \|y\|_{\mathscr{V}_2}^{-1}\varphi(y) - \|y\|_{\mathscr{V}_2}^{-1}\varphi(y_0) \to +\infty \text{ as } \|y\|_{\mathscr{V}_2} \to +\infty.$$

(c) λ_0*-pseudomonotony on* $\mathscr{V}_2$. Let $y_n \rightharpoonup y_0$ in $\mathscr{V}_2$, $\partial\varphi(y_n) \ni d_n \rightharpoonup d$ in $\mathscr{V}_2^*$ and the inequality (1.56) hold. Then, due to the monotony of $\partial\varphi$, for each $d_0 \in \partial\varphi(y_0)$ and for all $n \ge 1$

$$(d_n, y_n - y_0) = (d_n - d_0, y_n - y_0) + (d_0, y_n - y_0) \ge (d_0, y_n - y_0).$$

Hence

$$\varliminf_{n\to+\infty} (d_n, y_n - y_0) \ge \varliminf_{n\to+\infty} (d_0, y_n - y_0) = 0,$$

that together with (1.56) gives:

$$\lim_{n\to+\infty} (d_n, y_n - y_0) = 0.$$

Thus, for every $w \in \mathscr{V}_2$

$$\begin{aligned}\varliminf_{n\to+\infty} (d_n, y_n - w) &\ge \lim_{n\to+\infty} (d_n, y_n - y_0) + \varliminf_{n\to+\infty} (d_n, y_0 - w) \\ &= (d_0, y_0 - w).\end{aligned} \tag{3.45}$$

From another side,

$$\begin{aligned}(d_0, w - y_0) &\le \varlimsup_{n\to+\infty} (d_n, w - y_n) \le \varphi(w) - \varliminf_{n\to+\infty} \varphi(y_n) \\ &\le \varphi(w) - \varphi(y_0),\end{aligned} \tag{3.46}$$

because every convex lower semicontinuous functional is weakly lower semicontinuous. From (3.46) and from the definition of subdifferential map it follows that $d_0 \in \partial\varphi(y_0)$. From here, due to Proposition 1 and to the inequality (3.45), we will obtain the inequality (1.56) for $\mathscr{A} = \partial\varphi$ on $\mathscr{V}_2$.

So, due to Lemma 1.15, Lemma 1 and to Remark 1, all the assumptions (3.27) for the multivalued map

$$C(y) = A(y) + B(y), \quad y \in \mathscr{V}$$

are true. In order to finish the proof of the given statement it is enough to remark that problem (3.44) is equivalent to problem (3.29). Furthermore, problem (3.42)–(3.43) is equivalent to problem (3.30). The last one follows from the definition of subdifferential map, from Proposition 1 and from the formula:

$$D_+\varphi(u; v-u) := \lim_{t\to 0+} \frac{\varphi(u+t(v-u)) - \varphi(u)}{t} = [\partial\varphi(u), v-u]_+.$$

The Corollary is proved. □

3.3.1 *The Class of Multivalued Penalty Operators*

Let $K \subset \mathscr{V}$ be nonempty closed convex subset,

$$P_K(y) = \arg\min_{v\in K} \|y-v\|_{\mathscr{V}}, \quad y \in \mathscr{V}.$$

We consider the main convex (generally not strictly convex) lower semicontinuous functional

$$\varphi(y) = \|y - P_K y\|^2_{\mathscr{V}}, \quad y \in \mathscr{V}.$$

Let us put

$$\beta(y) = \partial\varphi(y) \in C_v(\mathscr{V}^*), \quad y \in \mathscr{V}.$$

In virtue of the properties of the subdifferential maps, the multivalued operator β is monotone, bounded, radially semicontinuous. So, it is enough to show that

$$K = \{y \in \mathscr{V} \mid \bar{0} \in \beta(y)\}.$$

"$\subset$" Let $y \in K$. Then $\varphi(y) = 0$ and for every $\omega \in \mathscr{V}, t > 0$

$$[\beta(y), \omega]_+ = [\partial\varphi(y), \omega]_+ \leftarrow \frac{\varphi(y+t\omega) - \varphi(y)}{t} = \frac{\varphi(y+t\omega)}{t} \geq 0,$$

as $t \searrow 0+$. Hence, $\bar{0} \in \beta(y)$.

"$\supset$" Let $\bar{0} \in \beta(y)$. Then for every $\omega \in \mathscr{V}$ (in particular for $\omega \in K$)

$$0 \leq [\beta(y), \omega - y]_+ = [\partial\varphi(y), \omega - y]_+ \leq \varphi(\omega) - \varphi(y).$$

Hence, $\varphi(y) \le 0$ and $y \in K$.

Let again X be the reflexive Banach space, X^* be its topologically adjoint,

$$\langle \cdot, \cdot \rangle_X : X^* \times X \to \mathbb{R} \text{ is canonical paring.}$$

We assume that for some interpolation pair of reflexive Banach spaces X_1, X_2, $X = X_1 \cap X_2$. Then due to Theorem 1.3 $X^* = X_1^* + X_2^*$. Remark also that

$$\langle f, y \rangle_X = \langle f_1, y \rangle_{X_1} + \langle f_2, y \rangle_{X_2} \quad \forall f \in X^* \, \forall y \in X,$$

where $f = f_1 + f_2$, $f_i \in X_i^*$, $i = 1, 2$.

Let $L : D(L) \subset X \to X^*$ be linear dense defined maximally monotone operator with definitional domain $D(L)$, $A : X_1 \rightrightarrows X_1^*$, $B : X_2 \rightrightarrows X_2^*$, $N : Y \rightrightarrows Y^*$ ($Y \subset X$ with compact embedding) be some multivalued maps. We consider the next problem:

$$\begin{cases} \langle Ly, \omega - y \rangle_X + [A(y), \omega - y]_+ \\ \quad + [B(y), \omega - y]_+ + [N(y), \omega - y]_+ \\ \quad\quad \ge \langle f, \omega - y \rangle_X \quad \forall \omega \in D(L) \cap K, \\ y \in D(L) \cap K, \end{cases} \tag{3.47}$$

for some fixed $f \in X^*$ and the bodily convex set $K \subset X$.

Our aim consists of proving the existence of solutions by the method of singular perturbation with the penalty method (see [LIO69]).

Remark 3.10. Further we consider that $D(L)$ is a reflexive Banach space with the graph norm

$$\|y\|_{D(L)} = \|y\|_X + \|Ly\|_{X^*} \quad \forall y \in D(L).$$

This condition assures the maximal monotony of L on $D(L)$ (Corollary 1.8).

Let us consider the operators A, L and the convex set K such that

$$\text{The operator } L : D(L) \subset X \to X^* \text{ is maximally monotone on } D(L), \text{ linear and dense defined;} \tag{3.48}$$

$$K \text{ is a convex closed subset from } X \text{ such that } \exists \beta_0 \in K \cap D(L) : \quad \bigcup_{t>0} t(K - \beta_0) = X; \tag{3.49}$$

$$\begin{gathered} \text{The multivalued map } A : X \to C_v(X^*) \text{ is } \lambda_0 - \text{pseudo monotone on } D(L), \text{ locally finite-dimensionally bounded,} \\ \text{it satisfies Condition } (\Pi) \text{ and for some} \\ y_0 \in K \cap D(L) \quad \frac{[A(y), y - y_0]_+}{\|y\|_X} \to +\infty \quad \text{as} \quad \|y\|_X \to \infty; \end{gathered} \tag{3.50}$$

$$\begin{gathered} \beta : X \to C_v(X^*) \text{ is the monotone, bounded, radially semicontinuous multivalued ``penalty'' operator, that} \\ \text{corresponds to the set } K, \text{ i.e. } K = \{y \in X \mid \beta(y) \ni \bar{0}\}; \end{gathered} \tag{3.51}$$

Remark 3.11. The sufficient condition for (3.49) is:
K is a convex closed subset from X such that $D(L) \cap \operatorname{int} K \neq \emptyset$.

Theorem 3.6. *Let the conditions (3.48)–(3.51) hold, $f \in X^*$ be arbitrary fixed. Then for each $\varepsilon > 0$ the problem*

$$\begin{cases} Ly_\varepsilon + A(y_\varepsilon) + \frac{1}{\varepsilon}\beta(y_\varepsilon) \ni f, \\ y_\varepsilon \in D(L) \end{cases} \tag{3.52}$$

has a solution. Moreover, there is a sequence $\{y_\varepsilon\}_\varepsilon \subset D(L)$ such that

(a) For every $\varepsilon > 0$ y_ε is the solution of problem (3.52).
(b) There exists a subsequence $\{y_\tau\}_\tau \subset \{y_\varepsilon\}_\varepsilon$ such that for some $y \in X$ $y_\tau \to y$ weakly in X.
(c) y is the solution of the next problem:

$$\begin{cases} \langle Ly, v-y\rangle_X + [A(y), v-y]_+ \geq \langle f, v-y\rangle_X \quad \forall v \in K \cap D(L), \\ y \in K \cap D(L). \end{cases} \tag{3.53}$$

Proof. By the analogy with [LIO69, c.396] without loss of generality we may consider that $y_0 = \bar{0} \in K$. Otherwise, the maps $\widetilde{A}(\cdot) = A(\cdot - y_0)$, $\tilde{f} = f - Ly_0$, $\widetilde{L} = L$, the set $\widetilde{K} = K - y_0$, $\tilde{y_0} = \bar{0}$ and $\widetilde{\beta_0} = \beta_0 - y_0$ satisfy the conditions (3.48)–(3.51).

For every $\varepsilon > 0$ let us enter a new multivalued map:

$$A_\varepsilon(y) := A(y) + \frac{1}{\varepsilon}\beta(y), \quad y \in X.$$

In virtue of Proposition 1.22 and of Lemma 1.15, $A_\varepsilon : X \to C_v(X^*)$ is λ_0-pseudomonotone on $D(L)$. Due to the boundness of β, thanks to Condition (Π) and to the locally finitedimensionally boundness for A it follows that A_ε is locally finitedimensionally bounded and it satisfies Condition (Π).

Now, let us use the coercivity condition. From (3.50) it follows the existence of $R > 0$ such that

$$[A(y) - f, y]_+ \geq 0 \quad \forall y \in X : \|y\|_X = R.$$

Then, for every $\varepsilon > 0$

$$\begin{aligned} [A_\varepsilon(y) - f, y]_+ &\geq [A(y) - f, y]_+ + \frac{1}{\varepsilon}[\beta(y), y - \bar{0}]_- \\ &\geq [A(y) - f, y]_+ + \frac{1}{\varepsilon}[\beta(\bar{0}), y]_+ \\ &= [A(y) - f, y]_+ \geq 0 \quad \forall \|y\|_X = R. \end{aligned}$$

Hence, we can apply Theorem 2.2 for

$$X_1 = X_2 = X,\ D(L) = D(L),\ L = L,\ \mathscr{A} \equiv \bar{0},\ \mathscr{B} = A_\varepsilon,\ f = f,\ R = R.$$

Then, we obtain that for every $\varepsilon > 0$ there exists $y_\varepsilon \in X$ such that

$$y_\varepsilon \text{ is the solution of (3.52)}, \qquad \|y_\varepsilon\|_X \le R. \tag{3.54}$$

We remark that the constant R does not depend on $\varepsilon > 0$.

From (3.54) it follows that there exist $d_\varepsilon \in A(y_\varepsilon)$, $b_\varepsilon \in \beta(y_\varepsilon)$ such that

$$Ly_\varepsilon + d_\varepsilon + \frac{1}{\varepsilon} b_\varepsilon = f. \tag{3.55}$$

Due to $\bar{0} \in K \cap D(L)$ and to the monotony of L and β we have:

$$\langle d_\varepsilon, y_\varepsilon \rangle_X \le -\langle Ly_\varepsilon, y_\varepsilon \rangle_X + \frac{1}{\varepsilon}\langle b_\varepsilon, \bar{0} - y_\varepsilon \rangle_X + \langle f, y_\varepsilon \rangle_X \le \|f\|_{X^*} R < +\infty.$$

In virtue of Property (Π) for A and from (3.54) it follows that there exists $c_1 > 0$ such that

$$\|d_\varepsilon\|_{X^*} \le c_1 \quad \forall \varepsilon > 0. \tag{3.56}$$

Moreover, from (3.55) and (3.56) it follows that

$$\begin{aligned} 0 &\le \langle b_\varepsilon, y_\varepsilon - \beta_0 \rangle_X = \varepsilon \langle f - d_\varepsilon - Ly_\varepsilon, y_\varepsilon - \beta_0 \rangle_X \\ &\le \varepsilon(\|f\|_{X^*} + c_1 + \|L\beta_0\|_{X^*})(R + \|\beta_0\|_X) =: c_2 \cdot \varepsilon \to 0 \quad \text{as} \quad \varepsilon \searrow 0+. \end{aligned} \tag{3.57}$$

From the monotony of β, from (3.57) and from (3.49) it follows that for every $\omega \in X$ $\exists t > 0$: $t\omega + \beta_0 \in K$ and

$$\begin{aligned} \frac{1}{\varepsilon}\langle b_\varepsilon, \omega \rangle_X &= \frac{1}{t\varepsilon}\langle b_\varepsilon, t\omega - y_\varepsilon \rangle_X + \frac{1}{t\varepsilon}\langle b_\varepsilon, y_\varepsilon \rangle_X \\ &\le \frac{1}{t\varepsilon}[\beta(y_\varepsilon), t\omega - y_\varepsilon]_+ + \frac{1}{t}c_2 \le \frac{1}{t\varepsilon}[\beta(\omega), t\omega - y_\varepsilon]_- + \frac{1}{t}c_2 \le \frac{1}{t}c_2. \end{aligned}$$

Hence, due to the Banach–Steinhauss Theorem there exists $c_3 > 0$ such that

$$\|b_\varepsilon\|_{X^*} \le \varepsilon c_3 \quad \forall \varepsilon \in (0, \varepsilon_0), \tag{3.58}$$

for some $\varepsilon_0 > 0$.

The conditions (3.48), (3.55) and (3.58) imply that for every $\omega \in X$

$$|\langle Ly_\varepsilon, \omega \rangle_X| \le (\|f\|_{X^*} + c_2 + c_3)\|w\|_X \quad \forall \varepsilon \in (0, \varepsilon_0).$$

Hence, there exists $c_4 > 0$ such that

$$\|Ly_\varepsilon\|_{X^*} \leq c_4 \quad \forall \varepsilon \in (0, \varepsilon_0). \tag{3.59}$$

Passing to limit. From the estimations (3.54), (3.56), (3.59), due to the Banach–Alaoglu Theorem it follows that there exists a subsequence $\{y_\tau\}_\tau$ from $\{y_\varepsilon\}_\varepsilon$ such, that for some $y \in D(L)$, $d \in X^*$

$$y_\tau \rightharpoonup y \text{ in } D(L),\ Ly_\tau \rightharpoonup y \text{ in } X,\ d_\tau \rightharpoonup d \text{ in } X^*,\ b_\tau \rightharpoonup \bar{0} \text{ in } X^* \text{ as } \tau \searrow 0+. \tag{3.60}$$

In virtue of Proposition 1.22, the map β is λ_0-pseudomonotone on X. Moreover, due to (3.57) and (3.60) we have:

$$\lim_{\tau \searrow 0+} \langle b_\tau, y_\tau - y\rangle_X = \overline{\lim_{\tau \searrow 0+}} \langle b_\tau, y_\tau - y\rangle_X \leq 0.$$

Hence, due to a subsequence, for every $\omega \in X$

$$0 = \underline{\lim_{\tau \searrow 0+}} \langle b_\tau, y_\tau - \omega\rangle_X \geq [\beta(y), y - \omega]_-.$$

The last relation is equivalent to $\bar{0} \in \beta(y)$. Hence, in virtue of (3.51), we obtain that $y \in K$.

Now, let us show that

$$\overline{\lim_{\tau \searrow 0+}} \langle d_\tau, y_\tau - y\rangle_X \leq 0. \tag{3.61}$$

Really, from (3.55) and from (3.51) it follows that for every $v \in D(L) \cap K$

$$\begin{aligned}
\langle d_\tau, y_\tau - v\rangle_X &= \frac{1}{\varepsilon}\langle b_\tau, v - y_\tau\rangle_X + \langle f, y_\tau - v\rangle_X + \langle Ly_\tau, v - y_\tau\rangle_X \\
&\leq \frac{1}{\varepsilon}[\beta(y_\tau), v - y_\tau]_+ + \langle f, y_\tau - v\rangle_X + \langle Ly_\tau, v - y_\tau\rangle_X \\
&\leq \frac{1}{\varepsilon}[\beta(v), v - y_\tau]_- + \langle f, y_\tau - v\rangle_X + \langle Ly_\tau, v - y_\tau\rangle_X \\
&\leq \langle f, y_\tau - v\rangle_X + \langle Ly_\tau, v - y\rangle_X + \langle Ly, y - y_\tau\rangle_X \\
&\leq \langle f, y_\tau - v\rangle_X + \langle Lv, v - y_\tau\rangle_X,
\end{aligned} \tag{3.62}$$

as $\bar{0} \in \beta(v)$. So,

$$\overline{\lim_{\tau \searrow 0+}} \langle d_\tau, y_\tau\rangle_X \leq \langle d, v\rangle_X + \langle f, y - v\rangle_X + \langle Lv, v - y\rangle_X \quad \forall v \in D(L) \cap K.$$

But in virtue of (3.60), if we put in the last relation $v = y$, we obtain:

$$\overline{\lim_{\tau \searrow 0+}} \langle d_\tau, y_\tau\rangle_X \leq \langle d, y\rangle_X.$$

Therefore, due to (3.60), the inequality (3.61) is true.

Let us use the λ_0-pseudomonotony of A. From (3.60) and (3.61) it follows that there exist the subsequences $\{y_\nu\}_\nu \subset \{y_\tau\}_\tau$ and $\{d_\nu\}_\nu \subset \{d_\tau\}_\tau$ such that

$$\varliminf_{\nu \searrow 0+} \langle d_\nu, y_\nu - v\rangle_X \ge [A(y), y - v]_- \quad \forall v \in X, \tag{3.63}$$

in particular, from the inequality (3.61) it follows

$$\lim_{\nu \searrow 0+} \langle d_\nu, y_\nu - y\rangle_X = 0.$$

Hence, due to (3.60), (3.62) and (3.63),

$$\langle Ly, v - y\rangle_X + [A(y), y - v]_- \le \langle f, y - v\rangle_X \quad \forall v \in K \cap D(L),$$

that is equivalent (due to Proposition 1) to (3.53).

The Theorem is proved. □

Corollary 3.2. *Let the conditions (3.48), (3.49), (3.51) hold, $A : X_1 \rightrightarrows X_1^*$ and $B : X_2 \rightrightarrows X_2^*$ are finite-dimensionally locally bounded, λ_0-pseudomonotone on $D(L)$ multivalued maps and satisfy Condition (Π). We consider, that the embedding of $D(L)$ in some Banach space Y is compact and dense, the embedding of X in Y is dense and continuous and let $N : Y \rightrightarrows Y^*$ be a locally bounded multivalued map, such that the graph of N is closed in $Y \times Y_w^*$ (i.e. with respect to the strong topology of Y and the weakly star one in Y^*) and which satisfies Condition (Π). Furthermore, for some $y_0 \in K \cap D(L)$*

$$\frac{[A(y), y - y_0]_+}{\|y\|_{X_1}} \to +\infty \quad as \quad \|y\|_{X_1} \to \infty,$$

$$\frac{[B(y), y - y_0]_+}{\|y\|_{X_2}} \to +\infty \quad as \quad \|y\|_{X_2} \to \infty,$$

$$\varliminf_{\|y\|_X \to \infty} \frac{[N(y), y - y_0]_+}{\|y\|_X} > -\infty, \tag{3.64}$$

$f \in X^*$ be arbitrary fixed. Then for each $\varepsilon > 0$ the problem

$$\left.\begin{aligned} &Ly_\varepsilon + A(y_\varepsilon) + B(y_\varepsilon) + N(y_\varepsilon) + \frac{1}{\varepsilon}\beta(y_\varepsilon) \ni f, \\ &y_\varepsilon \in D(L) \end{aligned}\right\} \tag{3.65}$$

has a solution. Moreover, there is a sequence $\{y_\varepsilon\}_\varepsilon \subset D(L)$ such that

(a) For every $\varepsilon > 0$ y_ε is the solution of problem (3.65).
(b) There exists a subsequence $\{y_\tau\}_\tau \subset \{y_\varepsilon\}_\varepsilon$ such that for some $y \in X$ $y_\tau \rightharpoonup y$ in X.
(c) y is the solution of problem (3.47).

Remark 3.12. The sufficient condition for (3.64) is:

$$\exists C_1, C_2 > 0: \qquad \|N(y)\|_+ \le C_1 + C_2\|y\|_X \qquad \forall y \in X.$$

Proof. Let us set $C(y) = A(y) + B(y) + N(y)$ for each $y \in X \subset Y$. In virtue of the continuous embedding $X \subset Y$ with Lemma 2 it follows, that C satisfies Condition (Π) on X. The finite-dimensionally locally boundness of C is clear. Due to Proposition 1.26 the map C is λ_0-pseudomonotone on $D(L)$. The +-coercivity of $C(\cdot + y_0)$ on X directly follows from Proposition 6, from Lemma 1 for A and B and from the condition (3.64). So, we apply Theorem 3.6 for C, L, K. Thus, problem (3.47) has at least one solution obtained by the penalty method. □

3.4 Evolution Variation Inequalities with Noncoercive Multivalued Maps

As before let $(V_1, \|\cdot\|_{V_1})$ and $(V_2, \|\cdot\|_{V_2})$ be reflexive Banach spaces continuously embedded in Hilbert space $(H, (\cdot,\cdot))$ such that for some numerable set $\Phi \subset V = V_1 \cap V_2$

$$\Phi \text{ is dense in } V,\ V_1,\ V_2 \text{ and in } H.$$

After identification $H \equiv H^*$ we obtain

$$V_1 \subset H \subset V_1^*, \qquad V_2 \subset H \subset V_2^*,$$

with continuous and dense embedding, $(V_i^*, \|\cdot\|_{V_i^*})$, $i = 1,2$ is topologically adjoint of V_i space with respect to the canonical bilinear form

$$\langle \cdot,\cdot \rangle_{V_i} : V_i^* \times V_i \to \mathbb{R},$$

which coincides on $H \times \Phi$ with the inner product $(\cdot,\cdot)$.

Let us consider the functional spaces $X_i = L_{r_i}(S; H) \cap L_{p_i}(S; V_i)$, where S is a finite time interval, $1 < p_i \le r_i < +\infty$. The spaces X_i are reflexive Banach spaces with the norms

$$\|y\|_{X_i} = \|y\|_{L_{p_i}(S;V_i)} + \|y\|_{L_{r_i}(S;H)},$$

$X = X_1 \cap X_2$, $\|y\|_X = \|y\|_{X_1} + \|y\|_{X_2}$. Let X_i^* $(i = 1,2)$ be topologically adjoint with X_i. Then,

$$X^* = X_1^* + X_2^* = L_{q_1}(S; V_1^*) + L_{q_2}(S; V_2^*) + L_{r_1'}(S; H) + L_{r_2'}(S; H),$$

where $r_i{}^{-1} + r_i'{}^{-1} = p_i{}^{-1} + q_i{}^{-1} = 1$ $(i = 1,2)$. Let us define the duality form on $X^* \times X$

$$\langle f, y\rangle = \int_S (f_{11}(\tau), y(\tau))_H d\tau + \int_S (f_{12}(\tau), y(\tau))_H d\tau + \int_S \langle f_{21}(\tau), y(\tau)\rangle_{V_1} d\tau$$
$$+ \int_S \langle f_{22}(\tau), y(\tau)\rangle_{V_2} d\tau = \int_S (f(\tau), y(\tau))\, d\tau,$$

where $f = f_{11} + f_{12} + f_{21} + f_{22}$, $f_{1i} \in L_{r_i'}(S; H)$, $f_{2i} \in L_{q_i}(S; V_i^*)$ $(i = 1, 2)$. Note that $\langle \cdot, \cdot\rangle$ coincides with the inner product in $\mathscr{H} = L_2(S; H)$ on $\mathscr{H}$.

Theorem 3.7. *Let $V_2 = H$, $r_1 \ge 2$, $p_2 = r_2 = 2$, $\lambda_0 > 0$, $A + \lambda_0 I : X_1 \rightrightarrows X_1^*$ is +-coercive, r.l.s.c. multivalued operator with (X_1, W)-s.b.v., $\varphi : X_2 \to \mathbb{R}$ is convex lower semicontinuous functional. Then for each $f \in X^*$ there is at least one solution $y \in W$ for the problem:*

$$\langle y', \xi - y\rangle + [Ay, \xi - y]_+ + \varphi(\xi) - \varphi(y) \ge \langle f, \xi - y\rangle \quad \forall \xi \in W, y(0) = \bar{0}, \quad (3.66)$$

under the condition, that A and $\partial\varphi$ are the Volterra operators that satisfy Condition (H).

Remark 3.13. In Theorem 3.7 we may change Condition (H) for A and $\partial\varphi$ and +-coercivity for $A + \lambda_0 I$ on −-coercivity for $A + \lambda_0 I$ on X_1.

Remark 3.14. At the last corollary we are not claim the coercivity for φ (resp. for $\partial\varphi$) on X_2.

Proof. Let us show that the operator $C = \lambda I + \lambda_0 I + A + \partial\varphi : X \rightrightarrows X^*$ satisfies all conditions of Corollary 2.2 for some $\lambda > 0$. For this purpose it is enough to show the same for multivalued map

$$B(y) = \lambda y + \partial\varphi(y) \quad \forall y \in X_2,$$

where $\lambda > 0$ is an arbitrary fixed. R.l.s.c. follows from u.s.c. of $\partial\varphi$ on X_2. In virtue of monotony of $\partial\varphi$ on X_2, for each $y \in X_2$

$$\begin{aligned}[B(y), y]_+ &= [(\lambda I + \partial\varphi)(y), y]_+ = \lambda\langle y, y\rangle + [\partial\varphi(y), y]_+ \\ &= \lambda(y, y)_{\mathscr{H}} + [\partial\varphi(y), y - \bar{0}]_+ \ge \lambda\|y\|_{\mathscr{H}}^2 + [\partial\varphi(\bar{0}), y - \bar{0}]_+ \\ &\ge \lambda\|y\|_{X_2}^2 - \|\partial\varphi(\bar{0})\|_+\|y\|_{X_2}.\end{aligned}$$

Thus, for each $y \in X_2$

$$\frac{[B(y), y]_+}{\|y\|_{X_2}} \ge \lambda\|y\|_{X_2} - \|\partial\varphi(\bar{0})\|_+ \to +\infty \text{ as } \|y\|_{X_2} \to +\infty.$$

So, B is +-coercive. From monotony of $\partial\varphi$, as $\lambda > 0$, it follows that B is also monotone and it has the semibounded variation on W. Thus, problem (3.66) has at least one solution $y \in W$ such that $y(0) = \bar{0}$. □

Example 3.1. Let $n \geq 1$, $\Omega \subset \mathbb{R}^n$ is bounded domain with the boundary $\partial\Omega$, $S = [0; T]$, $Q = S \times \Omega$, $p \in (1; 2]$: $W_0^{1,p}(\Omega) \subset L_2(\Omega)$; a functional $\Phi : \mathbb{R} \to \mathbb{R}$ is measurable and satisfies the conditions:

$$\exists C_1, C_2 > 0: \quad |\Phi(t)| \leq C_1|t| + C_2 \quad \forall t \in \mathbb{R};$$
$$\exists C_3 > 0: \quad (\Phi(t) - \Phi(s))(t - s) \geq -C_3(s - t)2 \quad \forall t, s \in \mathbb{R};$$

a functional $\psi : \mathbb{R} \to \mathbb{R}$ is convex, lower semicontinuous and satisfies the "growth condition":

$$\exists C_4, C_5 > 0: \quad |\psi(t)| \leq C_4|t| + C_5 \quad \forall t \in \mathbb{R}.$$

Let $q \geq 2$: $1/p + 1/q = 1$. Let us set

$$X = L_p(S; W_0^{1,p}(\Omega)) \cap L_2(S; L_2(\Omega)),$$

$$X^* = L_q(S; W_0^{-1,q}(\Omega)) + L_2(S; L_2(\Omega)).$$

We consider such problem:

$$\begin{aligned}
&\int_Q \frac{\partial y(t,x)}{\partial t}(v(t,x) - y(t,x))dtdx \\
&+ \sum_{i=1}^n \int_Q \left(\left| \frac{\partial y(t,x)}{\partial x_i} \right|^{p-2} \frac{\partial y(t,x)}{\partial x_i} \right) \left(\frac{\partial v(t,x)}{\partial x_i} - \frac{\partial y(t,x)}{\partial x_i} \right) dtdx \\
&+ \int_Q \Phi(y(t,x))(v(t,x) - y(t,x))dtdx \\
&+ \int_Q \psi(v(t,x))dtdx - \int_Q \psi(y(t,x))dtdx \\
&\geq \int_Q f(t,x)(v(t,x) - y(t,x))dtdx \quad \forall v \in X,
\end{aligned} \tag{3.67}$$

$$y(t,x)\big|_{\partial\Omega} = 0 \quad \text{for a.e. } t \in S, \tag{3.68}$$

$$y(t,x)\big|_{t=0} = 0 \quad \text{for a.e. } x \in \Omega, \tag{3.69}$$

where $f \in X^*$ is arbitrary fixed.

Let us set $X_1 = X$, $X_2 = L_2(Q)$,

$$A(y) = -\sum_{i=1}^n \frac{\partial}{\partial x_i} \left(\left| \frac{\partial y}{\partial x_i} \right|^{p-2} \frac{\partial y}{\partial x_i} \right) \quad \forall y \in X_1,$$

$$\varphi(y) = \int_Q \psi(y(t,x))dtdx, \quad y \in X_2.$$

After integration the inequality (3.67) by parts we will obtain the problem:

$$\langle y', v - y\rangle + \langle A(y), v - y\rangle + \varphi(v) - \varphi(y) \geq \langle f, v - y\rangle, \qquad y(0) = \bar{0}. \quad (3.70)$$

As generalized solution of (3.67)–(3.69) we understand the solution of (3.70) in the class $W = \{y \in X \mid y' \in X^*\}$. In virtue of Theorem 3.7, problem (3.67)–(3.69) has generalized solution $y \in W$.

3.5 On Solvability of the Class of Evolution Variation Inequalities with W_λ-Pseudomonotone Maps

Many important applied problems are reduced to so called problems with unilateral boundary conditions or to variation inequalities which generate differential-operator inclusions. The next problem is the most simple example of such type [1–5]: in the region Ω with the bound Γ we have to find a solution of the equation $\Delta y = f$ such that on Γ the conditions

$$u \geq 0, \qquad \frac{\partial u}{\partial n} \geq 0, \qquad u\frac{\partial u}{\partial n} = 0.$$

are fulfilled.

The generalized solution of such problem does not satisfy the integral identity (as, for example, in Dirichlet problem), but it satisfies some integral inequality, that is called the variation inequality.

Let $(V, \|\cdot\|_V)$ be the strictly normalized reflexive separable Banach space that continuously and densely embedded in the Hilbert space $(H, (\cdot,\cdot))$. The strict normalization of the space means the strict convexity of the norm $\|\cdot\|_V$ on V.

We identify the topologically adjoint space with H regarding bilinear form $(\cdot,\cdot)$ with H, $(V^*, \|\cdot\|_{V^*})$ is strictly normalized Banach space, that is topologically adjoint with V with regarding $(\cdot,\cdot)$. Then the next chain of continuous and dense embeddings takes place:

$$V \subset H \equiv H^* \subset V^* \quad (3.71)$$

Let $1 < p, q < +\infty$, $^1/_p + {^1/_q} = 1$, $S = [0, T]$, $T > 0$. Let us consider spaces

$$X = L_p(S; V), \quad X^* = L_q(S; V^*),$$

the pairing

$$\langle u, v\rangle_X = \int\limits_S (u(t), v(t))dt, \quad u \in X, \quad v \in X^*,$$

and

$$W = \{y \in X | y' \in X^*\}, \quad W_1 = \{y \in X | y' \in L_1(S; V^*)\},$$
$$W_2 = \{y \in W \mid y(0) = 0\}$$

with norms

$$\|u\|_W = \|u\|_X + \|u'\|_{X^*} \quad \forall u \in W$$
$$\|u\|_{W_1} = \|u\|_X + \|u'\|_{L_1(S;V^*)} \quad \forall u \in W_1,$$
$$\|u\|_{W_2} = \|u\|_W \quad \forall u \in W_2.$$

For the operator $A : X \to X^*$, convex closed set $K \subset X$ $(0 \in K)$ and fixed function $f \in X^*$ we consider the problem of searching of weak solution for the evolution variation inequality:

$$\begin{cases} \langle w', w - y\rangle_X + \langle A(y), w - y\rangle_X \geq \langle f, w - y\rangle_X \\ \forall w \in K \cap W_2, \\ y \in K. \end{cases} \tag{3.72}$$

Here w' is the derivative of the element $w \in X$ in the sense of $D^*(S; V_\sigma^*)$.

Let us remark that W with nature operations is Banach space that satisfies the next properties:

(a) W is continuously and densely embedded in $C(S; H)$.
(b) $\forall u, v \in W$ the next

$$\langle u', v\rangle_X + \langle v', u\rangle_X = (u(T), v(T)) - (u(0), v(0)) \tag{3.73}$$

is fulfilled; if $u = v$ we obtain:

$$\langle u', u\rangle_X = (\|u(T)\|_H^2 - \|u(0)\|_H^2)/2 \tag{3.74}$$

Let us improve conditions for parameters of problem (3.72), for which we will prove the weak solvability. In order to do this we will use the penalty method. As the penalty operator β and convex set K let us consider

$$\beta(y)(t) = \beta(y(t)), \quad K = K(t) \quad \text{for almost every} \quad t \in S,$$

where $\beta(v) = J(v - P_K v)$, $v \in V$, $J : V \to V^*$ is defined so:

$$\|J(v)\|_{V^*} \|v\|_V = (J(v), v), \quad \|J(v)\|_{V^*} = \|v\|_V^{p-1};$$

P_K is orthogonal projection operator from V on K. Let us remark that $\beta(v) = 0 \Leftrightarrow v \in K$.

Let us remark that $\beta : V \to V^*$ is bounded monotone demicontinuous operator, so $\beta : X \to X^*$ is λ -pseudomonotone on X bounded operator.

Theorem 3.8. *Let the next conditions*

(a) the operator $A : X \to X^*$ is bounded, λ–pseudomonotone on W_1 and coercive.
(b) $K \subset V$ is convex closed set, $0 \in intK \subset V$ are fulfilled.

Then, for an arbitrary $f \in X^*$ there exists the solution of problem (3.72).

Proof. For an arbitrary $\varepsilon > 0$ let us consider a new map

$$A_\varepsilon(y) := A(y) + \frac{1}{\varepsilon}\beta(y), \quad y \in X.$$

Lemma 3.1. *For an arbitrary $\varepsilon > 0$ the operator $A_\varepsilon(y)$ is bounded, λ–pseudomonotone on $W \subset W_1$ and coercive, and the problem*

$$\begin{cases} y'_\varepsilon + A(y_\varepsilon) + \frac{1}{\varepsilon}\beta(y_\varepsilon) = f, \\ y_\varepsilon \in W_2 \end{cases} \tag{3.75}$$

has the solution y_ε such that

$$\|y_\varepsilon\|_X \le c, \quad \|A(y_\varepsilon)\|_{X^*} \le c, \quad \|y'_\varepsilon\|_{L_1(S;V^*)} \le c \quad \forall \varepsilon > 0, \tag{3.76}$$

for some $c > 0$, that does not depend on $\varepsilon > 0$.

Proof. Let us use the coercivity condition for A. There exists such $R > 0$ that

$$\langle A(y) - f, y\rangle_X \ge 0 \quad \forall y \in X : \ \|y\|_X = R. \tag{3.77}$$

For every $\varepsilon > 0$

$$\begin{aligned} \langle A_\varepsilon(y) - f, y\rangle_X &\ge \langle A(y) - f, y\rangle_X + < \beta(y), y - 0\rangle_X/\varepsilon \\ &\ge \langle A(y) - f, y\rangle_X + \langle \beta(0), y\rangle_X/\varepsilon \\ &= \langle A(y) - f, y\rangle_X \ge 0 \quad \forall \|y\|_X = R. \end{aligned}$$

In particular, from here it follows the coercivity for A_ε on X. The boundness for A_ε is fulfilled, because operators A and β are bounded; λ–pseudomonotony for A_ε on $W \subset W_1$ follows from the same property for A and β. Let us consider the operator $L : W_2 \to X^*$, that is defined as:

$$Ly = y', \quad y \in W_2.$$

Let us remark that L is maximally monotone operator on W_2. $L^* : W_2 \to X$ is adjoint with L operator in the sense of the nonbounded operators theory.

So, from (3.77), from the boundness for A and from the previous results we have that for an arbitrary $\varepsilon > 0$ problem (3.75) has the solution y_ε such that

$$\|y_\varepsilon\|_X \le R, \quad \|A(y_\varepsilon)\|_{X^*} \le c_1.$$

for some $c_1 > 0$, that does not depend on $\varepsilon > 0$.

From the condition (b) it follows that for some $r > 0$ $\overline{B_r} = \{v \in V \mid \|v\|_V \leq r\} \subset K$. Then $\forall \varepsilon > 0$, for a.e. $t \in S$ we have

$$\|\beta(y_\varepsilon(t))\|_{V^*} \leq (\beta(y_\varepsilon(t)), y_\varepsilon(t))/r. \tag{3.78}$$

Indeed, $\forall \varepsilon > 0$, for a.e. $t \in S$, $\forall v \in V : \|v\|_V = 1$ we have

$$\begin{aligned}(\beta(y_\varepsilon(t)), v) &= (\beta(y_\varepsilon(t)),\ rv - y_\varepsilon(t)\ )/r + (\beta(y_\varepsilon(t)), y_\varepsilon(t)\ )/r \\ &\leq (\beta(rv), rv - y_\varepsilon(t)\ )/r + (\beta(y_\varepsilon(t)), y_\varepsilon(t))/r \\ &= (\beta(y_\varepsilon(t)), y_\varepsilon(t))/r.\end{aligned}$$

Now let us show that

$$\langle \beta(y_\varepsilon), y_\varepsilon \rangle_X \leq (\|f\|_{X^*} + c_1)\, R\, \varepsilon \quad \forall \varepsilon > 0 \tag{3.79}$$

As $\forall \varepsilon > 0$ $y_\varepsilon \in X$ is the solution of problem (3.75), then in view of (3.74),

$$\langle \beta(y_\varepsilon), y_\varepsilon \rangle_X / \varepsilon = \langle f - A(y_\varepsilon) - y'_\varepsilon, y_\varepsilon \rangle_X \ \leq\ (\|f\|_{X^*} + c_1)R.$$

From (3.78) to (3.79) $\forall \varepsilon > 0$ it follows the next inequality

$$\|\beta(y_\varepsilon)\|_{L_1(S;V^*)} \leq\ (\|f\|_{X^*} + c_1)R\,\varepsilon/r =: c_2\varepsilon \tag{3.80}$$

in particular, in view of (3.75), (3.76), (3.80) we have

$$\|y'_\varepsilon\|_{L_1(S;V^*)} \leq c_3 \quad \forall \varepsilon > 0,$$

where c_3 is the constant that does not depend on $\varepsilon > 0$.
The Lemma is proved. □

Let us continue the proof of the Theorem. From the monotony for β and from (3.79) to (3.80) it follows that for an arbitrary $w \in X$, $\varepsilon > 0$

$$\begin{aligned}0 \leq \langle \beta(y_\varepsilon), w \rangle_X \ &\leq\ \langle \beta(y_\varepsilon), w - y_\varepsilon \rangle_X \ + \langle \beta(y_\varepsilon), y_\varepsilon \rangle_X \\ &\leq \langle \beta(w), w - y_\varepsilon \rangle_X \ + c_2 r\varepsilon \to \langle \beta(w), w - y \rangle_X < +\infty \quad \text{as } \varepsilon \to 0+.\end{aligned}$$

So, in consequence of the Banach–Steingauss Theorem, there exists $c_4 > 0$ such that

$$\|\beta(y_\varepsilon)\|_{X^*} \leq c_4 \quad \forall \varepsilon \in (0, \varepsilon_0) \tag{3.81}$$

for some $\varepsilon_0 > 0$.

Let' remark also that for an arbitrary $\omega \in D(L^*)$

$$\langle y'_\varepsilon, \omega \rangle_X = \langle L^*\omega, y_\varepsilon \rangle_X \leq\ \|\omega\|_{D(L^*)} R \quad \forall \varepsilon > 0.$$

Therefore, there exists $c_5 > 0$ such that

$$\|y'_\varepsilon\|_{D(L^*)^*} \le c_5 \quad \forall \varepsilon > 0. \tag{3.82}$$

From here and from (3.75) it follows that

$$\beta(y_\varepsilon) \to 0 \quad \text{in} \quad D(L^*)^* \quad \text{as} \quad \varepsilon \to 0+\,. \tag{3.83}$$

From (3.75), (3.76) and from the monotony for L we have:

$$\begin{aligned}\langle y'_\varepsilon, \omega\rangle_X &= \langle y'_\varepsilon, \omega - y_\varepsilon\rangle_X + \langle y'_\varepsilon, y_\varepsilon\rangle_X \\ &\le \langle \omega', \omega - y_\varepsilon\rangle_X + (\|f\|_{X^*} + c)c \\ &\le \|\omega'\|_{X^*}(\|\omega\|_X + c) + (\|f\|_{X^*} + c)c \quad \forall \omega \in W_2.\end{aligned}$$

Therefore, there exists $c_6 > 0$ such that

$$\|y'_\varepsilon\|_{W_2^*} \le c_5 \quad \forall \varepsilon \in (0; \varepsilon_0).$$

The passage to the limit. From the estimations (3.76) and (3.81), from the convergence (3.83), in consequence of the Banach–Alaoglu theorem, it follows the existence of sequences $\{y_\tau\}_\tau$ from $\{y_\varepsilon\}_\varepsilon$ such that for some $y \in X$, $d \in X^*$

$$\begin{aligned}&y_\tau \xrightarrow{w} y \text{ in } X, \quad A(y_\tau) \xrightarrow{w} d \text{ in } X^*, \\ &\beta(y_\tau) \xrightarrow{w} \bar{0} \text{ in } X^* \quad \text{as } \tau \to 0+\,.\end{aligned} \tag{3.84}$$

From (3.78) and (3.84) we have:

$$\lim_{\tau\to 0+} \langle \beta(y_\tau), y_\tau - y\rangle_X = \overline{\lim_{\tau\to 0+}} \langle \beta(y_\tau), y_\tau - y\rangle_X \le 0.$$

So, for an arbitrary $\omega \in X$

$$0 = \lim_{\tau\to 0+} \langle \beta(y_\tau), y_\tau - \omega\rangle_X \ge \langle \beta(y), y - \omega\rangle_X.$$

It means that $\bar{0} = \beta(y)$. Therefore,

$$y \in K. \tag{3.85}$$

Let us now prove that

$$\overline{\lim_{\tau\to 0+}} \langle A(y_\tau), y_\tau - y\rangle_X \le 0. \tag{3.86}$$

Indeed, from (3.75) for an arbitrary $v \in W_2 \cap K$ it follows:

$$\begin{aligned}\langle A(y_\tau), y_\tau - v\rangle_X &= \frac{1}{\varepsilon}\langle \beta(y_\tau), v - y_\tau\rangle_X + \langle f, y_\tau - v\rangle_X + \langle y'_\tau, v - y_\tau\rangle_X \\ &\le \frac{1}{\varepsilon}\langle \beta(y_\tau), v - y_\tau\rangle_X + \langle f, y_\tau - v\rangle_X + \langle v', v - y_\tau\rangle_X \\ &\le \frac{1}{\varepsilon}\langle \beta(v), v - y_\tau\rangle_X + \langle f, y_\tau - v\rangle_X + \langle v', v - y_\tau\rangle_X \\ &\le \langle f, y_\tau - v\rangle_X + \langle v', v - y_\tau\rangle_X, \end{aligned} \tag{3.87}$$

as $0 = \beta(v)$. So,

$$\overline{\lim_{\tau\to 0+}}\, \langle A(y_\tau), y_\tau\rangle_X \le \langle d, v\rangle_X + \langle f, y - v\rangle_X + \langle v', v - y\rangle_X \quad \forall v \in W_2 \cap K.$$

But from [LIO69, p. 284], in consequence of (3.87), there exists $v_j \in K \cap W_2$ such that $v_j \to y$ in X and $\overline{\lim_{j\to\infty}}\, \langle v'_j, v_j - y\rangle_X \le 0$. If we set in the last relation $v = v_j$, we will obtain:

$$\overline{\lim_{\tau\to 0+}}\, \langle A(y_\tau), y_\tau\rangle_X \le \langle d, y\rangle_X.$$

So, with respect to (3.84), the inequality (3.86) is true.

Let us use the λ-pseudomonotony for A on W_1. From (3.76), (3.84) and (3.86) it follows the existence of sequence $\{y_\nu\}_\nu \subset \{y_\tau\}_\tau$ such that

$$\underline{\lim_{\nu\to 0+}}\, \langle A(y_\nu), y_\nu - v\rangle_X \ge \langle A(y), y - v\rangle_X \quad \forall v \in X, \tag{3.88}$$

in particular, from the inequality (3.86) it follows that

$$\lim_{\nu\to 0+} \langle A(y_\nu), y_\nu - y\rangle_X = 0.$$

In consequence of (3.84) and (3.87),

$$\langle A(y), y - v\rangle_X \le \langle f, y - v\rangle_X + \langle v', v - y\rangle_X \quad \forall v \in K \cap W_2.$$

The Theorem is proved. □

So, we can prove the solvability for the class of evolution variation inequalities with essentially nonlinear pseudomonotone on W_1 operators by the penalty method. As an example, we can consider the variation inequality with the operator that is introduced in the form of monotone and demicontinuous one. In view of perspectives of obtained results, we can justify the solvability for the classes of nonautonomous evolution problems with free bound, with nonlinear conditions on the bound of region, in particular, for boundary problems of Signorini type.

3.6 The Modelling of Unilateral Processes of Diffusion of Petroleum in Porous Mediums with the Limiting Pressure Gradient

The character of filtration of oil through porous mediums of oil deposits depends on interaction between the oil and the rigid skeleton of ground. Within the framework of the classical theory the liquid filtration in porous mediums obeys the linear law of Darsi filtration. That is why the problem of filtration proposes the variation definition and it can be realized within the theory of boundary problems for the partial differential equations [AS86].

The presence of impurities (lampwax, asphalts and others), temperature anomalies and other factors essentially change the physics–chemical properties of the oil and lead to the appearance of specific effects of hindrance inside considered region as well as effects of undirectional conduction of boundary [AS86]. The equations of filtration based on classical linear law of Darsi are not the best form of considered processes description in this case. Taking into account the additional conditions for the solution of problem on the stage of its variation definition is often more substantial way of considered processes development. Such approach leads to variation inequalities which are introduced and developed in [DL76, LK83].

Let us consider the anomalous process of oil movement which contain lampwaxes and resinous-asphalt substances in porous medium of oil deposit. The oils which contain such heavy components can form the rigid structure at low temperatures, that at certain limiting pressure gradient can gain the movability. Such behavior of oil has the nonlinear character and can't be described by linear law of Darsi. The consequence of such anomaly is the appearance of stagnation zone. Such processes were considered as the filtration of visco-plastic liquid with thelimiting pressure gradient. The models of such processes were considered, in particular, in [LK83]. Let us describe these processes in the form of variation inequalities [DL76], which take into account the nonlinear effects of one-sided hindrances with the limiting pressure gradient, propose the algorithm of their realization and adduce the results of computations.

Let $y(t,x)$ be the function of pressure of oil with taking into account its visco-plastic properties that is defined on the bounded opened set Ω of the space $\mathbb{R}^3$ with smooth bound Γ and in the interval of time $(0,t_k)$ for $t_k < \infty$, $Q = \Omega \times (0,t_k)$, $\sum = \Gamma \times (0,t_k)$ is the solution of the variation inequality [ZN96]

$$\left(n(x)\frac{\partial y}{\partial t}, \upsilon - y\right) + (A(\lambda)y, \upsilon - y) + \psi(\upsilon) - \psi(y) \geq (f, \upsilon - y) \text{ in } Q$$

$$\forall \; v \in H^1(\Omega) = V, \tag{3.89}$$

with initial condition

$$y|_{t=0} = y_0 \text{ in } \Omega, \tag{3.90}$$

where $n(x) = \beta_C + m^*(x)\beta_H$; β_C is the coefficient of compressibilities of a porous medium β_H is the coefficient of compressibilities of oil; $m^*(x)$ is the porosity of a medium; $A(\lambda)(\cdot) = -\sum_{i=1}^{3} \frac{\partial}{\partial x_i}\left(b(x)k(x)\frac{\partial}{\partial x_i}\right)$, $b(x) = \frac{\rho}{\mu}$; ρ is the denseness of oil; μ is the viscosity of oil; $k(x)$ is the coefficient of transmissivity $x = (x_1, x_2, x_3)$. The variable $f(t,x) = \frac{1}{h(x)} \sum_{j=1}^{k} q_j(t)\,\delta(x - x^j)$ is the forcing function of the processes; $q_j(x)$ are discharges of productive slits, operated in subsets $\Omega_j \in \Omega$, $j = 1, \ldots, K$ the number of slits; $\delta(x - x^j)$ is the characteristic function. Let us establish that the transmissivity $k(x)$ and discharges of slits q have technological restrictions. We can define the limiting gradient by the relation of the form:

$$\left|\frac{\partial y(t,x)}{\partial x}\right| \le y_{\lim}(x), \tag{3.91}$$

where $y_{lim}(x)$ is the known value of limiting gradient

Let us add the physical sense to the system (3.89), (3.90), defining the functional ψ, that would provide the realization of the next physical conditions in the region Ω: if in some points of the spatial domain Ω the condition $\left|\frac{\partial y(t,x)}{\partial x}\right| > y_{\lim}(x)$ is fulfilled, then we have classical physical processes of oil filtration through the porous medium, obeying Darsi law. In the points of spatial domain Ω, where the association (3.91) is true, the filtration of a fluid stops and the stagnation domains are formed. That is why ψ, that adds to system (3.89), (3.90) the mentioned physical properties, will have the form

$$\psi = \begin{cases} \frac{1}{2}\left[m(t,x)y(t,x)\right]^2, & \left|\partial y(t,x)/\partial x\right| \le y_{\lim}(x), \\ 0, & \left|\partial y(t,x)/\partial x\right| > y_{\lim}(x), \end{cases} \tag{3.92}$$

where $m(t,x)$ is the known coefficient of hindrance. Let us define $m(t,x)$ as $m(t,x) \in M = L^{\infty}(Q)$, where M is the space of parameters m with the norm $\|m\|_M = \|m\|_{L^{\infty}(Q)}$. The set of admissible parameters $M_{\text{adm}} = \{m \in M \,|\, m_{\max} \ge m \ge 0 \text{ a.e.}\}$.

Following the procedure of the solution of the problem adduced in Sect. 2.7 let us inject $\varphi(\upsilon) = \frac{d\psi_2(\upsilon)}{d\upsilon}$ defined in the space $\Upsilon = L^{\infty}(Q)$ with the norm $\varphi(y)\| = \|\varphi\|_{L^{\infty}(Q)}$

$$\psi(y) = \begin{cases} m(t,x)y(t,x), & \left|\partial y(t,x)/\partial x\right| \le y_{\lim}(x), \\ 0, & \left|\partial y(t,x)/\partial x\right| > y_{\lim}(x), \end{cases}$$

or $\varphi(y) = m(y;t,x)y(t,x)$, where

$$m(y;t,x) = \begin{cases} m(t,x), & \left| \partial y(t,x)/\partial x \right| \le y_{\lim}(x), \\ 0, & \left| \partial y(t,x)/\partial x \right| > y_{\lim}(x), \end{cases} \tag{3.93}$$

$m(t,\ x)$ is known. So, the relation (3.93) defines the coefficient $m(y;t,x)$ of known structure, spatio-temporal characteristics for that are unknown.

Problem (3.89), (3.90) can be reduced to the nonlinear problem with hindrance

$$n(x)\frac{\partial y}{\partial t} + A(\lambda)y + \varphi(m,y) = f \; in \, Q, \tag{3.94}$$

$$y|_{\Sigma} = 0, \tag{3.95}$$

$$y|_{t=0} = y_0 \;\; in \, \Omega. \tag{3.96}$$

As the solution of the system (3.94)–(3.96) we will consider the pair $\{\hat{y}(t,x), m(y;t,x)\}$. The problem of the search of spatio-temporal characteristics for $m(y;t,x)$ we will replace by the problem of search of $m(t,x)$, that is unknown. That is why the term $\varphi(y) = m(t,x)y(t,x)$ in (3.94) allows us to take into account the effect of missing of oil moveability, that appears at pressure gradients lower $y_{\lim}(x)$, i.e. $\left| \frac{\partial y(t,x)}{\partial x} \right| \le y_{\lim}(x)$.

Let us reduce the solution of considered problem to the problem of optimization of the search of unknown parameter $\widehat{m}(t,x)$, that satisfies the system (3.94)–(3.96) and provides the minimum of

$$J(m) = \int_0^{t_k}\int_{\Omega} \begin{Bmatrix} \left(\frac{\partial y(t,x)}{\partial t}\right)^2, & \left|\frac{\partial y(t,x)}{\partial x} \le y_{\lim}(x)\right| \\ \left(m\frac{\partial y(t,x)}{\partial t}\right)^2, & \left|\frac{\partial y(t,x)}{\partial x} > y_{\lim}(x)\right| \end{Bmatrix} dx\,dt \to \inf_{m \in M_{\text{adm}}} \tag{3.97}$$

where the upper branch of the functional $J(\cdot)$ fines the violation of missing of oil moveability condition, that appears at pressure gradients lower $y_{\lim}(x)$, else the lower branch of this functional provides the minimum of $m(t,x)$. It is possible to show that $J(\cdot)$ is continuously differentiable by m and y.

Let us solve the formulated minimization problem by the Lagrange method. In this case the system (3.94)–(3.97) will have the form

$$L(m,y,p) = J(m) + \left(n(x)\frac{\partial y}{\partial t} + A(\lambda)y + \varphi(m;y) - f, p \right)\Big|_{\Sigma} \to \inf_{m \in M_{\text{adm}}} \tag{3.98}$$

with boundary and initial conditions (3.95), (3.96), where $p(t,x)$ is unknown variable that will be defined later.

The necessary conditions of optimum for the formulated problem with regard to unknown parameter $m \in intM_{\text{adm}}$ will have the form

$$\delta L(m) = \frac{\partial L}{\partial m}\delta m = 0 \quad \forall m \in \text{int}M_{\text{adm}}, \tag{3.99}$$

where m is required solution.

For $m \notin int M_{\text{adm}}$ it is necessary to complete the conditions (3.99) with the conditions of complementary slackness

$$\frac{\partial L}{\partial m} = 0 \quad \forall m \notin M_{\text{adm}}. \tag{3.100}$$

Varying the functional (3.98), we can show that in (3.99)

$$\frac{\partial L}{\partial m} = yp + \left\{ \begin{array}{ll} 0, & \left| {\partial y(t,x)}/{\partial x} \right| \le y_{\lim}(x) \\ 2m\left(\frac{\partial y(t,x)}{\partial t}\right)^2, & \left| {\partial y(t,x)}/{\partial x} \right| > y_{\lim}(x) \end{array} \right\} \tag{3.101}$$

where $p(t,x)$ can be obtained from the solution of the adjoint system

$$\begin{aligned} & -n(x)\frac{\partial p}{\partial t} + A^*(\lambda)p + mp \\ & + \left\{ \begin{array}{ll} \left(\frac{\partial y(t,x)}{\partial t}\right)^2, & \left| {\partial y(t,x)}/{\partial x} \right| \le y_{\lim}(x) \\ 2m\left(\frac{\partial y(t,x)}{\partial t}\right)^2, & \left| {\partial y(t,x)}/{\partial x} \right| > y_{\lim}(x) \end{array} \right\} = 0 \end{aligned} \tag{3.102}$$

with boundary and terminal conditions

$$p|_{\Sigma} = 0, \tag{3.103}$$

$$p|_{t=t_k} = 0. \tag{3.104}$$

Here the operator $A^*(\lambda)$ has the form

$$A^*(\lambda)(\cdot) = -\left\{ \sum_{i=1}^{3} \left[\frac{\partial}{\partial x_i}\left(b(x)k(x)\frac{\partial}{\partial x_i} \right) + c_i(x)\frac{\partial}{\partial x_i} \right] - d(x) \right\} (\cdot).$$

The procedure of search of unknown parameter m is based on the relation of gradient in the form

$$m^{i+1} = \Pr\left\{ m^i - \lambda_m \left(\frac{\partial L}{\partial m} \right)^i \right\} \tag{3.105}$$

where i is the number of gradient cycle, m^0 and λ_m are given.

The search of unknown variable based on the gradient relation (3.105) is over when the criterion of finishing

$$\frac{\left| J^i - J^{i+1} \right|}{J^i} \le \varepsilon \tag{3.106}$$

is fulfilled, and unknown variable, that match to this criterion will have the value $\widehat{m}$.

Joining relations (3.94)–(3.96), (3.102)–(3.106) with (3.100), (3.101), we will obtain the algorithm of realization of considered problem:

1. To $i = 0$, where i is the index of current iteration we give the starting value m^0.
2. For the step $i + 1$, taking into account the known m^i in view of (3.100) and (3.101) we compute $\frac{\partial L}{\partial m}$, where y and p are defined by relations (3.94)–(3.96) and (3.102)–(3.104) respectively.
3. In view of (3.105) we define the value of m^{i+1}.
4. We compute (3.97) and check the condition (3.106). If it is fulfilled then the algorithm is over, else we pass to point 2.

The result of realization of un. 1–4 of algorithm is the totality $\{\hat{y}, m\}$, that defines the solution of the problem of modelling the process of oil filtration in porous mediums taking into account the additional effect of missing of liquid phase at gradients that are lower than marginal level defined by visco-plastic properties of oil.

3.7 On Analysis and Control of Second Order Hemivariational Inequality with +-Coercive Multivalued Damping

In order to investigate mathematical models of nonlinear processes and fields of nonlinearized theory of viscoelasticity and piezoelectric, to study waves of different nature such scheme is frequently used: the given model can be reduced to some differential-operator inclusion or multivariational inequality in infinite-dimensional space [ZME04, ZKM08, ZGM00, DM05]. Further, using that or another method of approximation, we prove the existence of generalized solution of such problem, validate constructive methods of search of approximate solutions, study functional-topological properties of resolving operator [ZKM08]. If mentioned process is of evolution nature then his mathematical model can be described by the second order differential-operator inclusion [DM05, ZK09, PAN85]. At that relations between determinative parameters of original problem provide certain properties for multi-valued (in general case) map in differential-operator scheme of investigation. It worth to notice that in the majority of woks concerning given direct of investigations rather strict conditions to "damping" concerning uniform coercivity, boundedness, generalized pseudo-monotony are required [DM05, PAN85]. Such conditions, as a general rule, provide not only the existence of solutions of such problems, but the dissipation of all solutions too. Sometimes such conditions provide the existence of global compact attractor, but this information is not always naturally expresses the real behavior of considered geophysical process or field [KMVY08]. Because of that it appears the necessity to investigate functional-topological properties of resolving operator for differential-operator inclusion, in particular, for those inclusions which describe new wider classes of nonlinear processes and fields of nonlinearized theory of viscoelasticity under valid sufficient weakness of upper mentioned properties of differential operators with suitable applications to concrete mathematical models.

In the given paper we consider problems of analysis and control of second order differential-operator inclusion with weakly-coercive, pseudo-monotone maps. We study the dependence of functional parameters of problem, consider the problem of optimal control. Obtained results can be applied to mathematical models of non-linearized theory of viscoelasticity.

3.7.1 Setting of the Problem

Let V_0, Z_0 are real reflexive separable Banach spaces with corresponding norms $\|\cdot\|_{V_0}$ and $\|\cdot\|_{Z_0}$, H_0 be real Hilbert space with the inner product $(\cdot,\cdot)$, identifying with its topologically conjugated space H_0^*. Let us suggest that the embedding $V_0 \subset Z_0$ is compact and dense, and the embedding $Z_0 \subset H_0$ is continuous and dense. We will have such chain of continuous and dense embeddings [ZKM08, GGZ74, KMP08] $V_0 \subset Z_0 \subset H_0 \subset Z_0^* \subset V_0^*$, where Z_0^* and V_0^* are corresponding topologically conjugated spaces with Z_0 and V_0 with corresponding norms $\|\cdot\|_{Z_0^*}$ and $\|\cdot\|_{V_0^*}$. Let us denote: $S = [\tau, T]$, $-\infty < \tau < T < +\infty$, $p \geq 2$, $q > 1$: $\frac{1}{p} + \frac{1}{q} = 1$,

$$\begin{aligned}
H &= L_2(S; H_0),\ Z = L_p(S; Z_0),\ V = L_p(S; V_0),\\
H^* &= L_2(S; H_0),\ Z^* = L_q(S; Z_0^*),\ V^* = L_q(S; V_0^*),\\
W &= \{y \in V \,|\, y' \in V^*\},
\end{aligned}$$

where y' is a derivative in the sense of $D^*(S; V_0^*)$ of element $y \in V$ [GGZ74]. Let us remark that the embeddings $V \subset Z \subset H \subset Z^* \subset V^*$ are continuous and dense. Moreover the embedding $W \subset Z$ is compact [LIO69, KMP08], and the embedding $W \subset C(S; H_0)$ is continuous [GGZ74, KMP08].

It worth to remark also that the canonical pairings $\langle\cdot,\cdot\rangle_{V_0} : V_0^* \times V_0 \to \mathbb{R}$ and $\langle\cdot,\cdot\rangle_{Z_0} : Z_0^* \times Z_0 \to R$ coincide on $H_0 \times V_0$ with the inner product in H_0. Then the pairing $\langle\cdot,\cdot\rangle_V : V^* \times V \to R$ and, respectively, $\langle\cdot,\cdot\rangle_Z : Z^* \times Z \to R$ coincides on $H \times V$ with inner product in H, namely

$$\langle f, u\rangle := \langle f, u\rangle_V = \langle f, u\rangle_Z = (f, u) = \int_S (f(s), u(s))\, ds,\ \ f \in H,\ u \in V.$$

Let $\hat{U}$, $\hat{K}$ are Hausdorff locally convex linear topological spaces (LTS), $U \subset \hat{U}$, $K \subset \hat{K}$ are some non-empty sets, $A : V \times U \to C_v(V^*)$, $C : Z \times K \to C_v(Z^*)$ are multi-valued maps with nonempty convex weakly compact values in corresponding spaces V^* and Z^*; $B : V \to V^*$ be linear operator; $f \in V^*$, $a \in H_0$, $b \in V_0$ are arbitrary fixed elements.

It is setting the problem about the studying of functional-topological properties of resolving operator $K(u, v, a, b, f)$ of the next problem

$$\begin{cases} y'' + A(y', u) + B(y) + C(y, v) \ni f, \\ y(\tau) = b, \quad y'(\tau) = a. \end{cases} \tag{3.107}$$

Here

$$K(u,v,a,b,f) = \{(y,y') \in C(S;V_0) \times W \mid y \text{ is the solution of (3.107)}\},$$

the derivative y' of the element y is considered in the sense of distribution space $D^*(S;V_0^*)$.

Let us remark that as the embedding $W \subset C(S;H_0)$ is continuous, then initial condition in (3.107) have the sense.

3.7.2 Classes of Parameterized Multi-Valued Maps

Let Y be some real reflexive Banach space, Y^* be its topologically conjugated, $\langle\cdot,\cdot\rangle_Y : Y^* \times Y \to R$ be the pairing, $A : Y \rightrightarrows Y^*$ be the strict multi-valued map, at that $A(y) \neq \emptyset\ \forall y \in Y$. Let $\hat{W}$ be some normalized space that is continuously embedded into Y, $\hat{X}$ be some Hausdorff LTS, $X \subset \hat{X}$ be some non-empty set. Let us consider the parameterized multi-valued map $A : Y \times X \rightrightarrows Y^*$.

Definition 3.5. A strict multi-valued map $A : Y \times X \rightrightarrows Y^*$ is said to be:

- λ_0-*quasi-monotone* on $\hat{W} \times X$, if for any sequence $\{y_n, a_n\}_{n\geq 0} \subset \hat{W} \times X$ such that $y_n \xrightarrow{w} y_0$ in $\hat{W}$, $a_n \to a_0$ in $\hat{X}$, $d_n \xrightarrow{w} d_0$ in Y^* as $n \to +\infty$, where $d_n \in \overline{co}\, A(y_n, a_n)\ \forall n \geq 1$, from the equality $\varlimsup_{n\to\infty} \langle d_n, y_n - y_0\rangle_Y \leq 0$ it follows the existence of subsequence $\{y_{n_k}, d_{n_k}, a_{n_k}\}_{k\geq 1}$ of $\{y_n, d_n, a_n\}_{n\geq 1}$, that the next inequality $\varliminf_{k\to\infty} \langle d_{n_k}, y_{n_k} - \omega\rangle_Y \geq [A(y_0, a_0), y_0 - \omega]_-\ \ \forall \omega \in Y$ holds true.
- *Bounded*, if for every $L > 0$ and for the bounded set $D \subset X$ in the topology of the space $\hat{X}$ there exists such $l > 0$, that

$$\|A(y,u)\|_+ \leq l \qquad \forall \{y,u\} \in Y \times D : \|y\|_Y \leq L;$$

- *Demi-closed*, if for an arbitrary sequence $\{y_n, u_n\}_{n\geq 0} \subset Y \times X$ such that $y_n \to y$ in Y, $u_n \to u_0$ in $\hat{X}$, $d_n \xrightarrow{w} d_0$ in Y^*, where $d_n \in \overline{co}\, A(y_n, u_n)\ \forall n \geq 1$, it follows that $d_0 \in \overline{co}\, A(y_0, u_0)$.

3.7.3 Abstract Results

Let us study functional-topological properties of resolving operator of problem (3.107) concerning closeness in certain topologies. The necessity of justification of such properties concerns problems of control of mathematical models of non-linear geophysics processes and fields that can be described by (3.107) as well as studying of dynamics of solutions of such problems.

Theorem 3.9. *Let $A : V \times U \to C_v(V^*)$ be λ_0-quasimonotone on $W \times U$, bounded operator, $B : V \to V^*$ be linear continuous operator, and a multi-map $C : Z \times K \to C_v(Z^*)$ be bounded and demi-closed. Let us consider a sequence $\{f_m, a_m, b_m, u_m, v_m\}_{m\geq 1} \subset V^* \times H_0 \times V_0 \times U \times K$. We suggest that for all $m \geq 1$ $(y_m, y'_m) \in K(u_m, v_m, a_m, b_m, f_m)$ and the next convergence take place:*

$$\begin{aligned} &f_m \to f_0 \text{ in } V^*, \ a_m \to a_0 \text{ in } H_0, \ b_m \to b_0 \text{ in } V_0, \\ &u_m \to u_0 \text{ in } \hat{U}, \ v_m \to v_0, \text{ in } \hat{K}, \ y'_m \xrightarrow{w} g \text{ in } V. \end{aligned} \tag{3.108}$$

Then there exists such $y \in C(S; V_0)$, that $y' \in W$, $y' = g$ and $(y, y') \in K(u_0, v_0, a_0, b_0, f_0)$. Moreover,

$$y_m \to y \text{ in } C(S; V_0), \ m \to +\infty, \tag{3.109}$$

$$y'_m \xrightarrow{w} y' \text{ in } W, \ m \to +\infty, \tag{3.110}$$

$$\forall t \in S \ \ y'_m(t) \xrightarrow{w} y'(t) \text{ in } H_0, \ m \to +\infty. \tag{3.111}$$

Proof. Let conditions of the theorem are hold true. For fixed $b \in V_0$ let us consider Lipschitz-continuous operator $R_b : Z \to Z$ ($V \to V$ respectively), that is well defined by the next relation

$$(R_b y)(t) = b + \int_\tau^t y(s)ds, \quad \forall y \in Z \text{ (respectively } \forall y \in V), \ \forall t \in S.$$

Let us consider a multi-valued operator $\bar{A} : V \times \bar{U} \to C_v(V^*)$

$$\begin{aligned} \bar{A}(y, \bar{u}) &= A(y, u) + B \circ R_b(y) + C(R_b y, v), \ y \in V, \ \bar{u} \\ &= (u, v, b) \in \bar{U} = U \times K \times V_0. \end{aligned}$$

If $(y, y') \in K(u, v, a, b, f)$, then $z = y'$ is the solution of such problem

$$\begin{cases} z' + \bar{A}(z, u, v, b) \ni f, \\ z(\tau) = a. \end{cases} \tag{3.112}$$

Vice versa, if $z \in W$ is the solution of problem (3.112), then $(y, y') = (R_b z, z) \in K(u, v, a, b, f)$.

Let $\{f_m, a_m, b_m, u_m, v_m\}_{m\geq 1} \subset V^* \times H_0 \times V_0 \times U \times K$, $(y_m, y'_m) \in K(u_m, v_m, a_m, b_m, f_m)$, $m \geq 1$ and (3.108) holds true. Let us set $z_m = y'_m \forall m \geq 1$. Then $y_m = R_{b_m} z_m \forall m \geq 1$. Let us remark that from inclusion (3.112) it follows that $\forall m \geq 1 \exists d_m \in \bar{A}(z_m, u_m, v_m, b_m)$ such that

$$d_m = f_m - z'_m \in \bar{A}(z_m, u_m, v_m, b_m). \tag{3.113}$$

From (3.108) it follows the boundedness of sequences $\{z_m\}$, $\{u_m\}$, $\{v_m\}$, $\{b_m\}$, $\{f_m\}$ in corresponding topologies of spaces V, $\hat{U}$, $\hat{K}$, V_0 and V^*. Then the boundedness of $\{d_m\}$ in V^* follows from the boundedness of maps $A : V \times U \to C_v(V^*)$, $B : V \to V^*$, $C : Z \times K \to C_v(Z^*)$, then continuity of embeddings $V \subset Z$, $Z^* \subset V^*$ and from estimations

$$\left\| R_{b_1} x_1 - R_{b_2} x_2 \right\|_V \leq \alpha \left[\|b_1 - b_2\|_{V_0} + \|x_1 - x_2\|_V \right] \ \forall x_1, x_2 \in V,\ b_1, b_2 \in V_0,$$

$$\left\| R_{b_1} x_1 - R_{b_2} x_2 \right\|_Z \leq \beta \left[\|b_1 - b_2\|_{V_0} + \|x_1 - x_2\|_Z \right] \ \forall x_1, x_2 \in Z,\ b_1, b_2 \in V_0,$$

where α, β are constants that are not depend on b_i, x_i.

So,

$$\exists c_1 > 0 : \ \forall m \geq 1 \ \ \|d_m\|_{V^*} \leq c_1. \tag{3.114}$$

The boundedness of $\{z'_m\}_{m\geq 1}$ in V^* follows from the boundedness $\{f_m\}$ in V^* and (3.114). Therefore,

$$\exists c_2 > 0 : \ \ \forall m \geq 1 \ \ \left\|z'_m\right\|_{V^*} \leq \|z_m\|_W \leq c_2. \tag{3.115}$$

As the embedding $W \subset C(S; H_0)$ is continuous (see. [ZKM08, GGZ74]), then through (3.115), we obtain

$$\exists c_3 > 0 : \ \ \forall m \geq 1, \ \ \forall t \in S \ \ \|z_m(t)\|_{H_0} \leq c_3. \tag{3.116}$$

In view of (3.108) and estimations (3.114)–(3.116), taking into account the continuity of map $y \mapsto y'$ in $\mathscr{D}^*(S; V_0^*)$, we have

$$\begin{aligned} &z_m \xrightarrow{w} g \ \ W,\ d_m \xrightarrow{w} d = f_0 - g' \ \text{in } V^*, \\ &\forall t \in S \ \ z_m(t) \xrightarrow{w} g(t) \ \text{ in } H_0 \ \text{as } m \to \infty. \end{aligned} \tag{3.117}$$

From here and from (3.108), particularly, it follows that

$$g \in W \ \text{ and } \ g(\tau) = a_0 \tag{3.118}$$

Let us show that g satisfies inclusion $g' + \bar{A}(g, \bar{u}_0) \ni f_0$, where $\bar{u}_0 = (u_0, v_0, b_0)$. As $g' + d = f_0$, then it is sufficiently to show that $d \in \overline{A}(g, \bar{u})$.

Firstly let us make sure that

$$\overline{\lim_{m\to\infty}} \ \langle d_m, z_m - g\rangle \leq 0. \tag{3.119}$$

Indeed, in view of (3.113), $\forall m \geq 1$ we have

$$\begin{aligned} \langle d_m, z_m - g\rangle &= \langle f_m, z_m\rangle - \left\langle z'_m, z_m\right\rangle - \langle d_m, g\rangle \\ &= \langle f_m, z_m\rangle - \langle d_m, g\rangle + \frac{1}{2}\left(\|z_m(\tau)\|^2_{H_0} - \|z_m(T)\|^2_{H_0}\right). \end{aligned} \tag{3.120}$$

Further, for left and for right parts of equality (3.120) we calculate an upper limit as $m \to \infty$:

$$\begin{aligned}\overline{\lim_{m\to\infty}} \langle d_m, z_m - g\rangle &\le \overline{\lim_{m\to\infty}} \langle f_m, z_m\rangle + \overline{\lim_{m\to\infty}} \langle d_m, -g\rangle \\ &+ \overline{\lim_{m\to\infty}} \frac{1}{2}\left(\|z_m(\tau)\|_H^2 - \|z_m(T)\|_H^2\right) \\ &\le \langle f_0, g\rangle - \langle d, g\rangle + \frac{1}{2}\left(\|g(\tau)\|_{H_0}^2 - \|g(T)\|_{H_0}^2\right) \\ &= \langle f_0 - d, g\rangle - \langle g', g\rangle = 0.\end{aligned}$$

The last is true in consequence of [GGZ74, Lemma I.5.3], and (3.117). Inequality (3.119) is checked.

From the definition of $\bar{A}$ it follows that $\forall m \ge 1\ \exists \zeta_m \in A(z_m, u_m),\ \exists \xi_m \in C(R_b z_m, v_m)$ such that

$$\forall m \ge 1\ \ d_m = \zeta_m + B \circ R_{b_m}(z_m) + \xi_m. \tag{3.121}$$

From (3.108), (3.117), the boundedness of $C : Z \times K \to C_v(Z^*)$ and the compactness of embedding $W \subset Z$ we have that

$$z_m \to g \text{ in } Z,\ \ R_{b_m} z_m \to R_{b_0} g \text{ in } Z,\ \ \xi_m \xrightarrow{w} \xi \text{ in } Z^*,\ \ m \to +\infty \tag{3.122}$$

From the demi-closeness of $C : Z \times K \to C_v(Z^*)$ it follows that

$$\xi \in C(R_{b_0} g, v_0). \tag{3.123}$$

As

$$\left\| \int_s^t z_m(s) ds \right\|_{V_0} \le |t-s|^{\frac{1}{q}} \|z_m\|_V \le c_4 |t-s|^{\frac{1}{q}} \quad \forall t, s \in S,\ \forall m \ge 1,$$

where $c_4 > 0$ is the constant that does not depend on $m \ge 1,\ s, t \in S$, then in consequence of (3.108)

$$R_{b_m} z_m \to R_{b_0} \text{ in } C(S; V_0),\ \ m \to +\infty, \tag{3.124}$$

particularly,

$$B \circ R_{b_m} z_m \to B \circ R_{b_0} g \text{ in } V^*,\ \ m \to +\infty. \tag{3.125}$$

So,

$$\left\langle B \circ R_{b_m}(z_m) + \xi_m, z_m - g\right\rangle \to 0,\ \ m \to +\infty \tag{3.126}$$

and

$$\begin{aligned}\lim_{m\to\infty} \left\langle B \circ R_{b_m}(z_m) + \xi_m, z_m - \omega\right\rangle &= \left\langle B \circ R_{b_0}(g) + \xi, g - \omega\right\rangle \\ &\ge \left\langle B \circ R_{b_0}(g), g - \omega\right\rangle + \left[C(R_{b_0} g, v_0), g - \omega\right]_- \quad \forall \omega \in V. \end{aligned} \tag{3.127}$$

From (3.119), (3.121) and (3.126) it follows that

$$\overline{\lim_{m\to\infty}} \langle \zeta_m, z_m - g\rangle \le 0. \tag{3.128}$$

From (3.121)–(3.124) and (3.117) it follows also that

$$\zeta_m \xrightarrow{w} \zeta = d - \xi - B \circ R_{b_0}(g) \text{ in } V^*, \quad m \to +\infty. \tag{3.129}$$

So, thanks to (3.117), (3.128), (3.129) and λ_0-quasimonotony A on $W \times U$ we have that up to subsequence $\{z_{m_k}, u_{m_k}, d_{m_k}\}_{k\ge1} \subset \{z_m, u_m, d_m\}_{m\ge1}$, $\underline{\lim}_{k\to+\infty} \langle \zeta_{m_k}, z_{m_k} - \omega\rangle \ge [A(g,u_0), g-\omega]_- \quad \forall \omega \in V$. From here and from (3.119), (3.126), particularly, it follows that $\lim_{k\to\infty} \langle \zeta_{m_k}, z_{m_k} - g\rangle = 0$ and $\langle \zeta, g-\omega\rangle = \lim_{k\to\infty} \langle \zeta_{m_k}, g - \omega\rangle \ge [A(g,u_0), g-\omega]_- \quad \forall \omega \in V$. The last and together with (3.127) provides that $\langle d, g-\omega\rangle \ge [\bar{A}(g,u_0,v_0,b_0), g-\omega]_- \quad \forall \omega \in V$. This means that $d \in \bar{A}(g,u_0,v_0,b_0)$. Setting $y = R_{b_0}g,\ y' = g$, we obtain $(y,y') \in K(u_0,v_0,a_0,b_0,f_0)$. Let us remark that (3.109) is the direct consequence of (3.125), and (3.110) and (3.111) follows from (3.117).

The theorem is proved. □

Let us suggest also that there exist real Hilbert spaces V_σ, V_{σ_1} such that embeddings $V_\sigma \subset V_0 \subset V_{\sigma_1} \subset H_0$ are continuous and dense. Then the embedding $V_\sigma \subset H_0$ is compact. Let us set $W_\sigma = \{y \in V \mid y' \in L_q(S; V_\sigma^*)\}$, where V_σ^* is topologically conjugated space with V_σ, y' is a derivative of the element $y \in V$ in the sense $D^*(S; V_\sigma^*)$ [GGZ74].

Theorem 3.10. *If for some $u \in U$, $v \in K$, $a \in H_0$, $b \in V_0$ the map $A(\cdot,u) : V \to C_v(V^*)$ is λ_0-pseudomonotone on W_σ and*

$$\begin{aligned} \exists c_1, c_2, c_3 > 0: \quad \forall y \in V,\ [A(y,u), y]_+ \ge c_1 \|y\|_V^p - c_2,\\ \|A(y,u)\|_+ \le c_3(1 + \|y\|_V^{p-1}); \end{aligned} \tag{3.130}$$

the map $B : L_2(S; V_{\sigma_1}) \to L_2(S; V_{\sigma_1}^*)$ satisfies such property:

$$\forall u \in V \quad (Bu)(t) = B_0 u(t) \ \textit{for a.e. } t \in S,$$

where $B_0 : V_{\sigma_1} \to V_{\sigma_1}^*$ is linear, bounded, self-conjugated, monotone operator; and the map $C(\cdot, v) : Z \to C_v(Z^*)$ is demi-closed and

$$\begin{aligned} \exists \varepsilon^* > 0: \quad \forall y \in Z \quad & \sup_{d \in C(y,v)} \|d\|_{Z^*} \\ & \le \left(c_1 \gamma^{-p}(T-\tau)^{-p/q} - \varepsilon^*\right)\left(1 + \|y\|_Z^{p-1}\right), \end{aligned} \tag{3.131}$$

where $\gamma \equiv$ const: $\|\cdot\|_{Z_0} \le \gamma \|\cdot\|_{V_0}$, then for an arbitrary $f \in V^*$ $K(u,v,a,b,f) \ne \emptyset$.

Proof. Let us consider such $\omega \in W$, that $\omega(\tau) = a$ and the map $\bar{A} : V \to C_v(V^*)$, $\bar{A}(y) = A(y+\omega, u)+C(R_b(y+\omega), v)$, $y \in V$. From conditions (3.130)–(3.131) it follows that (see [ZK09]) for some $\alpha_1, \alpha_2 > 0$ $[\bar{A}(y), y]_+ \geq \alpha_1 \|y\|_V^p - \alpha_2$, $y \in V$. Repeating suggestions from [ZK09, p. 207], taking into account that the embedding $W_\sigma \subset Z$ is compact, we obtain that the map $\bar{A}$ is bounded and λ_0-pseudomonotone on W_σ. Further, following proof of Theorem 3.9 from [ZK09], we obtain that the problem

$$\begin{cases} y'' + \bar{A}(y') + By \ni f - \omega' - B \circ R_{\bar{0}}\omega \\ y(\tau) = b, \;\; y'(\tau) = \bar{0} \end{cases} \tag{3.132}$$

has the solution $y \in C(S; V)$ such that $y' \in W$. Doing replacement $z = y + R_{\bar{0}}\omega$ in (3.132) we obtain the required statement.

The theorem is proved. □

Let us suggest also that $\hat{K} = (X^*, \sigma(X^*, X))$, where X is some real separable or reflexive Banach space, $K \subset \hat{K}$ is *-weakly compact non-empty subset, and the map $A : V \to C_v(V^*)$ does not depend on the parameter $u \in U$. Then we will write $K(v, a, b, f)$ instead $K(u, v, a, b, f)$. Let us fix an arbitrary $f \in X^*$, $a \in V$, $b \in H$. Let us set

$$G_{ad} = \{(y, y', v) \in C(S; V_0) \times W \times K | \, (y, y') \in K(v, a, b, f)\}.$$

Theorem 3.11. *Let* $L : C(S; V_0) \times (W; \sigma(W^*; W)) \times (X; \sigma(X^*; X))$ *be lower semi-continuous functional such that* $\forall \; u \in C(S; V_0)$, $v \in W$, $w \in X^*$ $L(u, v, w) \geq \varphi_1(\|v\|_V) + \varphi_2(\|w\|_{X^*})$, *where* $\varphi_i : R_+ \to R$ *such that* $\varphi_i(s) \to +\infty$, $s \to +\infty$, $i = 1, 2$. *Let us suggest that* $A : V \to C_v(V^*)$, $B : V \to V^*$, $C : Z \times K \to C_v(Z^*)$ *satisfy conditions of Theorems 3.9, 3.10. Then the problem*

$$\begin{cases} L(y, y', v) \to \inf, \\ (y, y', v) \in G_{ad} \end{cases} \tag{3.133}$$

has a solution.

Proof. Proof directly follows from statements of Theorems 3.9, 3.10 and from integrated Weierstrass theorem (see Lemma 3). □

3.7.4 Applications to Dynamical Contact Problems with Nonlinear Damping

Let us consider the viscoelastic body that in strain-free state fills up the bounded domain $\Omega \subset R^d$, $d = 2, 3$. Let us suggest that the boundary $\Gamma = \partial\Omega$ is regular (3.114) and Γ is divided into three pairwise disjoint measurable parts Γ_D, Γ_N and

Γ_C such that $meas(\Gamma_C) > 0$ [DM05]. The body is gripped on Γ_D such that the displacement field turns there to zero. Let us suggest also that the given vector of the volume power f_1 is distributed in Ω, and the surface force f_2 is distributed on Γ_D. The body can trace into contact with the foundation on potential connecting face Γ_C. Let us set $Q = \Omega \times (0, T)$ for $0 < T < +\infty$. Let us denote the displacement field by $u : Q \to R^d$, the stress tensor by $\sigma : Q \to S_d$, and the deformation tensor by $\varepsilon(u) = (\varepsilon_{ij}(u))$, $\varepsilon_{ij}(u) = \frac{1}{2}(u_{i,j} + u_{j,i})$, where $i, j = \overline{1, d}$, S_d is the space $R_s^{d\times d}$ of d-order symmetric matrices. Following [MIG05], let us consider the multi-valued analog of Kelvin–Voigt viscoelastic determining relation

$$\sigma(u, u') \in \aleph(\varepsilon(u')) + \Im(\varepsilon(u)),$$

where $\aleph$ is multi-valued non-linear map, and $\Im$ is simple-valued linear determining map. Let us remark that in classical linear viscoelasticity the upper defined rule has the form $\sigma_{i,j} = c_{ijkl}\varepsilon_{kl}(u') + g_{ijkl}\varepsilon_{kl}(u)$, where $\aleph = \{c_{ijkl}\}$ and $\Im = \{g_{ijkl}\}$, $i, j, k, l = \overline{1, d}$ are tensors of viscosity and elasticity respectively. Let us denote normal and tangential components of displacement u on Γ by u_N and u_T, $u_N = u \cdot n$, $u_T = u - u_N n$, where n is the unit vector of outer normal line to Γ. Similarly, normal and tangential components of stress field on Γ are defined by $\sigma_N = (\sigma n) \cdot n$ and $\sigma_T = \sigma n - \sigma_N n$ respectively. On the connecting face Γ_C let us consider boundary conditions. The normal stress σ_N and the normal displacement u_N satisfy non-monotony normal condition of form flexibility $-\sigma_N \in \partial j_N(x, t, u_N, \varsigma)$ on $\Gamma_C \times (0, T)$. The frictional law between the frictional force σ_T and the tangential displacement u_T on Γ_C has the form $-\sigma_T \in \partial j_T(x, t, u_T, \xi)$ on $\Gamma_C \times (0, T)$. Here $j_N(\cdot, \cdot, \cdot, \varsigma) : \Gamma_C \times (0, T) \times R^d \to R$ and $j_T(\cdot, \cdot, \cdot, \xi) : \Gamma_C \times (0, T) \times R^d \to R$ are locally Lipschitz by the last variables of function, ∂j_N, ∂j_T are Clarke subdifferentials of corresponding functionals $j_N(x, t, \cdot, \varsigma)$, $j_T(x, t, \cdot, \xi)$. Such boundary conditions in particular cases involve classical conditions on the boundary of domain (see for example [DM05, Chap. 2.3], [PAN85]). Let us denote the initial displacement and the initial speed by u_0 and u_1. The classical formulation of contact problem has the form: to search such $u : Q \to R^d$ and $\sigma : Q \to S_d$, that

$$\begin{cases} u'' - div\sigma = f_1 & \text{in } Q, \\ \sigma \in \aleph(\varepsilon(u')) + \Im(\varepsilon(u)) & \text{in } Q, \\ u = 0 & \text{on } \Gamma_D \times (0, T), \\ \sigma n = f_2 & \text{on } \Gamma_N \times (0, T), \\ -\sigma_N \in \partial j_N(x, t, u_N, \varsigma), & -\sigma_T \in \partial j_T(x, t, u_T, \xi) \quad \Gamma_C \times (0, T), \\ u(0) = u_0, u'(0) = u_1 & \text{in } \Omega. \end{cases} \tag{3.134}$$

For the variational setting of such problem let us set

$$H_0 = L_2(\Omega; R^d), \quad \bar{H}_0 = L_2(\Omega; S_d),$$
$$\bar{H}_1 = \{u \in H_0 |\, \varepsilon(u) \in \bar{H}_0\} = H^1(\Omega; R^d), \quad V_0 = \{v \in \bar{H}_1 |\, v = 0 \ = 0\ \Gamma_D\}.$$

Using Green's formula, the definition of Clarke subdifferential [CLA90, CHI97] if initial data are rather smooth (see for detail [DM05, MIG05]), we can obtain (see for detail [DM05, MIG05]) the variational formulation of problem (3.134) for searching of such $u : [0, T] \to V$ and $\sigma : [0, T] \to \bar{H}_0$, that

$$\begin{cases} \langle u''(t), v\rangle_{V_0} + (\sigma(t), \varepsilon(v))_{\bar{H}_0} + \int\limits_{\Gamma_C} (j_N^0(x,t,u_N;v_N;\varsigma) \\ \quad + j_T^0(x,t,u_T;v_T;\xi))\, d\Gamma(x) \\ \geq \langle f(t), v\rangle_{V_0} \text{ for all } v \in V_0 \text{ and for a.e. } t \in [0, T], \\ u(0) = u_0, \; u'(0) = u_1, \end{cases} \tag{3.135}$$

where

$$\langle f(t), v\rangle_{V_0} = (f_1(t), v)_{H_0} + (f_2(t), v)_{L_2(\Gamma_N;R^d)} \text{ for all } v \in V_0 \text{ for a.e. } t.$$

Let $V = L_2(0, T; V_0)$, $W = \{w \in V \,|\, w' \in V^*\}$ and $\bar{\gamma} : H^\delta(\Omega; R^d) =: Z_0 \to H^{1/2}(\Gamma; R^d) \subset L_2(\Gamma; R^d)$ is the trace operator, $\delta \in (1/2; 1)$. Maps $A : V \to 2^{V^*}$, $B_0 : V_0 \to V_0^*$ are defined by the next way:

$$\begin{aligned} [A_0(t,u), v]_+ &= \sup\{(d, \varepsilon(v))_{\bar{H}_0} \,|\, d \in V_0^*,\ d(\cdot) \in \aleph(\cdot, t, \varepsilon(u(\cdot)))\},\ t \in S,\ u, v \in V_0, \\ A(u) &= \{d \in V^* \mid d(t) \in A_0(t, y(t)) \text{ for a.e. } t \in S\},\ u \in V, \\ \langle B_0 u, v\rangle_{V_0} &= (\Im(x,t,\varepsilon(u)), \varepsilon(v))_{\bar{H}_0} \quad \forall u, v \in V_0,\ t \in [0, T]. \end{aligned}$$

Here $[A(t,u), v]_+$ is the upper support function of the set $A(t,u) \subset V_0^*$. The functional $J : [0, T] \times L_2(\Gamma_C; R^d) \times K \to R$ is defined by the next way:

$$J(t, v, \eta) = \int\limits_{\Gamma_C} (j_N(x,t,v_N(x),\varsigma) + j_T(x,t,v_T(x),\xi))\, d\Gamma(x),$$

for $t \in [0, T]$, $v \in L_2(\Gamma_C; R^d)$ and $\eta = (\varsigma, \xi) \in K$, and $C : Z \times K \to C_v(Z^*)$ by the next way:

$$C(u, \eta) = \{d \in Z^* \mid d(t) \in \bar{\gamma}^*(\partial J(t, \bar{\gamma} u(t), \eta)) \text{ for a.e. } t \in [0, T]\},$$

where $\bar{\gamma}^*$ is the conjugated operator to $\bar{\gamma}$. So, we obtained the problem of searching of such $u \in V$, $u' \in W$, that

$$\begin{cases} u'' + A(u') + Bu + C(u, \eta) \ni f, \\ u(0) = u_0, \; u'(0) = u_1, \end{cases} \tag{3.136}$$

It is possible to show (see for detail [DM05, MIG05]), that every solution of problem (3.136) is the solution of problem (3.135). So, imposing such conditions for parameters of problem (3.134), that maps A, B, C satisfy conditions of Theorems 3.9–3.11 (see for details [ZME04, ZKM08, ZK09, MIG05]), we will obtain results concerning some properties of the resolving operator of problem (3.136), particularly, of problem of optimal control (3.133). Let us remark that the introduced scheme of investigation in the given work (see for details [ZME04, ZKM08, ZK09, MIG05]), in comparison with existing approaches, allows us, for example, weaken the technical condition of uniform"–"-coercivity to "+"-coercivity, the generalized pseudo-monotony to w_{λ_0}-pseudo-monotony etc. It worth to remark that concrete classes of differential operators of pseudo-monotone type which appear in problem (3.134), considered in details in works [BAR76, GGZ74, PAN85, KMVY08, DM05, LIO69, MIG05, GLT81] (see works and citations there).

References

[AE84] Aubin J-P, Ekeland I (1984) Applied nonlinear analysis. Wiley-Interscience Publication, New York
[AS86] Aziz H, Settary E (1986) Mathematical modelling of stratum systems. Nedra, Moscow
[BAR76] Barbu V (1976) Nonlinear semigroups and differential equations in Banach spaces. Editura Academiei Noordhoff International Publishing, Bucuresti
[CHI97] Chikrii AA (1997) Conflict-controlled processes. Kluwer Academic, Boston
[CLA90] Clarke FH (1990) Optimization and nonsmooth analysis. SIAM, Philadelphia
[DM05] Denkowski Z, Migorski S (2005) Existence of solutions to evolution second order hemivariational inequalities with multivalued damping. Syst Model Optim. doi: 10.1007/0-387-23467-5_14
[DL76] Duvaut G, Lions J-L (1976) Inequalities in mechanics and physics [in Russian]. Springer, Berlin
[GGZ74] Gajewski H, Gröger K, Zacharias K (1974) Nichtlineare operatorgleichungen und operatordifferentialgleichungen. Akademie, Berlin
[GLT81] Glowinski R, Lions JL, Tremolieres R (1981) Numerical analysis of variational inequalities. North Holland, Amsterdam
[KMVY08] Kapustyan OV, Mel'nik VS, Valero J, Yasinsky VV (2008) Global attractors for multivalued dynamical systems. Naukova dumka, Kyiv
[KMP08] Kasyanov PO, Mel'nik VS, Piccirillo AM (2008) On some approximations and main topological descriptions for special classes of Banach spaces with integrable derivatives. Methods Funct Anal Topol 14:255–270
[LIO69] Lions JL (1969) Quelques methodes de resolution des problemes aux limites non lineaires. Dunod Gauthier-Villars, Paris
[LK83] Lyashko AD, Korchevsky MM (1983) Difference methods of resolution of nonlinear problems of filtration theory. Izv vuzov matematika 7:28–45
[MIG05] Migorski S (2005) Boundary hemivariational inequalities of hyperbolic type and applications. J Global Optim 31:505–533
[PAN85] Panagiotopoulos PD (1985) Inequality problems in mechanics and applications: convex and nonconvex energy functions. Birkhäuser, Boston
[ZK09] Zadoyanchuk NV, Kasyanov PO (2009) FaedoGalerkin method for second-order evolution inclusions with W_λ-pseudomonotone mappings. Ukrainian Math J doi: 10.1007/s11253-009-0207-z
[ZKM08] Zgurovsky MZ, Kasyanov PO, Mel'nik VS (2008) Differential-operator inclusions and variation inequalities in infinite dimensional spaces. Naukova Dumka, Kyiv
[ZGM00] Zgurovskii MZ, Mel'nik VS (2000) Penalty method for variational inequalities with multivalued mappings. Part I Cybern Syst Anal. doi: 10.1007/BF02667060
[ZME04] Zgurovsky MZ, Mel'nik VS (2004) Nonlinear analysis and control of physical processes and fields. Springer, Berlin
[ZN96] Zgurovsky MZ, Novikov AN (1996) Analysis and control of unilateral physical processes [in Russian]. Naukova Dumka, Kiev

Appendix A
Vortical Flow Pattern Past a Square Prism: Numerical Model and Control Algorithms

Abstract A coupled Lagrangian–Eulerian numerical scheme for modeling the laminar flow of viscous incompressible fluid past a square prism at moderate Reynolds numbers is developed. The two-dimensional Navier–Stokes equations are solved with the vorticity–velocity formulation. The convection step is simulated by motion of Lagrangian vortex elements and diffusion of vorticity is calculated on the multi-layered adaptive grid. To reduce the dynamic loads on the body, the passive control techniques using special thin plates are proposed. The plates are installed either on the windward side of prism or in its wake. In the first case, the installation of a pair of symmetrical plates produces substantial decreasing the intensity of the vortex sheets separating in the windward corners of prism. In the second case, the wake symmetrization is achieved with the help of a long plate abutting upon the leeward surface. Both the ways bring narrowing of the wake and, as a result, decrease of the dynamic loads. With optimal parameters of the control system, the drag reduction is shown to decrease considerably.

A.1 Introduction

Let us consider some hydrodynamical application to differential-operator equations in infinite-dimensional spaces. We develop a coupled Larangian–Eulerian numerical scheme for modeling the laminar flow of viscous incompressible fluid past a square prism at moderate Reynolds numbers. Then we solve the two-dimensional Navier–Stokes equations with the vorticity–velocity formulation, that can be reduced to differential-operator equation. The convection step will be simulated by motion of Lagrangian vortex elements and diffusion of vorticity will be calculated on the multi-layered adaptive grid. To reduce the dynamic loads on the body, the passive control techniques using special thin plates will be proposed. The plates will be installed either on the windward side of prism or in its wake. In the first case, the installation of a pair of symmetrical plates produces substantial decreasing the intensity of the vortex sheets separating in the windward corners of prism. In the second case, the wake symmetrization is achieved with the help of a long plate abutting upon the leeward surface. Both the ways bring narrowing of the wake and, as a result, decrease of the dynamic loads. With optimal parameters of the control system, the drag reduction will be shown to decrease considerably.

Hydrodynamic characteristics of a non-streamline body can be determined through vortex structure of the flow near it. Firstly they essentially depend on wall flow type, namely this flow is either separated or without separation. At large Reynolds numbers flows without separation are turbulent and a body drag can be determined through frequency and intensity of coherent vortex formations separations from the buffer zone of body boundary layer into the outward domain. In spite of small size these vortex structures essentially determine an energy interchange between the flow and the body. General pattern of flow past the body also depends on generation-diffusion correlation of vorticity in the wall area. In this case to construct control schemes providing the body drag decrease we should consider the dynamics of vortexes whose sizes are close either to the width of the buffer zone or to the boundary layer thickness. Great number of theoretical and experimental investigations [LN76, MO98, MIG62] show a principal possibility of the vortex structure control in the boundary flows. A number of schemes and constructions proposed for a boundary flow transformation, such as turbulence stimulators, vortex generators, vortex destructors, broaches, injectors of special flows etc. found their practical applications in aviation, naval technologies and hydraulic devices.

The the case of separated or detached flows past bodies of a non-streamline shape (namely the bodies with sharp edges or with large downstream pressure gradients over smooth surface areas). Such flows are accompanied by generation of vortex structures whose sizes are commensurate with geometrical parameters of the streamed body. Vortex generation causes irreversible energy consumption of the flow and brings unsymmetry into the patterns of flows past symmetrical constructions. In this case hydrodynamic characteristics of the body are determined by dynamics of the formed circulation flow. Instability and dynamic properties of a circulation zone are determined through motion and interaction of vortex structures in it.

It is well known that in a homogeneous incompressible fluid the vorticity is generated only by the body and be the domain boundaries where the fluid flows. An intensity of the vorticity generation depends on a local surface curvature of the body. On the smooth surface areas this intensity is much smaller than that of a flow past sharp edges. Hence sometimes it is enough to carry out dynamic analysis of vorticity separating from the body edges. Examples of such approach can be found in works devoted to the numerical analysis of a flow past a wing with a sharp trailing edge. Vortex sheets, which have been formed as a result of separation, under small perturbation decay into discrete vortexes, which, in their turn, after combining form into large circulation zones (Calvin–Gelmgholc instability). With the lapse of time as a result of viscous forces action diffusion processes start influencing upon the dynamics of these vortex formations, particularly, it is a viscous diffusion who causes the generation of a new vorticity on the smooth surface areas. Therefore considering a flow past a body we usually have several vorticity sources situated on the sharp edges and also on smooth surface areas. This sources are interacting both directly (mixing) and through hydrodynamic velocity field. The possibility of decreasing an intensity of one vorticity source under the influence of the other attracts the practical interest. For example changing relative positions of vorticity sources through the body shape transformation we can largely decrease the intensity of vortexes generated by the flow past the body. In this case the body drag decreases and the level of non-stationary lateral forces reduces as well.

An artificial modification of the separated flow structure is one of the modern directions of the flow control theory. This theory is based on the development of algorithms for generation of large vortex formations with prescribed properties near the body [GG98, CLD94, COR96, SAV98]. Methods of artificial vortex generation can either passive (when our aim is to suppress vorticity generation processes that can be reached by modification of the construction shape) or active (with usage of control elements, for example, fluid withdrawal through slots or permeable surface areas, fluid injection, impurity of polymer additions into wall area through electric and magnetic fields action). The first class of these method does not require additional energy expenditure. But among the advantages of active schemes there is adaptivity, development and application possibilities for control algorithms together with their feedbacks when the intensity and the direction of an external influence on the flow depend on flow conditions.

Here we consider structure control schemes of the flow past of body non-streamline shape directed on decrease of hydrodynamic drug and construction vibration in the flow. It is important for increase of building reliability design, decrease of acoustic noises and power inputs caused by body motion in fluid. For control algorithm development understanding of generation and evolution processes as well as vortex dynamics in the flow past the body is important. The approbation of new algorithms can be carried out by the examples of the flow past relatively simple bodies. In view the analysis of flows past a square prism gives wide opportunities of approbation of control algorithm for flows past a body of non-streamline shape.

Investigation of the flow structure around the square prism is important in view of many factors: fundamental study of regularities of physical processes concerned with generation and interaction of separated circulation zones, obtaining patterns of flows past a body of non-streamline shape, solving practical engineering dynamics problems and reliability design problems for constructions exploitable in water or wind flows (flow past bridge bearings, tower buildings, elements of oceanographic equipment, offshore constructions).

In most experimental works devoted to the investigation of flows past the square prism an analysis was carried out at large Reynolds Numbers [BO82, VIC66, KNI90]. Exactly this range is important for the majority of engineering problems. The investigation results indicate that the flow pattern past the square prism and its hydrodynamical characteristics depend on Reynolds number far less than in the case of a round cylinder, and at $\mathrm{Re} > 10^3$ these characteristics would hardly change a lot. This is a result by an existence of a fixed separation on the sharp edges causing approximate constancy of vortex separation frequency. Let us denote that at moderate Reynolds numbers $5 \cdot 10 \leq \mathrm{Re} \leq 5 \cdot 10^3$, the flow structure around the prism is complicated enough. This fact results in certain changes of Strouhal number and of the square prism drag coefficient [OKA82].

First numerical investigation of flows past a square prisms are associated with the discrete-vortex method [NNT82]. Obtained there drag coefficients and frequencies of vortex separation are close to the experimental values at Re $\to \infty$. When Reynolds number is decreasing the mismatch between computed and experimental values ia increasing.

Numerical models of a flow past a square prism, considering viscous effects, are based on the complete Navier–Stokes equation system. Majority of such investigations concern with two-dimensional problems at small and moderate Reynolds numbers. One of the first and most complete investigations of this direction is the work [DM82]. The proposed algorithm combines elements of the grid method and the method of finite volumes. Taking into account limited capacity of the computation engineering, all computations was carried out with the help of crude mesh. Obtained data for hydrodynamic drag coefficient and Strouhal number, namely the number characterizing a frequency of vortex separations, differ within 10–20 % from their experimental values.

In the work [MFR94] the study of a laminar flow past an oscillating square cylinder is carried out by the method of finite volumes. There are considered basic regimes occurring as a result of forced oscillations of the cylinder over its natural oscillations caused by vortex separation. In the work [SND98] this method had already been used for investigation of two-dimensional and three-dimensional flows past a square prism at moderate Reynolds numbers. Obtained results show that transition from two-dimensional regime up to three-dimensional one goes on in the following range of Reynolds numbers: from 150 to 200. At the same time it is shown that Strouhal number and body drag coefficient obtained through two-dimensional simulation are close enough to their experimental values.

Grid numerical algorithm applied in the work [ZSG94] for the analysis of flow past the square cylinder, revolving on longitudinal axis, is based on usage of Navier–Stokes equations with "vorticity-flow function" formulation. The advantage of this approach is absence of the pressure terms in the equations that enables to use explicit time computational schemes and avoids the choice boundary conditions for pressure.

In the majority of numerical simulation methods for viscous flows of incompressible fluid as an independent unknown magnitudes they take velocity and pressure. In view of numerical modeling, the form of Navier–Stokes equations that was proposed by Lighthill [LIG63] is more convenient. As an independent unknown magnitudes he considered velocity and vorticity. This gives an opportunity to separate a problem of fluid kinematics by applying the Bio-Savart law. Numerical models based on such approach can be reduced to the limit integral equation for the velocity. After that standing by themselves vorticity problems are considered which also can be reduced to limit integral equations [WW86]. The advantage of this approach is the following: when computing the velocity it is only vorticity distribution that is being taken into consideration. In the majority of hydromechanical problems the domain occupied by the whirling fluid is mach smaller than the domain of significant velocity perturbations. Therefore the dimensions of the computation domain within the considered approach can be much smaller than in the case when required functions are velocity and pressure. Moreover in the problems of the external flow past a body, velocity boundary conditions of infinity are precisely fulfilled by analytical choice of Green's function, entering into Bio-Savart equation. In some cases, for example in the case of a plane plate, the second boundary condition on the wall (non-percolation condition) can be similarly fulfilled.

Here for computation of the flow past the square cylinder and analysis some control schemes we propose generalized vortex algorithm. It is based on the complete Navier–Stokes equation system with the "vorticity–velocity formulation" and uses grids, Lagrangian vortex points and the discrete vortex method.

A.2 Problem Definition

Let us consider a two-dimension laminar flow of viscous incompressible fluid past a square cylinder. Assuming critical parameters are the remote flow velocity U_∞ and the length of the side of a square a, we obtain the following dimensionless Navier–Stokes equations:

$$\frac{\partial V}{\partial t} + (V \cdot \nabla)V = -\Delta p + \frac{1}{\mathrm{Re}}\nabla^2 V, \tag{A.1}$$

$$\nabla \cdot V = 0, \tag{A.2}$$

where $V(x, y, t)$ is the fluid velocity, $p(x, y, t)$ is the pressure, ν is the fluid viscosity, $\mathrm{Re} = U_\infty a/\nu$.

Performing the operation *rotor* with respect to each term of (A.1) and putting the vorticity $\omega = \nabla \times V$ in view of (A.2) we obtain the equation describing evolution and diffusion of vorticity in the considered domain. Particularly, for two-dimensional problems we have:

$$\frac{\partial \omega}{\partial t} + (V \cdot \nabla)\omega = \frac{1}{Re}\Delta\omega. \tag{A.3}$$

Equation (A.3) implies: if in the time point t velocity and vorticity are given, then the vorticity distribution in the next time point $t + \Delta t$ can be found. Thereafter, under obtained values ω, using Bio-Savart formula and taking into account boundary conditions on the body surface, we can find new velocity values in the domain. This computation cycle, which firstly was described by Lighthill [LIG63], is the foundation of the vortex method. The distinctive feature of numerical algorithms based on this cycle consists in the way how diffusion and vorticity convection are calculated, and also in different approaches to modeling of vorticity generation on the body wall.

To solve the diffusion problem for vorticity ω it is necessary to fulfill boundary conditions of the body and boundary conditions of infinity. For velocity $V(x, y, t) = V_n(x, y, t) + V_\tau(x, y, t)$ these are the standard non-percolation and adhesion conditions:

$$V_n(x, y, t)|_L = U_{n\,body}, \tag{A.4}$$

$$V_\tau(x, y, t)|_L = U_{\tau\,body}, \tag{A.5}$$

where $U_{body} = U_{n\,body} + U_{\tau\,body}$ is the body velocity consisting of translation and rotation velocities in general case, L is the boundary of the body.

The choice of a boundary condition on the surface of the body L for the function ω is a non-trivial problem. It is associated with the way one describes the vorticity generation on the body wall. Due to Lighthill's method, which is effectively used in the modeling of discrete vortexes, the body surface is replaced by vortex sheet. In this case the vorticity value on the wall ω_0 depends on intensity of the vortex sheet γ. There are different approaches to finding a function ω. One of them is based on the fact that a jump of tangential velocity (V_τ) in incompressible ideal fluid crossing the vortex sheet is equal to $\gamma/2$. Then, under the adhesion condition, on the surface L the following relationship must be fulfilled:

$$V_\tau^0 + \gamma/2 = 0.$$

Taking into account that $\omega_0 = \gamma/h$, (h is a given short distance from the wall along the normal line or an appropriate sampling interval for computational grid associated with the body), we have:

$$\omega_0 = -\frac{2V_\tau^0}{h}\bigg|_L. \tag{A.6}$$

The velocity V_τ in formula (A.6) is calculated directly on the wall.

In [WU76] the magnitude ω_0 was determined using the expansion of tangential velocity into Taylor's series near the body surface. In this case, taking for instance the horizontal wall, we have:

$$\omega_0 = -\frac{2V_\tau(x, h/2)}{h} + \frac{\partial^2 V_\tau}{\partial y^2}\bigg|_{y=0} h/4 + O(h^2) + \dots. \tag{A.7}$$

If in formula (A.7) only terms containing first order infinitesimals are left, we obtain an expression which is an analogue of well-known Thompson's formula in $\omega - \psi$ model. Expressions (A.6), (A.7) are the examples of Dirichlet condition on the function ω_0. In the work [KOU93] for vorticity flow Neumann condition is used. It should be noted that there is no rigorous mathematical substantiation of boundary conditions for the function ω which would correlate the intensity of the vortex sheet around the body with vorticity generated by its walls. The choice of boundary condition for the function ω essentially depends on the numerical method used for solving vortex transfer equation (A.3).

Let us consider an unbounded fluid flow. Then for fluid velocity perturbations caused by the body, the damping condition is satisfied:

$$V(x, y, t) \to U_\infty, \quad \textit{if} \;\; r = \sqrt{x^2 + y^2} \to \infty. \qquad \text{(A.8)}$$

The problem formulation is supplemented with initial conditions:

$$\begin{aligned} V(x, y, 0) &= U_\infty(x, y), \\ \omega(x, y, 0) &= \nabla \times U_\infty(x, y). \end{aligned} \qquad \text{(A.9)}$$

A.3 Numerical Algorithm

For modeling the fluid flow described by (A.3) with corresponding boundary and initial conditions we propose the generalized vortex method combining grids usage with Lagrangian vortex elements. This method relies on construction of numerical algorithms based on discrete-vortex approximations.

The configuration of the computation domain and axes related with the considered body are shown in Fig. A.1. Since the body surface is the only vortex source in the flow we may assume that an input flow is vortex-free:

$$\omega|_{AB} = 0, \;\; \mathrm{V}|_{AB} = U_\infty. \qquad \text{(A.10)}$$

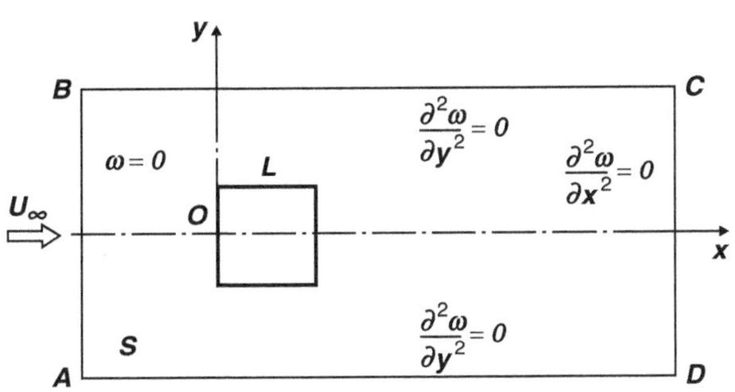

Fig. A.1 Computation domain

On other boundaries of this domain we suppose [GG05]:

$$\left.\frac{\partial^2\omega}{\partial y^2}\right|_{BC} = 0,$$
$$\left.\frac{\partial^2\omega}{\partial y^2}\right|_{AD} = 0,$$
$$\left.\frac{\partial^2\omega}{\partial x^2}\right|_{CD} = 0. \tag{A.11}$$

The boundary of the body L is simulated by continuous vortex sheet. Its intensity γ can be found using limit integral equations method. Taking into account the non-percolation condition (A.4) we obtain the following equation:

$$\int_L \gamma(\vec{r}\,',t)\,\frac{\partial G(\vec{r},\vec{r}\,')}{\partial n}\,dl(\vec{r}\,') + \int_S \omega(\vec{r}\,',t)\,\frac{\partial G(\vec{r},\vec{r}\,')}{\partial n}\,ds(\vec{r}\,') \;=\; V_{n\,body}, \tag{A.12}$$

where $\vec{r} = \vec{r}(t)$, $\vec{r}\,' = \vec{r}\,'(t)$ are position vectors, $G(\vec{r},\vec{r}\,')$ is Green function, which for two-dimensional problem and unbounded fluid flow is of the form:

$$G(\vec{r},\vec{r}\,') = \frac{1}{2\pi i}\ln(\vec{r} - \vec{r}\,').$$

Moreover, the following theorem on constancy of circulation in the domain must take place:

$$\int_L \gamma(\vec{r}\,',t)\,dl(\vec{r}\,') + \int_S \omega(\vec{r}\,',t)\,ds(\vec{r}\,') = 0. \tag{A.13}$$

The velocity field in the domain is defined due to Bio-Savart theorem:

$$V(\vec{r},t) = U_\infty + \int_L \gamma(\vec{r}\,',t)\,\frac{\partial G(\vec{r},\vec{r}\,')}{\partial n}\,dl(\vec{r}\,') + \int_S \omega(\vec{r}\,',t)\,\frac{\partial G(\vec{r},\vec{r}\,')}{\partial n}\,ds(\vec{r}\,'). \tag{A.14}$$

The vorticity field is approximated by the system of Lagrangian vortex elements, namely elements moving together with the fluid and circulating:

$$\omega(\vec{r},t) \approx \sum_k \Gamma_k f_\delta(\vec{r} - \vec{r}_k), \tag{A.15}$$

where Γ_k, $\vec{r}_k$ is a circulation and a position of k-th vortex, respectively, $f(\vec{r} - \vec{r}_k)$ is a vortex function which is equal to delta function for ideal fluid, δ is a radius of vortex core (we assume that inside vortex core viscous effects are essential but outside the velocity field is potential).

If over the flow field we put the grig such that the vorticity $\omega(x_k, y_k, t)$is uniformly distributed in every its cell with the number k, then for the circulation of the corresponding vortex element we have:

$$\Gamma_k(t) = \omega(x_k, y_k, t)\Delta s_k, \tag{A.16}$$

where Δs_k is grid element area.

Introduction of vortex core and the function f_δ is one of the ways of vortex motion regularization, and it is widely used in modern vortex methods [NMDK99, CK00]. This approach allows to get rid of singularity in the vortex point, otherwise incredible velocities induced by adjoint vortexes may occur. Instead of vortex points vortex blobs are considered with the core radius δ. For in-depth discussion concerning the choice of the function f_δ and the radius δ we refer to the monograph [CK00]. Here we use the second order Gaussian function [NMDK99]:

$$f_\delta(\bar{r} - \bar{r}_k) = \frac{\exp\left(-(\bar{r} - \bar{r}_k)^2\right)/\delta^2}{\pi\delta^2}.$$

Core parameter essentially depends on the size h of an element of the orthogonal analytical grid:

$$\delta = h^{0.9}.$$

The process is being time sampled with the step Δt. On each time step nonlinear (A.3) splits up on two equations, where the first one describes vorticity diffusion by means of viscous diffusion while the second one – by beans of convection:

$$\frac{\partial \omega}{\partial t} = \frac{1}{\mathrm{Re}}\Delta\omega, \tag{A.17}$$

$$\frac{\partial \omega}{\partial t} = -(V \cdot \nabla)\,\omega. \tag{A.18}$$

Let us note, that for numerical integrating of the equation system (A.17), (A.18) we use the explicit time integration scheme.

Over the flow field we consider an orthogonal analytical grid with the cell dimensions Δx, Δy. Then with every grid element we associate a vortex with intensity $\Gamma_{ij} = \omega_{ij}\Delta x \Delta y$.

As was mentioned before, the body surface is simulated by the vortex sheet. To find its intensity we apply the method of discrete vortexes [GG05] in the computational scheme. According to this method, sides of the square should be divided into equal vortex intervals, and each of them is replaced by a discrete vortex with circulation Γ_l^*, that is equal to sheet intensity along the interval:

$$\Gamma_l^* = \gamma^*(l)\Delta l, \quad l = 1, 2, \ldots, N,$$

where Δl is the length of the interval, N is the number of vortexes, disposed along the boundary (attached).

The reference points in which integral equations (A.12), (A.13) hold true are disposed in the middle between neighboring vortexes. Then finding the intensity of the vortex sheet modeling the body boundary can be reduced to solving (at each time step) of N linear algebraic equations with respect to unknown circulations of attached vortexes:

$$\sum_{l=1}^{N} \Gamma_l^* (v_n^*)_{ml} = -2\pi U_\infty - \sum_i \sum_j \Gamma_{ij} (v_n)_{mij}, \; m = 1, 2, \ldots, N-1,$$

$$\sum_{l=1}^{N} \Gamma_l^* = -\sum_i \sum_j \Gamma_{ij}, \tag{A.19}$$

where $(v_n^*)_{ml}$ is a normal speed, inducible in m-th reference point by l-th attached vector, $(v_n)_{mij}$ is a normal speed in m-th reference point, inducible by vortexes disposed in mesh points outside the body boundary.

Solving system (A.19) and using formula (A.6), we can find vorticity ω on the body wall. To provide the fulfillment of the Kutt–Gukovsky condition in the sharp edges of the square, we put the grid along its surface in such a way that vortexes were in corner points. This vortexes are considered free, namely they are moving with the local velocity of the fluid. This approach was proposed by S.M. Belocerkovsky and was successfully used for computation of the flow past wings and bluff bodies applying discrete-vortex method [GG05].

Taking into account discretization of the flow field and of the body surface, formula (A.14) for computation of velocity components u, v in the domain takes the following form:

$$u(x, y) = U_\infty - \sum_{l=1}^{N} \frac{\Gamma_l^*}{2\pi} \frac{y - y_l}{(x - x_l)^2 + (y - y_l)^2}$$

$$\times \left(1 - exp\left[(x - x_l)^2 + (y - y_l)^2\right] / \delta^2\right)$$

$$- \sum_i \sum_j \frac{\Gamma_{ij}^*}{2\pi} \frac{y - y_{ij}}{(x - x_{ij})^2 + (y - y_{ij})^2}$$

$$\times \left(1 - exp\left[(x - x_{ij})^2 + (y - y_{ij})^2\right] / \delta^2\right), \tag{A.20}$$

$$v(x, y) = \sum_{l=1}^{N} \frac{\Gamma_l^*}{2\pi} \frac{x - x_l}{(x - x_l)^2 + (y - y_l)^2}$$

$$\times \left(1 - exp\left[(x - x_l)^2 + (y - y_l)^2\right]/\delta^2\right)$$

$$+ \sum_i \sum_j \frac{\Gamma_{ij}^*}{2\pi} \frac{x - x_{ij}}{(x - x_{ij})^2 + (y - y_{ij})^2}$$

$$\times \left(1 - exp\left[(x - x_{ij})^2 + (y - y_{ij})^2\right]/\delta^2\right).$$

To solve viscous diffusion equation (A.17) for second order space derivatives we use centered difference approximation on the analytical grid. The time derivative can be approximated using first order explicit scheme. Hence, if we know vortex values in mesh points at time point t, its new values $\omega^{t+\Delta t}$ can be found from the formula:

$$\omega_{ij}^{t+\Delta t} = \omega_{ij}^t + \frac{\Delta t}{\mathrm{Re}} \left(\frac{\omega_{i+1\,j}^t - 2\omega_{i\,j}^t - \omega_{i-1\,j}^t}{\Delta x^2} + \frac{\omega_{i\,j+1}^t - 2\omega_{i\,j}^t - \omega_{i\,j-1}^t}{\Delta y^2} \right). \tag{A.21}$$

Let us note, that applying formula (A.21) results in calculating errors which are the source of artificial viscosity. This circuit viscosity depends on discretization steps, both time and space ones. It is related, specifically, with negligible diffusion on the first grid layer adjoining wall along the surface normal. Ti reduce the error and circuit viscosity there can be used the exact solution of (A.17), which is of the form [GG05]:

$$\omega^{t+\Delta t}(r) = \int_S \omega^t(\vec{r}\,')\, G(\vec{r}, \vec{r}\,')\, d\vec{r}\,', \tag{A.22}$$

where $G(\vec{r}, \vec{r}\,')$ is the Green function:

$$G(\vec{r}, \vec{r}\,') = \frac{\mathrm{Re}}{4\pi\Delta t} \exp\left[-\frac{1}{4\Delta t}\left(\vec{r} - \vec{r}\,'\right)^2\right].$$

Mathematical aspects of the scheme and problems arising when accounting a boundary effect on the diffusion process are considered in [BP86].

Approach based on the Green function (A.22) is more explicit though requires significant computer resources. Effectiveness of application of formulas (A.21), (A.22) depends on Reynolds number. Formula (A.21) is more effective for small Reynolds numbers, while the greater Reynolds numbers are the more explicit the second scheme is.

Equation (A.18) coincides with the corresponding equation describing vortex motion in the ideal incompressible fluid, where vortexes move together with fluid elements and their intensity does not change. Therefore instead of (A.18) we may consider the equation describing the motion of vortex elements x_v, y_v:

$$\frac{dx_v}{dt} = u_v,$$
$$\frac{dy_v}{dt} = v_v, \tag{A.23}$$

This approach is often used in traditional vortex algorithms which [CK00] require carrying out complex procedure of regridding, namely redistribution of circulation of free vortex on the mesh points. Let us note, that numerical integrating of (A.23) and application of approximate "*regridding*"–algorithms considerably increases circuit viscosity of computational scheme.

Here to integrate (A.18) we propose to consider the flow field as a collection of discrete volumes, for each of which the vortex conservation law holds true:

$$\int_{\Delta Q} \frac{\partial \omega}{\partial t} dq = -\int_{\Delta S} \omega \left(\vec{V} \cdot \vec{n}\right) dS, \tag{A.24}$$

where ΔQ is a discrete volume, on which flow field is divided, ΔS, $\vec{n}$ is a surface of this volume and its outer normal. To each discrete volume the mesh point (x_{ij}, y_{ij}) with vorticity ω_{ij} is related (Fig. A.2). In view of (A.24), the vorticity convection of this grid element in the given time point t is described by the following expression:

$$\frac{\Delta\omega_{ij}^t}{\Delta t} \Delta x \Delta y \approx \omega_{i-1j}^t u_{i-1j}^t \Delta y + \omega_{ij-1}^t v_{ij-1}^t \Delta x - \omega_{i+1j}^t u_{i+1j}^t \Delta y$$
$$- \omega_{ij+1}^t v_{ij+1}^t \Delta x \omega_{ij}^t u_{ij}^t \Delta y - \omega_{ij}^t v_{ij}^t \Delta x.$$

Hence we obtain the circulation of the considered vortex element in the next time point $t + \Delta t$:

$$\Gamma_{ij}^{t+\Delta t} = \Gamma_{ij}^t + \left(\omega_{i-1j}^t u_{i-1j}^t \Delta y + \omega_{ij-1}^t v_{ij-1}^t \Delta x - \omega_{i+1j}^t u_{i-+1j}^t \Delta y\right.$$
$$\left. - \omega_{ij+1}^t v_{ij+1}^t \Delta x - \omega_{ij}^t u_{ij}^t \Delta y - \omega_{ij}^t v_{ij}^t \Delta x\right) \Delta t. \tag{A.25}$$

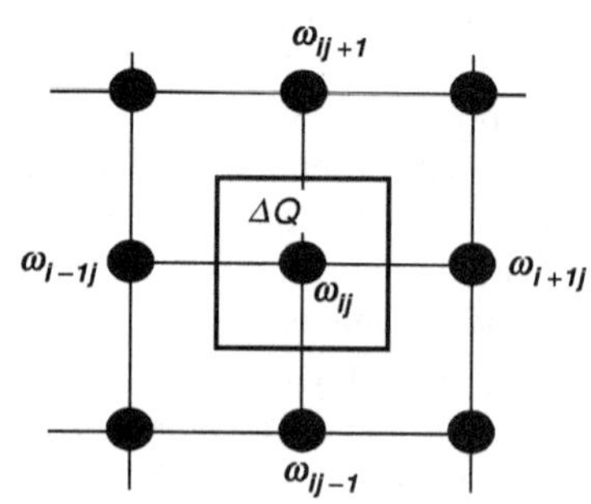

Fig. A.2 Grid element

Velocity components of the vortexes u, v can be found from expressions (A.20). It should be noted that the grid put over the flow field is adaptive, namely in each time point we consider only the mesh points with nonzero vorticity. This feature makes it possible to optimize calculations of velocity field.

Now let us consider the whole the computation algorithm in each time point:

1. Having known the vorticity values in the domain on the previous step, from the system of algebraic equations (A.19) we obtain the intensity of the attached vortex sheet and the intensity of vortexes on the sharp edges of the boundary.
2. From boundary condition (A.6) we obtain vorticity generated by the body surface.
3. Using boundary conditions (A.10), (A.11), we compute vorticity value on the boundaries of rated domain.
4. Using numerical integration of diffusion equation by formula (A.21) we calculate intermediate vorticity values ω_{ij}^* in the inner points of flow area.
5. Taking into account the obtained vorticity distribution we correct the intensity of attached vortex sheet in such a way that surface non-percolating condition was fulfilled.
6. Having obtained the vorticity field we define the velocity field in the domain using formulas (A.20).
7. Using formula (A.25) we compute the convection of vortex elements. Obtained values $\Gamma_{ij}^{t+\Delta t}$ in their turn define new values of vorticity $\omega_{ij}^{t+\Delta t}$.

A.4 Computation of Hydrodynamic Loads

To compute the hydrodynamic force acting on the body which is moving in the homogeneous fluid, we can use the theorem of impulses:

$$\vec{F} = -\frac{d}{dt}\int_S \vec{V}\, dr, \tag{A.26}$$

where $\vec{F} = \left(F_x, F_y\right)$, S is a fluid-filled domain, $\vec{V}, \vec{r}(x, y)$ is a velocity and a position vector of a fluid element, respectively.

From the continuity equation and from the definition of vorticity ω we obtain the following formula for fluid velocity:

$$\vec{V} = (\vec{r} \times \vec{\omega}) + \nabla \cdot (\vec{r}\,\vec{V}) - \nabla(\vec{r} \cdot \vec{V}). \tag{A.27}$$

Setting (A.27) in (A.26) and transforming integrals we obtain the force expression which depends only on the vorticity field characteristics [WU81, GG05]:

$$\vec{F} = -\frac{d}{dt}\int_S \vec{\omega} \times \vec{r}\, d\vec{r}. \tag{A.28}$$

This implies formulas for computation of resistance and lifting force:

$$F_x = -\frac{d}{dt}\int_S \omega y ds,$$
$$[3mm] F_y = \frac{d}{dt}\int_S \omega x ds. \tag{A.29}$$

Note that force component F_x in (A.29) includes form resistance as well as friction resistance.

Since the vorticity field is considered as superposition of vortex elements with intensities Γ_{ij} and coordinates $(x_{ij},\ y_{ij})$, we have formulas (A.29) in discrete form:

$$F_x = -\sum_i \sum_j \left(\frac{d\Gamma_{ij}}{dt} y_{ij} + \Gamma_{ij} v_{ij}\right),$$
$$[5mm] F_y = \sum_i \sum_j \left(\frac{d\Gamma_{ij}}{dt} x_{ij} + \Gamma_{ij} u_{ij}\right). \tag{A.30}$$

Dimensionless coefficients of hydrodynamic forces are defined in the following way:

$$C_x = \frac{2F_x}{U_\infty^2 a}, \quad C_y = \frac{2F_y}{U_\infty^2 a}. \tag{A.31}$$

A.5 Approbation of Numerical Scheme

Testing of the developed algorithm was carried out in several directions: analysis of numerical diffusion of computational scheme; accuracy evaluation of the algorithm with respect to description of vorticity generation on the body surface, error analysis when defining hydrodynamic forces acting on a streamline body.

A.5.1 Evolution of Vortex Point in Unbounded Flow

Suppose that in an unbounded domain there is a vortex point with intensity Γ_0. Its vorticity diffusion can be calculated by formulas (A.21), (A.23). Obtained vorticity distribution we compare with the known exact solution of Navier–Stokes equation which describes the development of two-dimensional vortex with the initial circulation Γ_0 in the viscous fluid [GG05]:

$$\omega(x, y, t) = \frac{Re}{4\pi t} \exp\left(-\frac{r^2 \mathrm{Re}}{4t}\right), \tag{A.32}$$

where $\mathrm{Re} = \Gamma_0/\nu,\ r = \sqrt{x^2, y^2},\ \nu$ is fluid viscosity.

In Fig. A.3 we can see time dependencies of vortex radius r_v, calculated on analytical grid $\Delta x = \Delta y = 0.01$ with the time step $\Delta t = 0.0025$ (markers), and the exact values r_v obtained from formula (A.32) (total line). The vortex radius is obtained from the relationship:

$$\omega|_{r>r_v} < 10^{-5}$$

From Fig. A.3 it follows that in the case $\mathrm{Re} = 10^2$ numerical diffusion is insignificant, and for $\mathrm{Re} = 10^3$ its value is within 10%. Increasing of numerical diffusion at large Reynolds numbers is essentially related with discretization parameters misfit: the steps Δx, Δy are too big for such Reynolds number. The best result is attained if $\Delta x, \Delta y \approx 1/\mathrm{Re}$. Comparison of curves in Fig. A.3 indicates that the numerical algorithm is convergent at moderate Reynolds numbers.

The choice of parameter value Δt is associated with the dimensions of grid elements and with local flow velocity. To provide the stability of the algorithm, vortex displacement within a time step must not exceed the minimal dimensions of a grid element Δx, Δy. In most cases it is enough that the step Δt was less by an order of the dimensions of grid elements.

A.5.2 *Flow Past a Flat Plate*

For accuracy evaluation of developed numerical algorithm describing vorticity generation process on the body surface we solved the problem of longitudinal flow past a flat plate. Characteristic parameters of this problem are the flow velocity U_∞ and the length of the plate L, such that: $Re_L = U_\infty L/\nu$, $\overline{x} = x/l$, $\overline{y} = y/l$. Calculation of vorticity generation based on the boundary condition (A.6). For modeling of vorticity diffusion formulas (A.21), (A.23) was used. Calculation results was compared with Blazius solution, which comes from boundary-layer theory [GG05]. In Figs. A.4 and A.5 we can see the stationary profile of longitudinal velocity $\overline{u} = u/U_\infty$ for the middle of the plate and the dimensionless friction factor

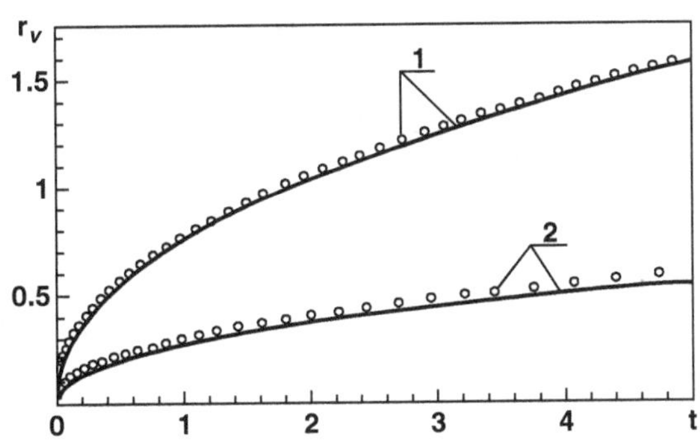

Fig. A.3 Vortex diffusion in unbounded domain. Comparison of calculation results *open circle* for vector r_v with the exact solution *solid line* : **1**, $\mathrm{Re} = 10^2$; **2**, $\mathrm{Re} = 10^3$

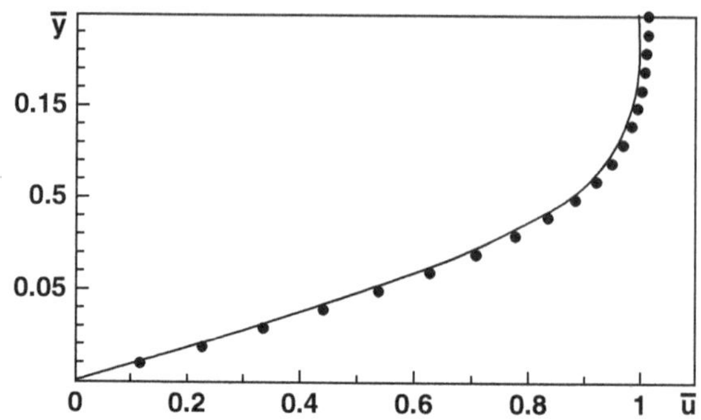

Fig. A.4 Longitudinal velocity profile in the attached layer for the middle of the plate at $\mathrm{Re}_L = 10^3$: *solid line* Blazius solution, *filled circle* computations

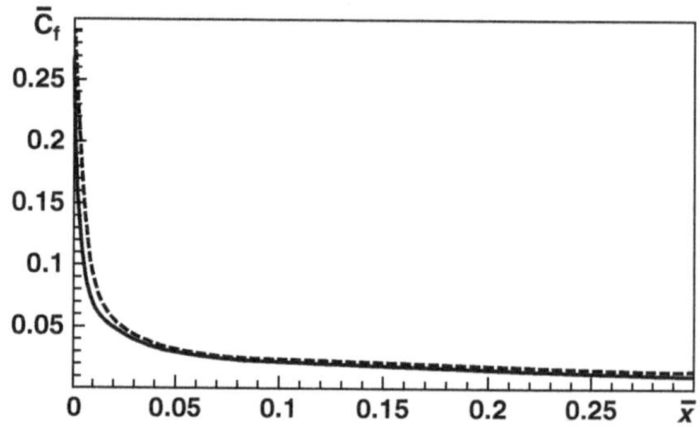

Fig. A.5 Plate's longitude coordinate dependence of the friction coefficient at $\mathrm{Re}_L = 10^3$: *solid line* Blazius solution, *dashed line* computations

$$\overline{C}_f = \frac{2}{\mathrm{Re}_L} \frac{\partial u}{\partial y} \bigg|_{wall} = -\frac{2}{\mathrm{Re}_L} \omega \,|_{wall} \tag{A.33}$$

on the plate, numerically obtained and obtained due to Blazius solution at $\mathrm{Re}_L = 10^3$. The comparison of these results indicates of high accuracy of considered scheme for modeling velocity and vorticity fields and for definition of hydrodynamic friction drag force of the plate.

A.5.3 Flow Past Square Prism

Developed numerical algorithm was used for modeling of two-dimensional laminar flow past a square prism at moderate Reynolds numbers. Carried out computations revealed some regularities of separated flow forming as well as supplemented the approbation of the numerical scheme by comparing obtained characteristics with well-known experimental data and similar numerical results of other researchers.

Computations was carried out in the range of Reynolds numbers from $\mathrm{Re} = 70$ to $\mathrm{Re} = 5 \cdot 10^3$. Further increase of Reynolds number needs considerable computational forces, due to decreasing dimensions of grid elements, as well as application of turbulence models. The grid which was used for calculations had square elements

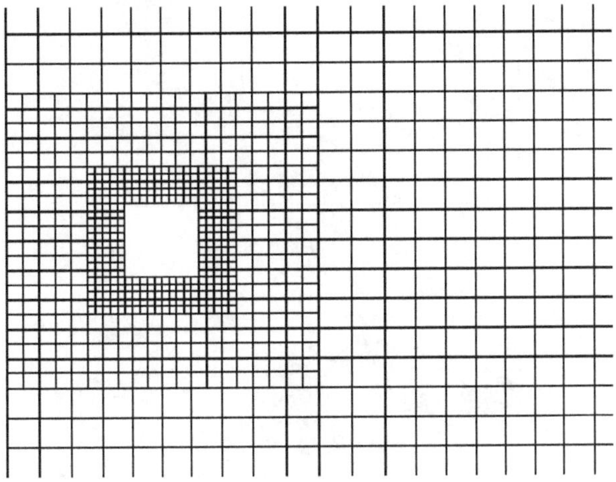

Fig. A.6 Configuration of the grid

and consisted of tree levels (Fig. A.6) with different densities. On the first level that is adjacent to the body, the dimensions of grid elements is associated with the number N of attached vortexes located along the prism walls (on the square side). On the next levels the dimensions of elements enlarge (double). Most of computations are done at $N = 100$.

In Fig. A.7 we see the computed instantaneous vorticity distributions in the wake past the square cylinder at different Reynolds numbers and at $\tau = 25$ ($\tau = tU_\infty/a$). This results indicate that the pattern of flow past the square cylinder considerably depends on Re. At Re ≤ 100 directly adjacent to the body circulation zones are of prolate form (Fig. A.7a). Intensity of vortex structures separating from this zones is comparatively small. Structures interactions in the wake lead to forming of the regular vortex street. When the Reynolds number increases the length of circulation zones decreases while the intensity of separating vortex beams proportionally goes up. Respectively vortex interaction in the wake amplifies and the width of the wake considerably increases. Since there is more vortexes entering the wake, the cylinder drag increases as well. At Re $\geq$ 1,000 the origin of the vortex sheet comes nearer to the body (Fig. A.7d–e). Increase of vortex circulation results in intensification of their interaction which in its turn brings increase of the street width as well as considerable deformations of separated vortex structures (up to their breakdown).

Mentioned features of the flow correlate with dependencies of instantaneous drag coefficient C_x and lifting force coefficient C_y on Reynolds numbers (Fig. A.8). Cited diagrams indicate that after short transient time phase in the wake past the body at $\tau > 10$ the structure of the flow approximates a periodic one that is characterized by Strouhal number St $= fa/U_\infty$, where f is a frequency of vortex separation. At Re $>$ 1,000 an additional mode appears. This fact is corroborated by the vortex distribution data in the wake (Fig. A.7e).

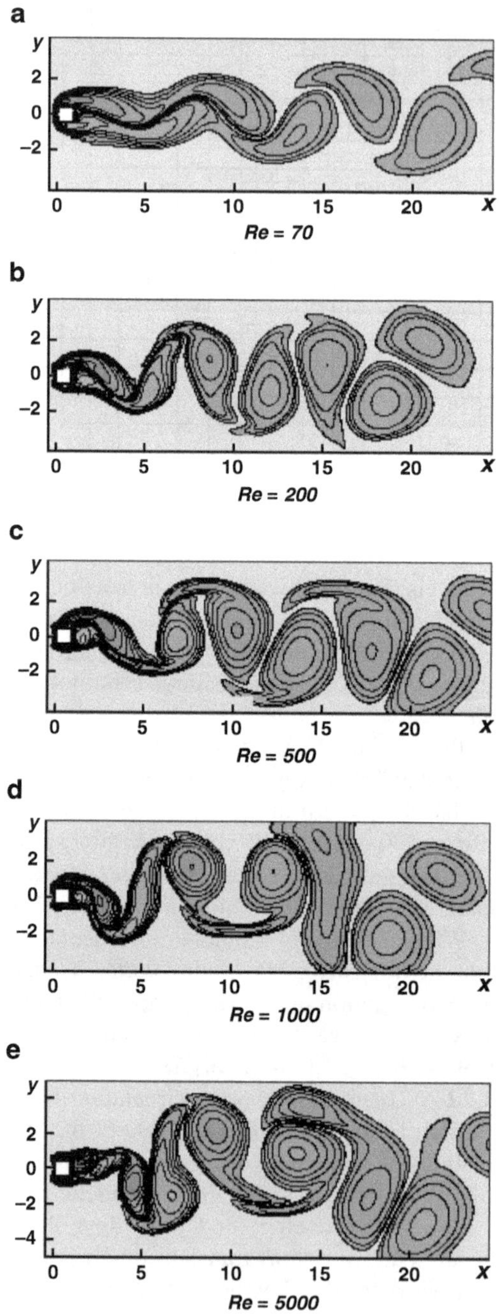

Fig. A.7 Vorticity distribution in the wake past the square cylinder at different Reynolds numbers, $\tau = 25$

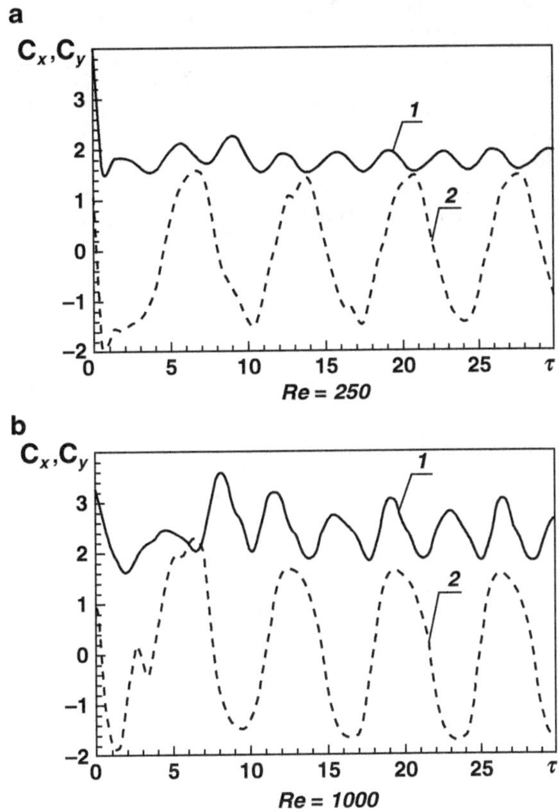

Fig. A.8 Time dependence of drag coefficient C_x **1** and of lifting force C_y, **2** of the square prism at different Reynolds numbers

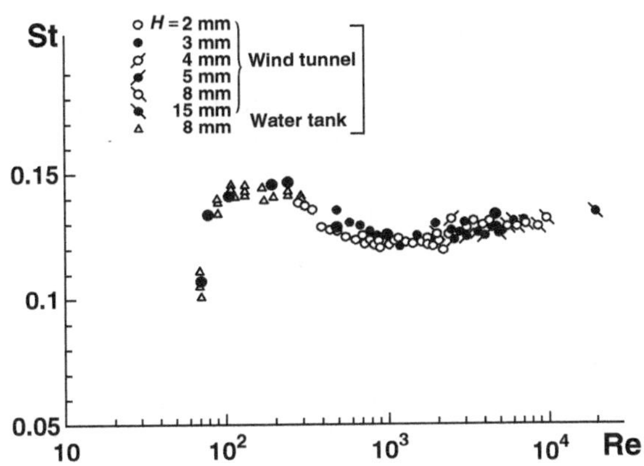

Fig. A.9 Dependence of a Strouhal number on a Reynolds number for the flow past the square prism and comparison of the obtained results with the experimental data; [OKA82] *filled circle* computations, *open circle*, *open triangle*, ... experiment

In Fig. A.9 the computation of Strouhal numbers are compared with the data from the work [OKA82] that is notable for broad experimental investigations of the flow past square prism at moderate Reynolds numbers. This comparative diagram indicates that the introduced numerical scheme high-precisely describes forming of the vortex wake past the body in the range of moderate Reynolds numbers.

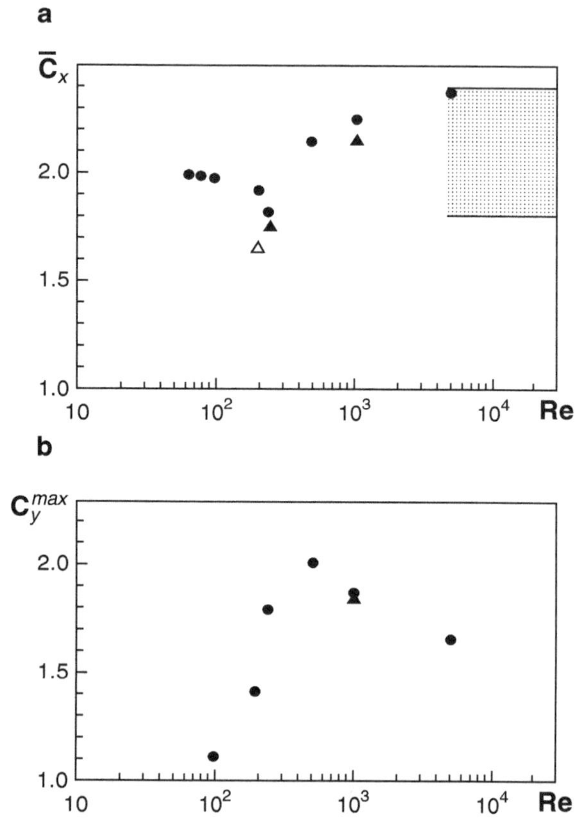

Fig. A.10 Dependencies of period average drag coefficient $\overline{C}_x$ (**a**) and the lifting force coefficient oscillation amplitude C_y^{max} (**b**) of the square prism on Reynolds numbers. Computation: *filled circle* this work, *filled triangle* [DM82], *open triangle* [MFR94]. Experiment: by the results from [BO82, VIC66]

In Fig. A.10 we see computed by formulas (A.31) dependencies of the period average drag coefficient $\overline{C}_x$ and the lifting force oscillation amplitude C_y^{max} of the square prism on Reynolds numbers. Let us denote that dependencies of the drag coefficient $\overline{C}_x$ and the frequency of vortex separation (of the number St) on Reynolds number are correlating: increase of Strouhal number up to St $= 1.45$ at Re $= 250$ is implied by narrowing of the wake directly after the square. This fact in its turn leads to the drag reduction (Fig. A.10a). Narrowing of the wake is essentially associated with a change of the flow pattern around the prism. At Re > 200 the vortical circulation zone arising near a side wall of the prism decreases in both crosswise and lengthwise directions (though, experiments indicate that the reattachment of the circulation flow to the walls in the case of a square prism does not take place). Further increase of $\overline{C}_x$ is implied by wake expansion due to decrease of viscous diffusion and of vortex circulation in the wake. The obtained data (Fig. A.10) for the magnitudes $\overline{C}_x$, C_y^{max} is close to the relevant results introduced in the works [DM82, MFR94], where computations were based on finite volume method. Experimental investigations of dynamic loads on the square cylinder substantially deal with large Reynolds numbers $\mathrm{Re} \geq 5 \cdot 10^3$. In this case the flow is essentially tree-dimensional and its characteristics depend on the rate of flow turbulence, velocity profile and also on parameters of experimental facility. This fact explains significant variation of values $\overline{C}_x$ obtained in different experiments.

In general the obtained results corroborate that the introduced numerical scheme is an effective tool for modeling of the viscous flow and it can be used for computation of flows past the bodies of complex configuration.

A.6 Algorithms for Flow Past a Square Cylinder Structure Control

Theoretical and experimental investigations of laminar flow past a square (namely two-dimensional case of flow past a square prism) at moderate Reynolds numbers indicated that wake structure and hydrodynamic drag coefficient C_x and lifting force coefficient C_y essentially depend on dimensions and interaction character of separated circulation zones, forming after the body.

Two following cases are qualitatively different: – when a separated zone over the side wall is local (bounded); – when a separated circulation zone generated on the windward corner of the prism is unclosed while its leeward corner is situated in the middle of this zone. In the first case the interaction between primary and quarter zones is weak and the wake progress is determined mostly by separation on the leeward corner. In the second case it is separation on the windward corner that is of most importance for the wake progress. The following algorithm for structure control of the flow past a square prism is based on physical properties of interaction between separated zones generated on windward and leeward corners of the prism.

The control objective is drag reduction of dynamic lateral forces producing hydroelastic oscillation of the flow structure. The strategy of such control is reduction of total vortex intensity, reduction of vortex flow past the body and minimization of the wake size. We propose to achieve an improvement of the flow structure around the square prism through artificial generation of special local separated circulation zones (standing vortexes) near it. Physical experiments carried out in Institute of hydromechanics of National Academy of Sciences of Ukraine with non-streamline bodies demonstrated an efficiency of this approach for flow structure change as well as for improvement of hydrodynamic characteristics [KR95]. To achieve the necessary positive effect an artificial separated zone must be stable with respect to external perturbations. Otherwise these perturbations may cause vortex emission into the flow. Moreover, a separated zone must be controllable (namely when its dimensions change with respect to the change of the flow conditions). In the optimal case the structure control system must have a feedback.

A.6.1 Two Symmetric Plates on the Windward Side of the Body

Here we consider the control technique of the flow past a square prism using special thin plates which are installed on the windward side of the prism (Fig. A.11). In the space between such plate and the corner of the prism a local separated zone is forming. Experiments indicate that the control objective is achieved when the flow line being separated from the plate edge A is attaching gradually to the corner B.

To study the flow past the body with control plates (Fig. A.11) we propose a few theoretical models. Certain general properties of detached flow can be described by the simplified model where the fluid supposed ideal and the circulation zone is simulated by one point vortex. For the justification of such model, it is associated with presence of general dynamic characteristics of separated zones which are the same for laminar, turbulent and potential flows (where viscous effects are insignificant). The analysis of dynamics of the point vortex describing separated circulation flow gives an information about topological characteristics of the flow: presence, disposition and type of critical points, their stability and reaction with respect to external perturbations [GG98, GG96].

Taking in view symmetry and stability of the flow in front of the body we consider only its upper part $y > 0$ (Fig. A.11). As before the boundary of the body is simulated by continuous vortex sheet which in its turn is replaced by a system of discrete vortexes in the numerical scheme. Then the complex potential of the flow is of the form:

$$\Phi(z) = U_\infty z + \frac{1}{2\pi i} \sum_{k=1}^{N} \Gamma_k \ln \frac{z - z_k}{z - \bar{z}_k} + \frac{1}{2\pi i} \Gamma_0 \ln \frac{z - z_0}{z - \bar{z}_0}, \tag{A.34}$$

where Γ_k and $z_k = x_k + i y_k$ is a circulation and complex coordinate of k-th attached (situated on the boundary) discrete vortex, respectively, Γ_0, $z_0 = x_0 + i y_0$ are parameters of a standing (immobile) vortex describing the circulation flow, N is a number of attached vortexes.

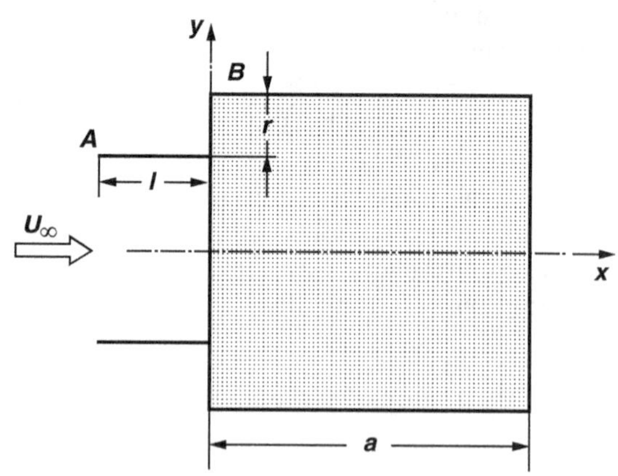

Fig. A.11 Configuration of the body with control plates

The problem is to find the parameters of the standing vortex $\overline{\Gamma}_0 = \Gamma_0/U_\infty a$, $\overline{x}_0 = x_0/a$, $\overline{y}_0 = y_0/a$ and parameters of the control plate $\overline{r} = r/a$ $\overline{l} = l/a$, (see Fig. A.11) such that there is a lack of vorticity generation in the points A and B (further we omit the dash indicating that the magnitude is dimensionless).

For solving such problem we have four nonlinear equations. The first two of them follow from vortex immobility condition:

$$\left.\frac{d\Phi}{dz}\right|_{z=z_0} = 0 \tag{A.35}$$

or

$$v_x(z_0) = 0, \ \ v_y(z_0) = 0.$$

Another two equations describe the Kutt–Gukovsky condition about velocity boundedness in the corner points A and B. The flow is optimal if

$$\Gamma_k\,|_A = 0\,, \ \ \Gamma_k\,|_B = 0, \tag{A.36}$$

where $\Gamma_k\,|_A$ and $\Gamma_k\,|_B$ are circulations of discrete vortexes, situated in the corner points A and B.

Taking in view the expression for the potential (A.34), the conditions (A.36) can be reduced to the following form:

$$v_x(z_0) - i v_y(z_0) = U_\infty + \frac{\Gamma_0}{4\pi y_0} + \frac{1}{2\pi i}\sum_{k=1}^{N}\Gamma_k\left(\frac{1}{z_0 - z_k} - \frac{1}{z_0 - \overline{z}_k}\right) = 0. \tag{A.37}$$

Unknown circulations of the attached vortexes Γ_k can be found from the system of linear equations that follows from non-percolation condition on the body surface and is similar with the system (A.19) where instead of the collection of vortexes we consider only one circulation vortex Γ_0.

The equations for Γ_0, x_0, y_0, r are essentially nonlinear. To solve them we used Broyden's numerical algorithm. Computations indicated that given one of control plate characteristics, for instance, the length of the plate l, the problem (A.35)–(A.36) has a unique solution; in the space between the plate and the prism side at $0.15 \le l \le 0.65$ there is a point with coordinates $x_0, \ y_0$ where the vortex with circulation Γ_0 is equilibrious (namely it is immobile or standing); exactly such vortex provides null vortex generation in the corner points A and B. The numerical behavior analysis of this vortex in the neighborhood of the equilibrium position showed that small flow perturbations result in precessional movement of the vortex about the point $x_0, \ y_0$. In view of dynamic system theory it means that such critical point is an elliptic one and the corresponding vortex is stable. Figure A.12 illustrates dependencies of the standing vortex coordinates x_0, y_0, circulation Γ_0 and the parameter r on the plate length l. These dependencies are close to linear ones. From the graph $r(l)$ we can define an optimal configuration of the control plates. The pattern of flow lines in front of the long body for the case $l = 0.3, r = 0.22$ is shown in Fig. A.13.

Fig. A.12 Dependencies of standing vortex circulation, its disposition and the parameter r on the plate length l: ***1***, $x_0(l)$; ***2***, $y_0(l)$; ***3***, $\Gamma_0(l)$; ***4***, $r(l)$

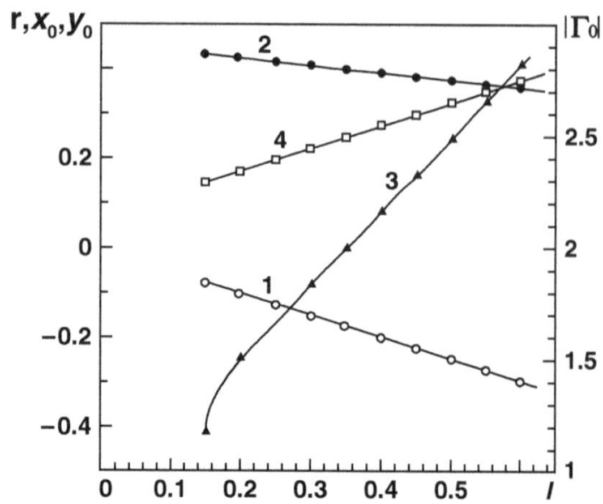

Fig. A.13 The pattern of flow lines with forming of the standing vortex in front of the long body with the control plates on the windward side

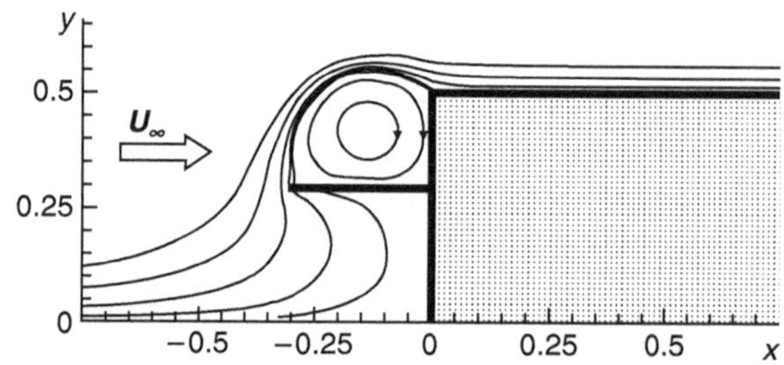

Computations carried out with respect to the simplified model, namely when the separated circulation flow is described by one point vortex, demonstrated principal possibility of existence of a standing (immobile) vortex in front of the body with special (control) thin plates on its windward side, and also this computation make it possible to define optimal parameters of the plates. For this way obtained optimal configuration of a square prism with control plates the flow past this prism was simulated basing on the complete Navier–Stokes equation system due to the scheme defined in the Sect. A.2. Figure A.14 illustrates instantaneous vorticity distribution in the wake past the square cylinder with control plates at $Re = 500$, when the characteristics of the plates are close to optimal ones: $l = 0.2, r = 0.16$.

Comparison of computation results shown in Figs. A.14 and A.7c indicates that of the wake past the prism changes after the installation of the plates. This change is expressed in:

- Narrowing of the vortex region both near the body and in the wake.
- Increasing of frequency of vortex separation and, respectively, extension of the wake; Strouhal number characterizing this process increases from $\mathrm{St} = 0.125$ for an ordinary square prism up to $\mathrm{St} = 0.143$ for a prism with control plates.
- Decreasing intensity of vortexes in the wake.

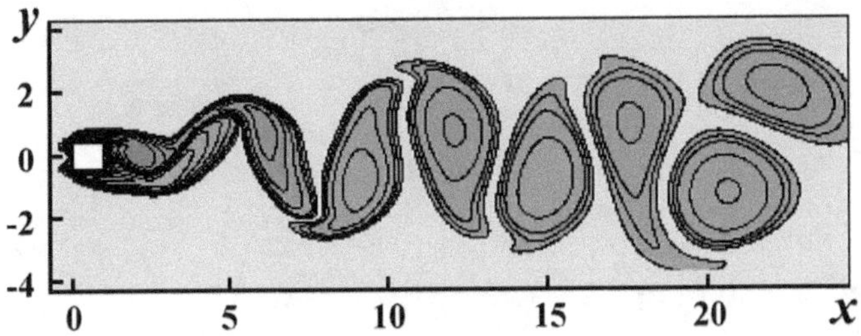

Fig. A.14 Vorticity distribution in the wake of the square cylinder with control plates in the optimal regime ($l = 0.2$, $r = 0.16$) at Re $= 500$, $\tau = 25$

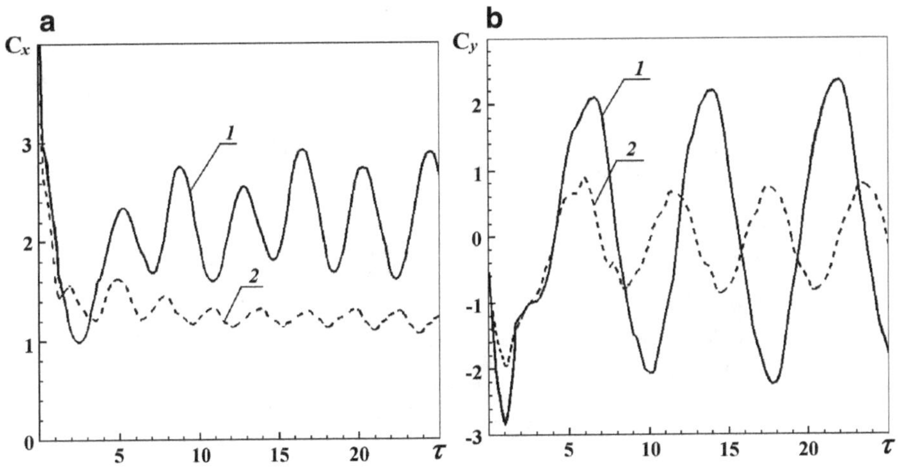

Fig. A.15 Time dependencies of drag coefficient C_x (**a**) and lifting force coefficient C_y (**b**) for the square cylinder at Re $= 500$: *1* with control, *2* with optimal control ($l = 0.2, r = 0.16$)

Specified changes have a positive effect on hydrodynamic characteristics of the prism (Fig. A.15, Re $=$ 500). The value of period average drag coefficient $\overline{C}_x$ at moderate Reynolds numbers decreases from 2.15 (for an "ordinary" square prism) down to $\overline{C}_x \approx 1.2$, that is about 55% of the previous value (Fig. A.15a). The lifting (or lateral) force oscillation amplitude respectively decreases from $C_y^{max} = 2$ down to $C_y^{max} = 0.8$ (Fig. A.15b). If the body is elastically fixed in the flow then the amplitude of the body oscillations caused by vortex separation substantially decreases.

Importance of parameter choice for control plates is demonstrated in Fig. A.16, where we present dependencies of the coefficients $C_x(\tau), C_y(\tau)$ at different plate–prism corner distances. Graphs $C_x(\tau), C_y(\tau)$, and the vorticity distribution in the wake (Fig. A.17) indicate occurrence of additional modes caused by vortex deformations in the wake.

Conclusions concerning optimal flow structure can be generalized over the range of large Reynolds numbers. This fact is supplemented by the results of numerical

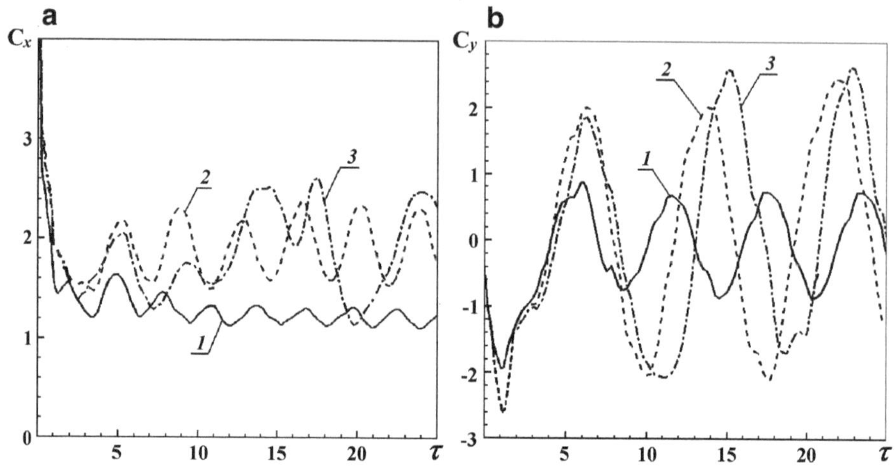

Fig. A.16 Time dependencies of drag coefficient C_x (**a**) and lifting force coefficient C_y (**b**) for the square cylinder at different parameters of control plates: *1* $l = 0.2, r = 0.16$, *2* $l = 0.2, r = 0.4$ *3* $l = 0.2, r = 0.075$, $\text{Re} = 500$

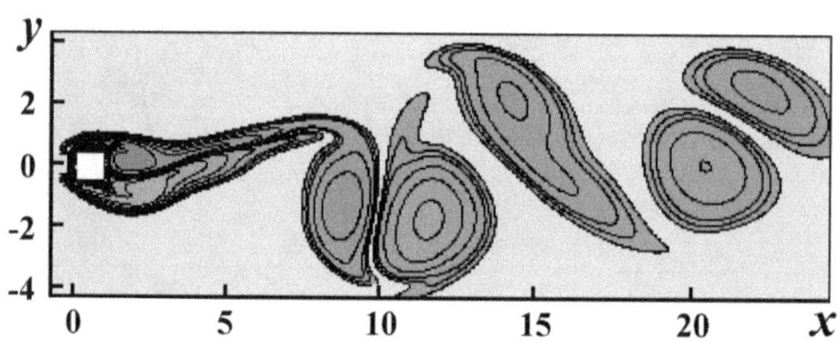

Fig. A.17 Vorticity distribution in the wake of the square cylinder with control plates in non-optimal regime ($l = 0.2, r = 0.075$) at $\text{Re} = 500$, $\tau = 25$

modeling for detached flow past a prism and a prism with plates based on the method of discrete vortexes. We supposed that from the corner points vortex sheets come down into the flow.

The computation results indicate three typical regimes of a flow past the body with control plates. If the plate is installed far from the corner then the vortex sheet and the corresponding flow line which has descended from the plate edge lie on the windward side of the square (Fig.A.18c). In this case formed circulation zone becomes weak and is of little influence upon the flow development around the body. The flow attaches to the body at the point lying substantially lower that the corner B. The vortex generated in front of the body weakly influence on vortex generation processes in the corner point B (on the edge of the body), and hence it cannot prevent a global flow separation in the neighborhood of this corner, and in its this turn causes a development of an intense vortex zone over the body wall.

The second type of the flow occurs when the control plate is very close to the corner (Fig. A.18b). In this case the flow does not attach to the body, and the detached circulation zone dos not localize. On the contrary, an intense detached zone is developing, as it had already been observed in the case of a body without plates. Let us note also, that approximated one-vortex model, describing a circulation flow through the behavior of the one vortex, did not show up any standing vortex in the space between the plate and the corner. When the plate is very close to the corner B the separation zone descending from the plate edge A overlays on the flow generated by the prism edge. The flow in the neighborhood of the corner B becomes very unstable. In this situation the body characteristics become worse compared with the case without the controls.

In the optimal case the flow line descending from the plate edge is gradually attaching to the body surface near the corner B (Fig. A.18a). Further this attached flow without any relevant perturbations moving along the surface. Vorticity generation on the corner B is suppressed by influence of the circulation zone in front of the body. In this case we can see the decrease of vortex intensity in the flow round the corner B that causes substantial drag decrease. It has to be mentioned that the optimal control plate parameters computed by applying each of the following schemes: the "viscous" model, the method of discrete vortexes and the model of the standing vortex, are almost coincident.

A.6.2 Stabilization of a Vortex Flow Past the Body with a Separating Plate

Substantial influence on the body drag has a structure of a flow in the near wake. It depends on intensity of vortex formation and floe pattern in the wake. Two following situations are possible:

- The flow past the body is stable; here two symmetrical circulation zones are forming; the process of whirling liquid carrying away into encircling flow is insignificant, hence the body drag is relatively small as well.
- Circulation zones past the body are unstable with respect to small perturbations; from their interaction area attaching directly to the leeward side of the body large beams of whirling fluid periodically join to the flow; hydrodynamic forces in this case are much greater that in the previous case.

A stable symmetrical flow pattern in the wake occurs in the experiments only ar small Reynolds numbers $\mathrm{Re} < 50$, when perturbations are suppressed by viscous force action. The stability analysis of the flow in the wake showed that non-symmetrical (with respect to the axis Ox – Fig. A.1 in our case) perturbations are most dangerous (destabilizing). To suppress such perturbations and stabilize the flow is possible using a separating plate situated behind the body (Fig. A.19). To find out regularities of vortex structure forming for a flow past a square prism with a plate we simulated the flow using numerical scheme defined in the Sect. A.2.

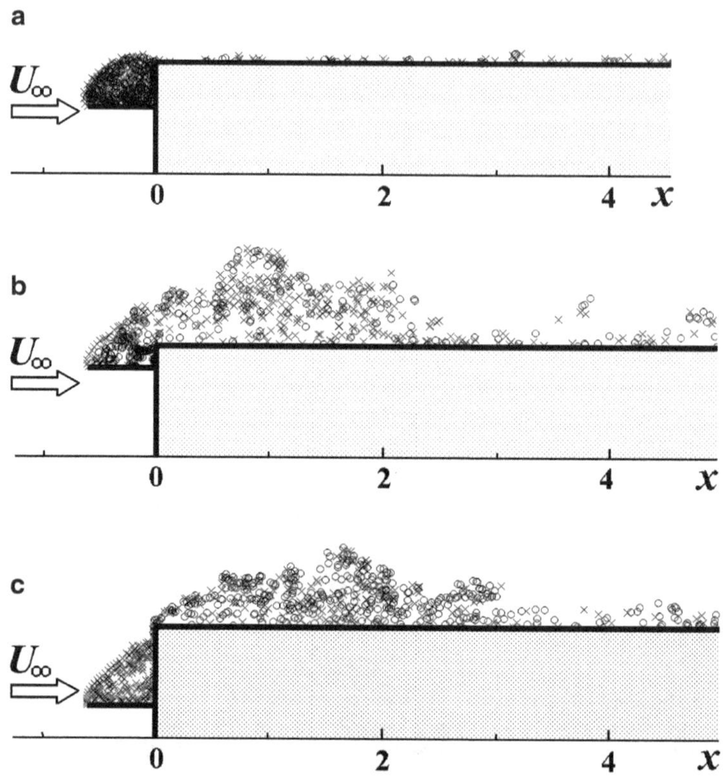

Fig. A.18 Discrete-vortex model of the flow in front of the body of non-streamline form with control plates on the windward side: optimal (**a**) and non-optimal (**b**, **c**) plate parameters

We considered laminar flow at $l = 2$. Obtained patterns of vortex distribution (Fig. A.20) indicate useful decrease of oscillatory motion of circulations zones in the wake. Installation of the plate causes narrowing of the wake and decreasing of intensity of the vortex beams attaching to the flow from the nearest wake. Respectively hydrodynamic loads decrease [GG05]. Experiments indicate that at increasing of Reynolds number this positive effect decreases though at large numbers Re installation of the separating plate results in noticeable drag decrease (at 10–20%).

A.7 Conclusions

The numerical algorithm for modeling the laminar flows of viscous incompressible fluid, particularly for non-stationary detached flow past the non-streamline bodies, is developed. The algorithm is based on usage of Navier–Stokes equation system with the "vorticity–velocity" formulation. This algorithm has the following advantages: the possibility of the significant decrease of dimensions of the computation

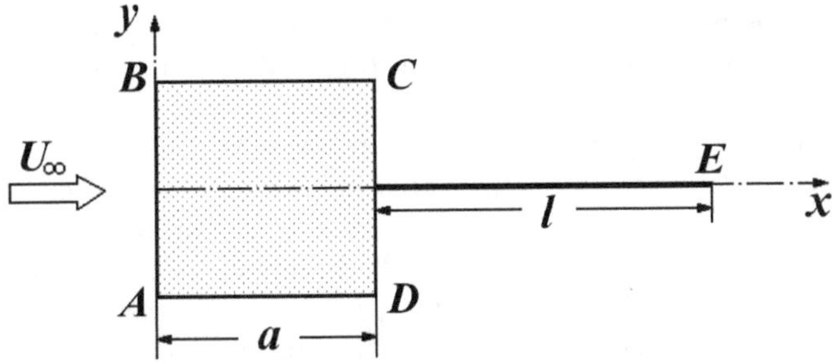

Fig. A.19 Diagram of the flow past the square prism with the separating plate

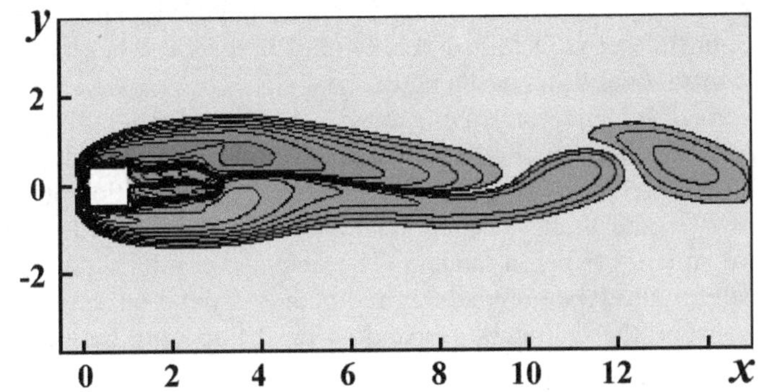

Fig. A.20 Pattern of vorticity isolines at the flow past the square prism with the separating plate

domain (we carry out the computations within the grid elements with nonzero vorticity values), high accuracy of the boundary conditions fulfillment in the corner points of the body surface and also high-accuracy of vortex value determination in the neighborhood of such points.

Algorithm was approbated on the two-dimensional diffusion problems for a vortex point in a viscous fluid, problems of longitudinal flow past a plate and problems of cross flow past a square prism. Computation result was compared with well-known analytical solutions, numerical simulation results carried out by other authors and also with experimental data. This computations showed that to provide the accuracy of numerical modeling grid spatial steps Δx, Δy and a time step Δt must fulfill certain conditions, for example, near the body surface

$$\Delta x, \Delta y \approx (1 \div 10)\frac{1}{\text{Re}}, \quad \frac{\Delta x}{\Delta t} \le V_{max},$$

(where V_{max} is the maximal local fluid velocity in the domain). Analysis of obtained results indicate effectiveness of the developed numerical algorithm for modeling the separated flows at moderate Reynolds numbers, particularly, non-stationary processes of vortex structure generation near the body surface and their dynamics in the wake past a non-streamline body.

Performed computations showed that forming of the wake past the body and the body drag essentially depend on the interaction of the separated circulation zones attaching directly to the body surface. Decreasing of non-stationary motion of such zones is accompanied by correspondent decrease of hydrodynamic loads. Basing on this investigation results two control algorithms for a flow past the square prism were considered. In the first case on the front (windward) side of the prism two symmetrical (control) plates were installed. In the second case on the back (leeward) side of the prism one plate was installed (separating vortex sheets with circulations opposite in direction). The analysis of computation results showed that at certain (optimal) parameters of the plates separated circulation zones near the body surface are stable. Non-stationary motion of such zones is decreasing, vorticity generation processes on the body slow down causing decrease of hydrodynamic drag and lateral force acting on the body. Obtained results may be useful when developing new methods of control over flows past a body.

The numerical scheme for solving Navier–Stokes equations represented and validated above has been applied to calculation fluid flows in multiply-connected domains. That allowed deriving regularities referring to generation of vortex wakes near systems of some bodies. Paper [GG08] deals with modelling the laminar flow past two square prisms in tandem. The obtained results show that evolution of the flow and hydrodynamic loads on the bodies depend on the flow pattern in the region between the cylinders. Depending on the spacing between the bodies the flow pattern may be symmetrical, when a stable vortex pare is there formed; non-symmetrical, with separation of large vortices from the front cylinder; and bifurcational, when a sudden jump from the first flow regime to another is possible. Special attention is paid to the effect of external disturbances on flow evolution. Wake behavior behind a parallel pair of square prisms, placed side-by-side normal to a uniform flow, is considered in paper [GG09]. The wake patterns are shown to depend strongly on the gap size between the bodies. When the space is small enough comparatively to the prism side, the wake resembles that seen behind a single body. At gap widths closed to the prism side the flip-flopping vortex shedding pattern where the wake flip-flops randomly between two states is observed. In this regime, the narrow and wide wakes with separate vortex shedding frequencies are formed behind the body system. For a large spacing, two synchronized vortex shedding patterns, antiphase and in-phase, are possible. At very large spaces two Karman vortex streets are generated behind each cylinder. The results obtained are used to ground the scheme of flow control near a square prism that utilizes the small gap in the center. The optimal characteristics of the control scheme are derived. In paper [VVGG09], the kinematics of flow in the system of ten staggered circular cylinders is calculated. The results show that interactions between the wake, which forms behind the entire construction, and the vortices generated near the cylinders in the system cause bifurcation in the vortical flow and appearance of the structures that are convected downstream in the gaps between the cylinders. The calculation results are compared with data of the physical experiments.

Acknowledgements This chapter prepared jointly with N.V. Gorban, I.N. Gorban and V.A. Gorban.

References

[BO82] Bearman PW, Obasaju ED (1982) An experimental study of pressure fluctuations on fixed and oscillating square-section cylinders. J Fluid Mech 119:297–321

[BP86] Benfatto G, Pulvirenti M (1986) Convergence of Chorin-Mardsen product formula in the half-plane. Commun Math Phys 106:427–458

[CLD94] Cortelezzi L, Leonard A, Doyle JC (1994) An example of active circulation control of the unsteady separated flow past a semi-infinite plate. J Fluid Mech 260:127–154

[COR96] Cortelezzi L (1996) Nonlinear feedback control of the wake past a plate with a suction point on the downstream wall. J Fluid Mech 327:303–324

[CK00] Cottet G-H, Koumoutsakos P (2000) Vortex methods: theory and practice. Cambridge University Press, London

[DM82] Davis RW, Moore EF (1982) A numerical study of vortex shedding from rectangles. J Fluid Mech 116:475–506

[GG96] Gorban VA, Gorban IN (1996) On resonance properties for vortexes by the inequality on boundary. Reports NAS of Ukraine 2:44–47

[GG98] Gorban V, Gorban I (1998) Dynamics of vortices in near-wall flows: eigenfrequencies, resonant properties, algorithms of control. AGARD Report. 827:15-1–15-11

[GG05] Gorban VA, Gorban IN (2005) Vortical flow pattern past a square prism: numerical model and control algorithms. Appl Hydromechan 6:23–48

[GG08] Gorban VA, Gorban IN (2008) On interaction studing for square cillinders, ordered by thandem. Appl Hydromechan 9:18–32

[GG09] Gorban VA, Gorban IN (2008) On analysis of two square cillinders disposed side by side system streamline. Appl Hydromechan 10:36–47

[KNI90] Knisely CW (1990) Strouhal number of rectangular cylinders at incidence: a review and new data. J Fluids Struct 4:371–393

[KR95] Koenig K, Roshko A (1995) An experimental study of geometrical effects on the drag and flow field of two bluff bodies separated by a gap. J Fluid Mechan 156:167–204

[KOU93] Koumoutsakos PD (1993) Direct numerical simulations of unsteady separated flows using vortex methods. PhD Thesis. California Institute of Technology, Pasadena

[LIG63] Lighthill MJ (1963) Laminar boundary layers. Oxford University Press, London

[LN76] Loehrke RI, Nagib HM (1976) Control of free stream turbulence by means of honeycombs: a balance between suppression and generation. Trans ASME Fluid Eng 98:342–353

[MFR94] Minewitsch S, Franke R, Rodi W (1994) Numerical investigation of laminar vortex-shedding flow past a square cylinder oscillating in line with the mean flow. J Fluids Struct 8:787–802

[MIG62] Migai VK (1962) Aerodynamic efficiency of irregular surface. J Eng Phys Thermophys 5:20–23

[MO98] Mochizuki S, Osaka H (1998) Drag reduction with submerged ribs and its mechanism in a turbulent boundary layer over D-type roughness. Proceedings of the International Simposium on Seawater Drag Reduction. pp 121–126

[NNT82] Nagano S, Naito M, Takata H (1982) A numerical analysis of two-dimensional flow past a rectangular prism by a discrete vortex model. Comput Fluids 10:243–259

[NMDK99] Naim HN, Milne RB, Devine KD, Kmpka SN (1999) A coupled Lagrangian-Eulerian scheme for reacting flow modeling. Vortex Flows and Related Numerical Methods III, ESAIM: Proceedings, Vol. 7, pp 304–313

[OKA82] Okajima A (1982) Strouhal number of rectangular cylinders. J Fluid Mechan 123: 379–398

[SAV98] Savchenko YuN (1998) Drag reduction through the nearwall vortex system management. Proceedings of the International Simposium on Seawater Drag Reduction. Newport, USA. pp 249–255

[SND98] Sohankar A, Norberg C, Davidson L (1998) Numerical simulation of flow past a square cylinder. Proceedings of FEDSM99 3rd ASME/JSME Joint Fluids Engineering Conference. July 18–23. San-Francisco, California, USA. pp 1–6

[VIC66] Vickery BJ (1966) Fluctuating lift and drag on a long cylinder of square cross-section in a smooth and in a turbulent stream. J Fluid Mechan 25:481–494

[VVGG09] Voskobiynik AV, Voskobiynik VA, Gorban VA, Gorban IN (2009) On vortex feature in conjugated flow between pale group of thrice-line bridge pier. Appl Hydromechan 11:16–29

[WW86] Wang CM, Wu JC (1986) Numerical solution of steady Navier-Stokes problems using integral representations. AIAA J 8:1305–1312

[WU76] Wu JC (1976) Numerical boundary conditions for viscous flow problems. AIAA J 14:1042–1049

[WU81] Wu JC (1981) Theory of aerodynamics force and moments in viscous flows. AIAA J 19:432–441

[ZSG94] Zaki TG, Sen M, Gad-el-Hak M (1994) Numerical and experimental investigation of flow past a freely rotatable square cylinder. J Fluids Struct 8:555–582

Index

FSC
www.fsc.org
MIX
Papier | Fördert
gute Waldnutzung
FSC® C083411

Kolibri 360 GmbH
Ferdinand-Jühlke-Straße 7
99095 Erfurt, Deutschland
produktsicherheit@kolibri360.de